SpringerWienNewYork

CISM COURSES AND LECTURES

The series presents lecture notes, monographs, edited works and proceedings in the field of Mechanics, Engineering, Computer Science and Applied Mathematics.
Purpose of the series is to make known in the international scientific and technical community results obtained in some of the activities organized by CISM, the International Centre for Mechanical Sciences.

INTERNATIONAL CENTRE FOR MECHANICAL SCIENCES

COURSES AND LECTURES - No. 537

GENERALIZED CONTINUA AND DISLOCATION THEORY

THEORETICAL CONCEPTS, COMPUTATIONAL METHODS AND EXPERIMENTAL VERIFICATION

EDITED BY

CARLO SANSOUR
UNIVERSITY OF NOTTINGHAM, NOTTINGHAM, GREAT BRITAIN

SEBASTIAN SKATULLA
CERECAM, DEPARTMENT OF CIVIL ENGINEERING,
UNIVERSITY OF CAPE TOWN, SOUTH AFRICA

SpringerWienNewYork

This volume contains 76 illustrations

SPIN 86085123

All contributions have been typeset by the authors.

ISBN 978-3-7091-1221-2 SpringerWienNewYork

PREFACE

The term generalised continua was coined in the proceedings of the IUTAM conference held in Freudenstadt in 1967; a remarkable gathering of some of the most influential names in mechanics at the time. The term generalisation was meant as an extension of the concept of the Cosserat continuum. In fact some hundred years ago, the Cosserat brothers introduced their concept of a continuum which not only was revolutionary, it made complete the ideas itself involved in the definition of a continuum. The angular momentum equation was treated on equal footing as that of linear momentum. It was explicitly used instead of being reduced to a symmetry statement of the stress tensor. Notwithstanding the appeal of the concept it was only after a paper by Ericksen and Truesdell in 1957 that the ideas of the Cosserat brothers were revived. In this paper the original concept was modified in two ways. On the one hand the idea of directors, the rotation of which provided the characteristic new degree of freedom, was introduced and on the other hand these directors where also allowed to deform to describe a micro deformation as well. Later, Eringen and associates coined the term micromorphic for such continua. They also used the term micropolar to describe the Cosserat continua. Moreover, unlike Ericksen and Truesdell and other early contributors in the field, they derived the strain measures based on the notion of the group element itself. That is the group of rotation tensors was directly introduced in the case of the Cosserat continuum and that of the linear transformations in the case of the micromorphic continuum. In spite of the quite advanced state of this kind of theories, in the following decades applications of generalised formulations were more or less restricted to rod and shell theories as, with few exceptions such as the dispersion relation in wave propagation, no real understanding of scale effects incorporated in these theories was available. Further, these theories were restricted to elastic material behaviour. With the progress in material sciences in understanding the material behaviour at micro and nano scales and beyond, scale effects became of common interest and so the interest in generalised formulations resurfaced. Also the availability of computer power meant that solutions of problems based on these theories became accessible albeit numerical ones. Moreover, computations at smaller scales allows to address the source of scale ef-

fects and provides much needed understanding as to the background of such generalised theories. In fact much of these sources lie in defects and dislocations at a much smaller scale which mannifest themselves in scale effects and inelastic deformations.

This book contains some of the lectures presented at the CISM Centre in Udine on generalised continua and dislocation theory.

The first contribution by Gérard Maugin deals with the driving forces on different types of "defects" such as, dislocations, disclinations, point defects, cracks, phase-transition fronts and shock waves in micropolar and micromorphic materials. Exploiting modern notions of mathematical physics (Noether's theorem, Lie groups, Cartan geometry) these so-called material or configurational *forces are related to the* Eshelby stress tensor *and a corresponding dissipative energy approach.*

The following chapter by Carlo Sansour and Sebastian Skatulla starts with a compact treatment of linear transformation groups (Lie Groups), associated Lie algebras and the role of the exponential map. Rules of differentiations and variations are discussed and many relevant formulas for the rotation group are derived. The subsequent excursion into the continuum theory of generalised continua sets out motivating the inclusion of the angular momentum equation within a framework of classical deformations. The Cosserat formulation comes as a natural extension of the classical formulation. After a critical assessment of the micromorphic theory by Eringen and associates, a unified framework of generalised continua based on the notion of fibre bundle is presented. It puts all formulations including higher gradient ones on a common ground and allows for a rational treatment of inelastic and non-linear material behaviour. Numerical results based on finite elements and meshfree methods illustrate the relevance of the framework.

The next contribution by Samuel Forest gives an account on continuum crystal plasticity starting from the classical theory and providing the transition to Cosserat, strain-gradient and micromorphic continuum-based formulations where the material hardening and evolution of dislocations and damage are related to the lattice curvature captured in the higher-order strains. The generalised theories are embedded in an energy balance equation and an entropy principle to derive the state laws and residual dissipation. The application to dam-

ages mechanics addresses the deficiency of classical crystal plasticity to realisticly predict crack tip behaviour, especially in fatigue. Moreover, plastic strain heterogeneities and scale effects can be addressed, e.g. found in thin polycrystal coatings. The resulting properties of the polycrystal material are obtained by homogenization scheme making use of Hill-Mandel's approach periodic Dirichlet boundary conditions.

The final chapter by Hussein Zbib is devoted to dislocation dynamics and its fundamental importance to describe irreversible deformations of crystalline materials linked to the motion of the constituent defects. The complex governing mechanisms located at micro- and nano scales and the associated local quantities such as plastic distortion and internal stresses are exhaustively treated. Moreover, their embedment into continuum mechanical frameworks which can be exploited computationally are given particular attention. It is demonstrated that the coupling of continuum mechanical analysis with suitable computational algorithms represents a powerful technique in material engineering analysis.

Finally, a word of clarification and gratitude. Regrettably there have been serious delays in publication of the book. However, the present contributions have been brought up to date and reflect state of the art in the field, also regarding the literature. In spite of the delay, Generalised Continua have lost nothing of their relevance to today's research in the broader field of solid mechanics and so we hope that this book provides a valuable contribution to introduce researchers to this active field of research. In this regard the editor is indebted to the authors for their patience and to Prof. Serafini for his encouragement and support to bring this book to a completion.

Carlo Sansour and Sebastian Skatulla

CONTENTS

Defects, Dislocations and the General Theory of Material Inhomogeneity

G. A. Maugin
Institut Jean Le Rond d'Alembert
Université Pierre et Marie Curie, Paris, France

Abstract The present lecture notes have for main purpose to introduce the reader to the notion of driving forces acting on defects in various classes of materials. These classes include elasticity, the standard case in its pure homogeneous form, and more complex behaviors including inhomogeneous and dissipative materials. A typical such driving force is the Peach–Koehler force acting on a dislocation line. More generally, these forces of a non-Newtonian nature are so-called *material or configurational* forces which are contributors to the canonical equation of momentum, here the momentum equation completely, canonically projected onto the material manifold. The latter indeed is the arena of all material defects and the essential ingredient then becomes the so-called material *Eshelby stress tensor*. This stress is the driving force behind various types of local matter rearrangements such as plasticity, damage, growth, and phase transformations. Its material divergence provides the sought driving force on different types of "defects" such as, dislocations, disclinations, point defects, cracks, phase-transition fronts and shock waves. Here the emphasis is placed on defects more particularly related to materials science and for materials presenting a microstructure such as polar materials and micromorphic ones. Of importance is the fact that the concept of driving force is always accompanied by a parallel energy approach, so that the dissipation (energy release rate) occurring during the progress of a defect is exactly the non-negative product of the driving force by the velocity of progress. Modern notions of mathematical physics (Noether's theorem, Lie groups, Cartan geometry) as well as efficiently adapted mathematical tools (*e.g.*, generalized functions or "distributions") are exploited where necessary. The three great heroes of the reported story are J. D. Eshelby, E. Kroener and J. Mandel.

1 Lecture 1: The Eshelby–Kroener View of Defects and their Driving Force: Peach–Koehler Force, Incompatibility, Eshelby Stress

1.1 The Case of a Dislocation Line (Peach–Koehler Force; Eshelby's Derivation)

We start with the evaluation of a material force acting on one singular line. This was first developed by Peach and Koehler (1950)) in a celebrated paper, a true landmark in dislocation theory. A dislocation line L is seen in continuum physics as a line along which the displacement vector of elasticity suffers, in a certain sense, a finite discontinuity, called the *Burgers vector*, that we shall note $\tilde{\mathbf{b}}$ in order to avoid any confusion with the Eshelby stress (although there exists a relation between these two notions). The magnitude and direction of $\tilde{\mathbf{b}}$ characterize the different types of dislocations (see Lardner, 1974). In discrete crystals $\tilde{\mathbf{b}}$ can only be equal to a finite number of the vectors of the lattice. What exactly occurs is that in the presence of a dislocation line L the displacement vector $\mathbf{u}$ is no longer a single-valued function of the coordinates: it receives a finite increment $\tilde{\mathbf{b}}$ in going along a circuit S around the dislocation line L. With a definite choice of sign, this is expressed by

$$\oint_S \mathrm{d}\mathbf{u} = \oint_S \frac{\mathrm{d}\mathbf{u}}{\mathrm{d}s}\,\mathrm{d}s = -\tilde{\mathbf{b}}, \tag{1}$$

where s is a line coordinate along the circuit S. A dislocation is called a screw dislocation when $\tilde{\mathbf{b}}$ is parallel to the unit tangent τ to L. It follows from this a singularity in the distortion $\beta = \nabla\mathbf{u}$. On subjecting an elastic crystal containing dislocations to an appropriate external loading, some of the atoms in the discrete view will move and the dislocation line will seem to move in the opposite direction. The dislocation is thus subjected to a "displacement" and the true mechanician will associate with that motion a "force". This is the *driving force* on the dislocation, not a force in the classical Newtonian sense, since the dislocation line is not a massive object but a mathematical notion (a singularity of the field). This force is called the *Peach–Koehler force* after its creators, the qualification of *creators – not* discoverers – being here justified since this force does not belong to the physical space. However, this "force" can be computed if we know the field solution of the problem at hand outside the singularity. Its original derivation goes as follows.

In the continuum elastic description of a defective crystal the interaction energy between a dislocation D characterized by an elastic displacement field $\mathbf{u}^D$ and an applied stress field σ^A is given by

$$E(D,A) = -\int_S \mathbf{n}\cdot\boldsymbol{\sigma}^A\cdot\mathbf{u}^D\,da = -\int_S \mathbf{u}^D\cdot\mathbf{T}^A\,da, \tag{2}$$

where $\mathbf{T}^A$ is the traction associated with stress $\boldsymbol{\sigma}^A$ and S is the boundary of the region containing the dislocation line L. Equation (2) is the expression of a *potential energy*. The dislocation itself is supposed *not* to produce any traction at S. Thus we may add the vanishing contribution $-\mathbf{n}\cdot\boldsymbol{\sigma}^D\cdot\mathbf{u}^A = -\mathbf{u}^A\cdot\mathbf{T}^D$ to the integrand in (2), and we obtain

$$E(D,A) = -\int_S (\mathbf{u}^D\cdot\mathbf{T}^A - \mathbf{u}^A\cdot\mathbf{T}^D)\,da. \tag{3}$$

According to the statement of the well-known *Rayleigh–Betti reciprocity theorem* of linear (isotropic or anisotropic) elasticity – *cf.* Maugin, 1992, p. 87, eqn. (A.16) – for any closed surface $\partial\Omega$ that does not embrace any body force or singularity, we have

$$\int_{\partial\Omega} (\mathbf{u}^1\cdot\mathbf{T}^2 - \mathbf{u}^2\cdot\mathbf{T}^1)\,da = 0, \tag{4}$$

for two elastic solutions labelled 1 and 2. Accordingly, the vector

$$\mathbf{g} := \mathbf{u}^1\cdot\boldsymbol{\sigma}^2 - \mathbf{u}^2\cdot\boldsymbol{\sigma}^1, \tag{5}$$

is divergence free in that enclosed surface $\partial\Omega$. Applying this reasoning to our dislocation case, the surface S can be replaced by any other surface enclosing the line L as the difference between (3) and the new integral will be zero. In particular, we may choose a surface such as in Figure 1 made up of an open tube enclosing L and the top Σ_+ and bottom Σ_- of an arbitrary discontinuity surface Σ leaning on the dislocation line L. Here Eshelby (1982, p. 211) proposes an ingenious argument that is often bypassed (compared to the treatments of Nabarro, 1967, p. 83, or Kosevich, 1979, p. 63). Suppose, to simplify the reasoning (but this is really immaterial) that the cross-section of the small tube that provides a jacket for the loop, is everywhere along L a circle of radius a. Then divide the tube into a large number of short cylindrical beads threaded by the dislocation (*i.e.*, somewhat like a necklace). While integrating the second contribution in (4) over one bead we may take the comparatively slowly spatially varying field $\mathbf{u}^A$ outside the integral. The stress field $\boldsymbol{\sigma}^D$ of a dislocation is of the order of the inverse, $1/r$, of the distance r from the dislocation (for this, see Nabarro, 1967, or Kosevich, 1979). Thus, the total traction exerted on the curved surface of the cylinder will be of the order of $a^2\,a^{-1} = a$, while each bead is in static

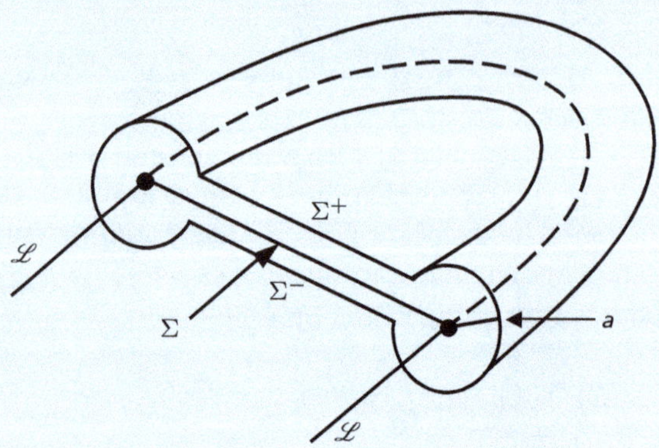

Figure 2.3 *Shrinking of* $\mathscr{S}$ *to a surface englobing* $\mathscr{L}$ *and a singularity surface* Σ.

Figure 1. Shrinking of $\mathcal{S}$ to a surface englobing $\mathcal{L}$ and a singularity Σ (Maugin, 1993).

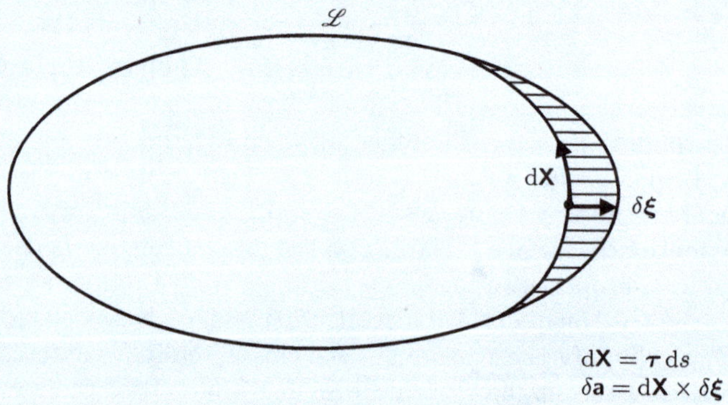

Figure 2.4 *Slight change in the dislocation loop* $\mathscr{L}$ *generating the Peach–Koehler force.*

Figure 2. Slight change in the dislocation loop $\mathcal{L}$ generating the Peach-Köhler force(Maugin, 1993).

equilibrium. Accordingly, we are now sure that the second contribution in (4) will go to zero with a. As to the first contribution, according to a classical evaluation of dislocation theory (see Nabarro, 1967, or Kosevich, 1979), we have $|\mathbf{u}^D| = \mathcal{O}(\ln(r))$ for small r, so that this contribution causes no trouble. As a result, we shall be left in (4) with the integral over the two sides Σ_+ and Σ_- of the cut Σ. The second contribution from (4) is continuous across Σ while $\mathbf{u}^D$ has a discontinuity of a value equal to the Burgers vector $\tilde{\mathbf{b}}$. Therefore, expression (4) is now reduced to

$$E(D,A) = -\tilde{\mathbf{b}} \int_{\Sigma=\Sigma_+} \mathbf{T}^A \,\mathrm{d}a, \tag{6}$$

where Σ_+ is oriented as Σ. To complete the proof, we consider that when the dislocation loop L suffers a little change of shape, say by an infinitesimal vectorial displacement $\delta\xi$ (*cf.* Figure 2), then the corresponding change in (6), δE is just the value of the integral in the right-hand side of (6) taken over the freshly formed portion of the cut. This additional infinitesimal surface element can be written in vector form as

$$\delta\mathbf{a} = \mathrm{d}\mathbf{X} \times \delta\boldsymbol{\xi}, \quad \mathrm{d}\mathbf{X} \equiv \boldsymbol{\tau}\,\mathrm{d}s, \tag{7}$$

where $\boldsymbol{\tau}$ is the unit tangent vector to L.

We can now write δE as

$$\delta E = -\int_L \mathbf{f}^{(PK)} \cdot \delta\boldsymbol{\xi}\,\mathrm{d}s, \tag{8}$$

where the Peach–Koehler "force" acting on the dislocation line L per unit length, due to an applied stress field $\boldsymbol{\sigma}^A$ is given by

$$\mathbf{f}^{(PK)} = (\tilde{\mathbf{b}} \cdot \boldsymbol{\sigma}^A) \times \tau. \tag{9}$$

Like other "configurational" material forces in other paragraphs, the Peach–Koehler force is generated in a *thought experiment* by a displacement of the defect, the dislocation loop L, in material space.

Several remarks are in order. First, there exists a discussion whether it is the whole of $\boldsymbol{\sigma}^A$ or just its deviatoric part that should be involved in the computation of $\mathbf{f}^{(PK)}$ – *cf.* Nabarro, 1967, p. 84. Second, expression (9) is very similar to that of the force acting on a current-carrying wire in applied electromagnetism. Third, a generalization of (9) based on *nonlinear elasticity* was given by Zorski (1981). Finally, a dynamical equation was derived when velocities are involved. This was achieved by Kosevich (1962, 1964) by using an analogy with Lorentz's derivation of the equation of motion of an electron accounting for its self-action (*cf.* Lorentz, 1952). The resulting equation reads

$$\mathbf{f}^{(PK)} = (\mathbf{p} \cdot \tilde{\mathbf{b}})\,(\mathbf{V} \times \tau), \tag{10}$$

where the left-hand side is given by expression (9), $\mathbf{p} = \rho_0\,\mathbf{v}$ is the linear momentum corresponding to a displacement rate due to both the external field and the self-field of the dislocation, and $\mathbf{V}$ is the velocity field of the position of the dislocation loop L. A field-theoretic derivation of (10) was given by Rogula (1977, p. 709). The right-hand side of (10) can be further transformed to give it the appearance of the product of a "mass" (so-called effective mass of a dislocation) and an "acceleration", so that (10) takes a "Newtonian" form – see Kosevich (1979, pp. 104–109) for this. Remarkably, the right-hand side of (10) is of second order jointly in the physical velocity of the material and the dislocation velocity. More recently a connection between $\mathbf{f}^{(PK)}$ and the *Eshelby stress* has been established.

1.2 Kroener's Approach to Incompatibility

In small strains the compatibility equations for integrating a displacement gradient (in general nine independent components) into a displacement vector (three components) are written as the set of six celebrated Saint-Venant conditions (ε_{ajk} is the Levi–Civita completely skewsymmetric alternating symbol)

$$S_{ab} \equiv -\varepsilon_{ajk}\,\varepsilon_{bli}\,\frac{\partial^2 \varepsilon_{ki}}{\partial x^j\,\partial x^l} = 0, \tag{11}$$

where $\varepsilon_{ki} = u_{(k,i)}$ or $\varepsilon = (\nabla \mathbf{u})_S$ in direct notation. Kroener (1958) introduced the source of elastic incompatibility $\boldsymbol{\eta}$ as the negative of the quantity defined in the first part of (11), *i.e.*,

$$S_{ab} + \eta_{ab} = 0. \tag{12}$$

Accordingly, in the absence of elastic incompatibility, we recover the integrability condition given by the second part of (11). We may conceive of (12) as a *balance* equation between the Einstein curvature tensor (this is what is S_{ab} in three dimensions) and Kroener's incompatibility tensor. But this is in fact related to a "defect of closure" and the concept of Burgers vector in a theory of continuous distributions of dislocations.

Indeed, equation (1) can also be written as

$$\oint_S \mathrm{d}x_j\,\beta_{ji} = -\tilde{b}_i, \tag{13}$$

where $\beta_{ji} = u_{i,j}$ is the distortion (displacement gradient). Using Stokes' theorem for a surface A leaning on S, we can rewrite this as

$$\int_A \mathrm{d}a_p\, \varepsilon_{pmi}\, \beta_{ij,m} = -\int_A \mathrm{d}a_p\, \tau_p\, \tilde{b}_j\, \delta(\xi), \tag{14}$$

where $\delta(\xi)$ is Dirac's distribution and ξ is the two-dimensional radius vector taken from the axis of the dislocation in the plane perpendicular to the unit vector τ at the given point. For arbitrary contour S and surface A, (13) yields

$$\varepsilon_{pmi}\, \beta_{ij,m} = -\tau_p\, \tilde{b}_j\, \delta(\xi). \tag{15}$$

As ξ goes to zero this relation becomes meaningless due to the obvious singularity arising in the limit. Passing now to the *continuous theory of dislocations*, (13) is generalized by introducing a tensor of dislocation density $\boldsymbol{\alpha}$, of components α_{ik}, such that

$$\int_A \mathrm{d}a_i\, \alpha_{ik} = \tilde{b}_k, \tag{16}$$

so that (1.15) is replaced by

$$\varepsilon_{ilm}\, \beta_{mk,l} = -\alpha_{ik}. \tag{17}$$

From this there follows immediately a **conservation law**:

$$\operatorname{div} \alpha = \mathbf{0}, \quad \text{or} \quad \frac{\partial \alpha_{ik}}{\partial x_i} = 0. \tag{18}$$

For a single dislocation this would be equivalent to the statement: "The Burgers vector is constant along the dislocation line". Furthermore, applying the operator $\varepsilon_{jpk}\, \partial/\partial x_p$ to (17) and symmetrizing with respect to i and j we obtain (12) in the form

$$\eta_{ij} = \frac{1}{2}\left(\varepsilon_{ipl}\, \frac{\partial \alpha_{jl}}{\partial x_p} + \varepsilon_{jpl}\, \frac{\partial \alpha_{il}}{\partial x_p}\right). \tag{19}$$

1.3 Passing to the Material Framework

For reasons to become clear later on we are interested in formulating some of the above-given elements in the material framework. We remind the reader of a few notations. Here a nondefective material body is a simply connected region B of a three-dimensional Euclidean manifold M^3, or simply M, called the *material manifold.* The elements of this manifold are so-called material points X. In (possibly but not necessarily) curvilinear coordinates X^K, $K = 1, 2, 3$, this point is simply represented by the boldface letter $\mathbf{X}$. To each point X on M is attached a density, the matter density ρ_0, which is the density of matter at the reference configuration K_R. This may be a function of X, as is the case in materially inhomogeneous bodies, and perhaps, but rarely, a function of Newtonian time t itself. The latter scalar parameter belongs to an ordered one-dimensional continuum, the positive real line R which presents no defects. *I.e.*, time itself cannot be "fractured". With this we have introduced the *basic space-time parametrization* of the classical mechanics of deformable solids, the set $(\mathbf{X}, t)$. The motion (or deformation) of the material body B of M is the time ordered sequence of the positions, sometimes called *placements*, occupied by the point X in Euclidean physical space E^3, the arena of classical phenomenological physics. This is expressed by the sufficiently (as needed) regular space-time parametrized mapping

$$\mathbf{x} = \chi(\mathbf{X}, t). \tag{20}$$

This is often (but not necessarily) reported to a Cartesian system of coordinates x^i, $i = 1, 2, 3$.

Note that physical space here is always Euclidean, since we work in Newtonian physics, while M could be non-Euclidean (as would be the case in a *defective* material body). The set of geometrical points $\mathbf{x}(B, t$ fixed) constitutes the *actual* or *current configuration* K_t of the body at time t. Usually, an origin of time, say t_0, is chosen such that $t_0 < t$, and (20) then reads $\mathbf{x}_0 = \chi(\mathbf{X}, t_0)$. When this one and (20) are sufficiently smooth, and in particular, invertible, we can rewrite (20) as

$$\mathbf{x} = \chi\big(\chi^{-1}(\mathbf{x}_0, t_0), t\big) = \bar{\chi}(\mathbf{x}_0, t; t_0) = \tilde{\chi}(\mathbf{x}_0, t). \tag{21}$$

This representation of the direct motion is called Lagrangian, the $\mathbf{x}_0$ being Lagrangian coordinates. The configuration $K_0 = K_t(t = t_0)$ of the body, the *initial configuration* at $t = t_0$, belongs to the sequence of "actual" configurations. This is the motion description preferred in fluid mechanics. Many authors identify the two representations (20) and (21) by identifying $\mathbf{X}_0$ and $\mathbf{x}_0$. But the motion representation (20) is somewhat more abstract and is essentially due to Gabrio Piola (1848) in a paper of far reaching insight. Indeed, the consideration of the material configuration K_R that corresponds to an ideally unstrained and unloaded configuration, corresponding usually to a minimizer of the energy (*cf.* Lardner, 1974), is essential in studying the material symmetry of solid bodies and defining material properties in a general manner. While (20) is called *the direct motion mapping* $K_R \to K_t$ *at* t, in the same smoothness conditions as above, the *inverse* motion is given by

$$\mathbf{X} = \chi^{-1}(\mathbf{x}, t). \tag{22}$$

The direct $\mathbf{F}$ and inverse $\mathbf{F}^{-1}$ motion **gradients** are defined thus

$$\mathbf{F} := \nabla_R \chi \equiv \frac{\partial \chi}{\partial \mathbf{X}}, \quad \mathbf{F}^{-1} := \nabla_R \chi^{-1} \equiv \frac{\partial \chi^{-1}}{\partial \mathbf{x}}. \tag{23}$$

It is immediately checked that

$$\mathbf{F} \cdot \mathbf{F}^{-1} = \mathbf{1}, \quad \mathbf{F}^{-1} \cdot \mathbf{F} = \mathbf{1}_R, \tag{24}$$

where the symbols $\mathbf{1}$ and $\mathbf{1}_R$ represent the unit dyadics in E^3 and on M, respectively. It must be emphasized that both $\mathbf{F}$ and $\mathbf{F}^{-1}$ are *not* tensors in the traditional sense because they are geometric objects defined on two different manifolds simultaneously. In a picturesque language, we can say that they have one foot in K_t and another in K_R. Such objects are so-called *two-point tensor fields.* They have components

$$\mathbf{F} = \left\{ F^i_K \equiv F_{iK} \right\}, \quad \mathbf{F}^{-1} = \left\{ (\mathbf{F}^{-1})^K_i = (\mathbf{F}^{-1})^{Ki} \right\}, \tag{25}$$

where the upward or downward position of lower Latin indices is irrelevant by virtue of the Cartesian representation chosen in K_t. Speaking of an *a priori* symmetry of $\mathbf{F}$ and $\mathbf{F}^{-1}$ is a mathematical nonsense since one must specify with respect to what metric is tensorial symmetry defined. The Jacobian determinant of $\mathbf{F}$ is noted

$$J_F = \det \mathbf{F}. \tag{26}$$

Of course, $J_{F^{-1}} = \det \mathbf{F}^{-1} = (J_F)^{-1}$. If ρ_0, the *matter density* at $\mathbf{X}$, does not depend on time, the actual mass density ρ is related to ρ_0 by the change of volume between configurations, *i.e.*,

$$\rho(\mathbf{x}, t) = \rho_0(\mathbf{X})\, J_F^{-1}. \tag{27}$$

Since densities are always positive, only deformation mappings such that J_F *is positive and never vanishes*, are considered. In physical terms this signifies the *impenetrability* of matter.

Deformation measures are typical "metrics" (truly symmetric tensors). Some of them can be defined thus (here the superscript T denotes the operation of *transposition*, δ's are Kronecker symbols):

$$\mathbf{C}(\mathbf{X}, t) := \mathbf{F}^T \mathbf{F} = \left\{ C_{KL} = F^i_K\, \delta_{ij}\, F^j_L \right\}, \tag{28}$$

and

$$\mathbf{C}^{-1} := (\mathbf{F}^{-1})\, (\mathbf{F}^{-1})^T = \left\{ (\mathbf{C}^{-1})^{KL} = (\mathbf{F}^{-1})^K_i\, \delta^{ij}\, (\mathbf{F}^{-1})^L_j \right\}. \tag{29}$$

These are defined over M, and are called the *Cauchy–Green* finite (material) strain tensor and the *Piola* finite (material) strain tensor, respectively. They are inverse to one another. These two measures are *absolute* ones. They are not compared to an undeformed metric. A natural *relative* strain measure is given by

$$\mathbf{E} := \frac{1}{2}\left(\mathbf{C} - \mathbf{1}_R\right). \tag{30}$$

A more general definition than this allows one to introduce a series of material strain measures such that, $m = \ldots, -2, -1, +1, +2, \ldots,$

$$\mathbf{E}^{(m)} := \frac{1}{m}\left(\mathbf{U}^m - \mathbf{1}_R\right). \tag{31}$$

Here $\mathbf{U}$ is the *right stretch (material) tensor* introduced in the polar decomposition of any non singular $\mathbf{F}$ – non vanishing $\det \mathbf{F}$ – according to a theorem due to Cauchy:

$$\mathbf{F} = \mathbf{R}\,\mathbf{U} = \mathbf{V}\,\mathbf{R}, \tag{32}$$

where $\mathbf{V}$ is the left stretch (spatial) tensor, and $\mathbf{R}$ is a rotation that belong to $SO(3)$, *i.e.* $\mathbf{R}^T = \mathbf{R}^{-1}$, $\det \mathbf{R} = +1$. Both $\mathbf{U}$ and $\mathbf{V}$ are positive definite. We immediately check that $\mathbf{C} = \mathbf{U}^2$, $J_F = \det \mathbf{U}$, and thus from (31), in particular,

$$\mathbf{E}^{(2)} \equiv \mathbf{E}, \quad \mathbf{E}^{(-2)} = \frac{1}{2}\left(\mathbf{1}_R - \mathbf{C}^{-1}\right). \tag{33}$$

Note that finding $\mathbf{U}$ from $\mathbf{C}$ is an awkward operation (finding the square root of a tensor).

The *displacement field* is the field $\mathbf{u}(\mathbf{X}, t)$ or $\bar{\mathbf{u}}(\mathbf{x}, t)$ defined by

$$\mathbf{x} = \mathbf{X} + \mathbf{u}(\mathbf{X}, t) \quad \text{or } \mathbf{x} = \mathbf{X} + \bar{\mathbf{u}}(\mathbf{x}, t), \tag{34}$$

On taking the material gradient ∇_R of the first of these and the spatial gradient ∇ of the second we obtain with (23),

$$\mathbf{F} = \mathbf{1} + \mathbf{H}, \ \ \mathbf{H} \equiv \nabla_R \mathbf{u} \quad ; \quad \mathbf{F}^{-1} = \mathbf{1} - \mathbf{h}, \ \ \mathbf{h} \equiv \nabla \bar{\mathbf{u}}. \tag{35}$$

It follows that we have the following *exact* formulas:

$$\mathbf{E} = \frac{1}{2}\left(\mathbf{H} + \mathbf{H}^T + \mathbf{H}^T\,\mathbf{H}\right), \quad \mathbf{e} = \frac{1}{2}\left(\mathbf{h} + \mathbf{h}^T - \mathbf{h}^T\,\mathbf{h}\right). \tag{36}$$

In *small strain theory* for which $\mathbf{H}$ and $\mathbf{h}$ are small in the sense that $|\mathbf{H}| \equiv (\mathit{trace}\,\mathbf{H}^T\mathbf{H})^{1/2}$ or $|\mathbf{h}| \equiv (\mathit{trace}\,\mathbf{h}^T\,\mathbf{h})^{1/2}$ is considered as an infinitesimal quantity of the first order, neglecting terms of second order in the "small" displacement gradients, we obtain the following approximation

$$\begin{aligned} \mathbf{E} = \mathbf{e} = \varepsilon = (\nabla \mathbf{u})_S \equiv \frac{1}{2}\left(\nabla \mathbf{u} + (\nabla \mathbf{u})^T\right), \\ \mathbf{R} - \mathbf{1} = \omega = (\nabla \mathbf{u})_A \equiv \frac{1}{2}\left(\nabla \mathbf{u} - (\nabla \mathbf{u})^T\right), \end{aligned} \tag{37}$$

where the subscripts S and A denote the operations of symmetrization and skew (anti)- symmetrization, respectively. The true tensors ε and ω are called the infinitesimal strain and rotation. Furthermore, to the same degree of approximation (tr = trace),

$$J_F \cong 1 + \operatorname{tr} \varepsilon, \quad J_F^{-1} \cong 1 - \operatorname{tr} \varepsilon. \tag{38}$$

Then we are back to the notion of **integrability** of the displacementh field. **Question**: Find a unique displacement $\mathbf{u}$ corresponding to a given deformed metric $\mathbf{C}$. Of course there must exist six so-called compatibility conditions in order to extract the three components of $\mathbf{u}$ from the nine components of $\mathbf{C}$ or $\mathbf{E}$. These were originally derived in the 19th century by Navier and Saint–Venant in small strain theory. For finite strain theory, it is noticed that in the absence of defects, the material manifold is *flat* (in the sense of Riemannian differential geometry) and must remain so in the course of the deformation. Accordingly, the Riemann curvature associated with $\mathbf{C}$ or $\mathbf{E}$ must always vanish (see, for instance, Maugin, 1993, pp. 54–57, for these developments). In three-dimensional space which is our concern, the Riemann curvature tensor reduces to the so-called *Einstein tensor.* For small strains this tensor, in term of the deformed metric ε, is given by

$$S_{ab} = -\varepsilon_{ajk}\, \varepsilon_{bli}\, \varepsilon_{ki,jl}, \tag{39}$$

which is none other than $(11)_1$.

Among the important operations involving deformation, we have those of convection, of a tensorial object, called *pull back* or *push forward* depending on whether the operation carries a tensorial object from the actual configuration to the reference one, or from the latter to the former. They are tensorial transformations effected with help of the motion mapping itself since these operations are conducted between two different manifolds. Historically first, but also endowed with a definite relevance in continuum mechanics, is the convection operation introduced by Piola, the *Piola transformation.* Let $\mathbf{A}$ be a vector field in the actual configuration. Then the material contravector defined by

$$\overline{\mathbf{A}} = J_F\, \mathbf{F}^{-1}\, \mathbf{A} = \left\{ \overline{A}^K = J_F (\mathbf{F}^{-1})^{Ki} A_i \right\}, \tag{40}$$

is the *Piola transform* of **A**. Conversely,

$$\mathbf{A} = J_F^{-1}\,\mathbf{F}\,\overline{\mathbf{A}} = \left\{ A = J_F^{-1}\, F^i_{\ K}\, \overline{A}^K \right\}. \tag{41}$$

We easily verify the following identities

$$\nabla_R (J_F\,\mathbf{F}^{-1}) = \mathbf{0}, \quad \nabla (J_F^{-1}\,\mathbf{F}) = \mathbf{0}, \tag{42}$$

from which there follows that

$$\nabla_R \cdot \overline{\mathbf{A}} = J_F\,\mathbf{F} \cdot (\nabla_R \mathbf{A}) = J_F\,\nabla \cdot \mathbf{A}. \tag{43}$$

This reminds us of the formula for the change of elementary volume, dv and dV, between the actual and reference configurations:

$$dv = J_F\, dV, \tag{44}$$

so that

$$(\nabla_R \cdot \overline{\mathbf{A}})\, dV = (\nabla \cdot \mathbf{A})\, dv. \tag{45}$$

By the same token it is salient to remind the reader of the so-called Nanson's formula for the change between oriented surface elements $\mathbf{n}\, ds$ and $\mathbf{N}\, dS$ of the same surface with respective unit normals $\mathbf{n}$ and $\mathbf{N}$ in the actual and reference configurations:

$$\mathbf{n}\, ds = J_F\, \mathbf{N} \cdot \mathbf{F}^{-1}\, dS, \quad \mathbf{N}\, dS = J_F^{-1}\, \mathbf{n} \cdot \mathbf{F}\, ds. \tag{46}$$

If $\boldsymbol{\sigma}$ is a spatial tensor defined per unit area in K_t, then we readily check that

$$\mathbf{n} \cdot \boldsymbol{\sigma}\, ds = \mathbf{N} \cdot \mathbf{T}\, dS, \tag{47}$$

where the two-point tensor field $\mathbf{T}$ is such that

$$\mathbf{T} = J_F\, \mathbf{F}^{-1}\, \boldsymbol{\sigma} = \left\{ T_i^K = J_F\, (\mathbf{F}^{-1})^{Kj}\, \sigma_{ji} \right\}, \quad \boldsymbol{\sigma} = J_F^{-1}\, \mathbf{F}\, \mathbf{T}. \tag{48}$$

Note that eq. (47) still has a foot in the actual configuration (physical space). The object $\mathbf{T}$ is the Piola transformed of $\boldsymbol{\sigma}$, but on the first index only. Of course, if one takes the material divergence to the left, noted div_R, of $\mathbf{T}$, one gets immediately (compare to (45))

$$\mathrm{div}_R\, \mathbf{T} = J_F\, \mathrm{div}\, \boldsymbol{\sigma} = J_F\, \nabla \cdot \boldsymbol{\sigma}, \tag{49}$$

where the symbol ∇ denotes the spatial divergence taken at the left of a tensorial object. Equation (49), just like (46), still has a foot in the actual configuration. Now we can revisit some of the notions introduced in Paragraphs 1.1 and 1.2.

1.4 Local Rearrangement of Matter

For the sake of illustration we consider the case of finite-strain pure elasticity for which the equilibrium equations in the *Cauchy format* in the *actual configuration* read

$$\operatorname{div} \boldsymbol{\sigma} = \mathbf{0} \text{ in } B, \quad \mathbf{n} \cdot \boldsymbol{\sigma} = \mathbf{T}^d \text{ at } \partial B, \tag{50}$$

if $\boldsymbol{\sigma}$ is the Cauchy stress tensor. Applying the operations (40) through (49) we also have the so-called *Piola–Kirchhoff* format of these equations

$$\operatorname{div}_R \mathbf{T} = 0 \text{ in } B_R, \quad \mathbf{N} \cdot \mathbf{T} = \overline{\mathbf{T}}^d \text{ at } \partial B_R, \tag{51}$$

where $\mathbf{T}$, a two-point tensor field, is called the first *Piola–Kirchhoff stress.* The two equations in (51) still have components in the actual configuration and, therefore, are *not completely material.* In pure, possibly anisotropic and inhomogeneous, elasticity in finite-strains, $\mathbf{T}$ is derived from an elastic energy per unit of the reference configuration $W = \overline{W}(\mathbf{F}; \mathbf{X})$ by

$$\mathbf{T} = \frac{\partial \overline{W}}{\partial \mathbf{F}}. \tag{52}$$

In the case when $\mathbf{T}$ is function of $\mathbf{F}$ and $\mathbf{F}$ *only*, where $\mathbf{F}$ is a true gradient, (52) represents the essence of *pure homogeneous elasticity* – a paradigmatic case as we shall see herein after – with

$$\mathbf{T} = \overline{\mathbf{T}}(\mathbf{F}) = \frac{\partial W(\mathbf{F})}{\partial \mathbf{F}}. \tag{53}$$

This plays the role of a *standard* with which any other situation in a solid is compared (case of more complicated functional dependences of W). Indeed, as soon as W becomes an explicit function of additional arguments, we are no longer in this ideal framework. This happens whether the additional argument is another field variable such as temperature in thermoelasticity, or electric polarization or magnetization in electro-magneto-elasticity (*cf.* Maugin, 1988), or else any variables such as so-called internal variables of state supposed to account for the hidden complexity of microscopic processes which have a macroscopic manifestation in the form of thermodynamic irreversibility (*i.e.*, dissipation; *cf.* Maugin, 1999a). Another frequent possibility (such as above) is that the energy W depends explicitly on the material particle X, in which case $W = \overline{W}(\mathbf{F}; \mathbf{X})$ and the elastic material is said to be *materially inhomogeneous.* We have called *material force of inhomogeneity* the material co-vector defined by

$$\mathbf{f}^{inh} := - \left. \frac{\partial \overline{W}}{\partial \mathbf{X}} \right|_{expl}, \tag{54}$$

if $\overline{W}$ is a sufficiently smooth function of $\mathbf{X}$, and where the subscript *expl* means that the material gradient is taken at fixed field (here $\mathbf{F}$). In composite materials where inhomogeneities manifest abruptly by jumps in material properties, (54) must be replaced by a distributional (generalized function) definition. The *force* $\mathbf{f}^{inh}$ belongs in the world of **material forces** (*cf.* Maugin, 1993, 1995) since it is a co-vector on the material manifold. It is a *directional indicator* of the changes of elastic properties as it is oriented opposite to the direct explicit gradient of W.

Now we can exploit the *thought experiment* of Epstein and Maugin (1990a,b). To that purpose, imagine that at *each material point* $\mathbf{X}$ we can give to the material deformation energy the appearance of that of a pure homogeneous elastic body (dependence on one deformation only and nothing else) by applying the appropriate *local* (at $\mathbf{X}$) *change*, here noted $\mathbf{K}$, of reference configuration, a so-called *uniformity map* in the language of Noll (1967) and Wang (1967). We consider this along with the concomitant change of volume and write

$$W = \overline{W}(\mathbf{F};\mathbf{X}) = J_K^{-1}\,W(\mathbf{F}\,\mathbf{K}(\mathbf{X})) = \widetilde{W}(\mathbf{F},\mathbf{K}). \tag{55}$$

We can compute the partial derivatives of the last function by obtaining thus

$$\mathbf{T} = \frac{\partial \overline{W}(\mathbf{F};\mathbf{X})}{\partial \mathbf{F}}, \quad \mathbf{b} = -\frac{\partial \widetilde{W}(\mathbf{F},\mathbf{K})}{\partial \mathbf{K}}\,\mathbf{K}^T = W\,\mathbf{1}_R - \mathbf{T}\,\mathbf{F}. \tag{56}$$

Thus there exits a relationship between the notion of material inhomogeneity and that of *configurational or Eshelby material (here quasi-static) stress tensor* $\mathbf{b}$. This is made more visible by applying the definition (54):

$$\begin{aligned}\mathbf{f}^{inh} &= -\left.\frac{\partial \overline{W}(\mathbf{F};\mathbf{K})}{\partial \mathbf{X}}\right|_{expl} = -\frac{\partial \widetilde{W}(\mathbf{F},\mathbf{K})}{\partial \mathbf{K}}\cdot\frac{\partial \mathbf{K}}{\partial \mathbf{X}} \\ &= \mathbf{b}\cdot\mathbf{K}^{-T}\cdot\frac{\partial \mathbf{K}}{\partial \mathbf{X}} = \mathbf{b}\cdot\left(\mathbf{K}^{-T}\cdot(\nabla_R\mathbf{K})^T\right)\end{aligned} \tag{57}$$

On the other hand, if we compute the material divergence of $\mathbf{b}$ in the case of quasi-statics in the absence of body force, for which the equilibrium at $\mathbf{X}$ is simply given by $(51)_1$, we have

$$\begin{aligned}\mathrm{div}_R\,\mathbf{b} &= \nabla_R W - (\mathrm{div}_R\,\mathbf{T})\cdot\mathbf{F} - \mathbf{T}\cdot(\nabla_R\mathbf{F})^T \\ &= \left(\frac{\partial W}{\partial \mathbf{F}} - \mathbf{T}\right)\cdot(\nabla_R\mathbf{F})^T + \left.\frac{\partial W}{\partial \mathbf{X}}\right|_{expl},\end{aligned} \tag{58}$$

or, on account of $(56)_1$ and (57),

$$\mathrm{div}_R\,\mathbf{b} = -\mathbf{f}^{inh}. \tag{59}$$

Here the material force of inhomogeneity is deduced from (or balanced by) the material divergence of the configurational stress. It is justified to call **configurational forces** these forces that are deduced through an operation acting on the configurational stress, whether by differentiation or integration (*e.g.*, over a material surface, along a material contour in 2D). If we combine the results of (57) and (59), we also obtain an equation for $\mathbf{b}$ which involves the local transformation $\mathbf{K}$ in a source term, that is,

$$\mathrm{div}_R\, \mathbf{b} + \mathbf{b} \cdot \Gamma = \mathbf{0}, \tag{60}$$

where we have defined a *material connection* $\Gamma(\mathbf{K})$ by

$$\Gamma(\mathbf{K}) = (\nabla_R \mathbf{K}^{-1}) \cdot \mathbf{K} = -\mathbf{K}^{-1} \cdot (\nabla_R \mathbf{K})^T. \tag{61}$$

The result (60) is due to Epstein and Maugin (1990a,b). If $\mathbf{K}$ is the same for all points $\mathbf{X}$, then $\nabla_R \mathbf{K} = \mathbf{0}$, and (60) reduces to the strict conservation law

$$\mathrm{div}_R\, \mathbf{b} = \mathbf{0}, \tag{62}$$

in the case (we remind the reader) of the absence of body force and neglect of inertia (quasi-statics). Otherwise, we can write

$$\mathbf{f}^{inh} = \mathbf{b} \cdot \Gamma(\mathbf{K}), \tag{63}$$

Then the above-reported intellectual construct means that the operation carried out (introduction of $\mathbf{K}$) brings the neighborhood of each material point $\mathbf{X}$ into a *prototypical situation* of the pure elastic type which allows one to compare the response of different points. Since this is *point-like*, the operation will not result in an overall smooth manifold, but in a collection of non-fitting neighborhoods or infinitesimal chunks of materials, and $\mathbf{K}$ will not, accordingly, be itself a gradient. It may at most be a *Pfaffian form*. Of course, if $\mathbf{K}$ is not integrable, so is the case of $\bar{\mathbf{F}} = \mathbf{F}\mathbf{K}$. With eqns. (60)–(61) we enter the *geometrization of continuum mechanics* that was started in the mid 1950s by scientists such as Kondo (1952), Kroener (1958), Noll (1967) and Wang (1967) among others. This was thoroughly reviewed in Maugin (2003a,b). This ambitious program belongs in the Hilbertian–Einsteinian tradition of geometrization of physics. Of course, in this line of thought, the writing (60) does not fulfill the whole program because the two sides of (60) contain contributions of a different nature, the non-Riemannian geometry being contained only in the right-hand side through the notion of connection based on $\mathbf{K}$. However, there is progress here compared to other approaches in the sense that the whole equation (60) is written on the material manifold, which indeed is the arena of what may happen to the material in its intimacy (*e.g.*, defects). Let us examine the source term (60) more thoroughly and discuss more deeply the geometrical connotations.

First of all, in components, (60) reads

$$b^{J}_{.I,J} + b^{K}_{.J}(\mathbf{K}^{-1})^{\alpha}_{.K} K^{J}_{.\alpha,I} = 0. \tag{64}$$

This is a first-order differential equation which is identically satisfied by the tensor **b** associated with a solution of an elastic boundary-value problem. But generally speaking the **K** transformation creates a so-called *distant parallelism* (called in the past *absolute parallelism or Fernparallelismus*), and thus a (generally non-metric) *connection* (*cf.* Choquet–Bruhat, 1968 or Lichnerowicz, 1976) as defined by (61). In words and following Elie Cartan (probably the main contributor to this field of geometry), distant parallelism in a Riemannian space is materialized by the fact that, if we attach to each point in space a reference frame, and this in some arbitrary manner, then it is sufficient to agree that two vectors of any origin, $\boldsymbol{A}$ and $\boldsymbol{B}$, are parallel or equipollent if they have equal projections (components) on the rectangular frames at $\boldsymbol{A}$ and $\boldsymbol{B}$. Then the reference frames themselves are parallel to each other in that sense! In this process it is clear that the metric of the relevant space and the parallelism are dependent on one another, but for each given metric there is an infinity of distant parallelisms compatible with that metric and, conversely, given a distant parallelism there exists an infinity of metrics compatible with it. Of course, in a Riemannian space, the notion of *Riemannian curvature* plays a fundamental role: it is related to the deviation undergone by a vector when the latter is transported in a *parallel* manner around a closed circuit. This notion disappears in the condition of distant parallelism, *i.e.*, a Riemannian space with distant parallelism has *no* curvature. Still, something distinguishes it from a Euclidean space, and that is *torsion.* As a consequence all the intrinsic geometric properties which characterize a Riemannian space with distant parallelism derive from its torsion (E. Cartan, in some uncontrolled enthusiasm, once said that "*if physics can be geometrized at all, then all physical laws must be expressible in terms of partial differential equations governing the torsion of the relevant space*" (Cartan, 1931, translation from the French by GAM)). This was very far-sighted in so far as unified gravitational theories are concerned. But in a way it also applies to continuum mechanics on the material manifold, our present concern.

To be more specific, define a *moving crystallographic frame* over the material body B by

$$\mathbf{E}_{\alpha} = K^{K}_{.\alpha} \frac{\partial}{\partial X^{K}}. \tag{65}$$

Two vectors at different points in the reference configuration are, by definition, **K**-parallel if they have the same components in their respective crystallographic bases. This leads to the introduction of a covariant derivative (here denoted ";") of a vector **V** of components V^I:

$$V^I_{.;J} \frac{\partial}{\partial X^I} \otimes dX^J = V^\alpha_{.,J} \mathbf{E}_\alpha \otimes dX^J = \left((\mathbf{K}^{-1})^\alpha_{.I} V^I\right)_{,J} K^K_{.\alpha} \frac{\partial}{\partial X^K} \otimes dX^J, \tag{66}$$

or

$$V^I_{.;J} = V^I_{.,J} + \Gamma^I_{KJ} V^K, \tag{67}$$

with a connection $\mathbf{\Gamma}$ defined by (61).

Similarly, for a *one-form* (co-vector) **W**, working in the dual basis, we classically obtain

$$W_{I;J} = W_{I,J} - \Gamma^K_{IJ} W_K, \tag{68}$$

while for a mixed tensor **b**, we will have

$$b^J_{.I;K} = b^J_{.I,K} - \Gamma^L_{IK} b^J_{.L} + \Gamma^J_{LK} b^L_{.I}. \tag{69}$$

The connection symbol $\mathbf{\Gamma}$ is *not necessarily* symmetric (as is, in contrast, the Christoffell symbol based on a metric). It is the skew part of this connection which defines the *torsion* $\widetilde{\mathbf{T}}$ by

$$\widetilde{T}^I_{JK} := \Gamma^I_{JK} - \Gamma^I_{KJ}. \tag{70}$$

This allows one (Epstein and Maugin, 1990a) to show that (64) can be rewritten in the following remarkable form

$$\bar{b}^J_{.I;J} = \bar{b}^J_{.K} \widetilde{T}^K_{JI} + \bar{b}^J_{.I} \widetilde{T}^K_{JK}, \quad \bar{\mathbf{b}} \equiv J_K \, \mathbf{b}, \tag{71}$$

or

$$\mathrm{div}_{\mathbf{K}} \bar{\mathbf{b}} = B(\bar{\mathbf{b}}, \widetilde{\mathbf{T}}), \tag{72}$$

where $\mathrm{div}_{\mathbf{K}}$ denotes the covariant divergence based on **K**, and $B(.,.)$ denotes the specific bi-linear form introduced in the first of (71). The formula for the divergence of a determinant has been employed in writing this on account of the introduction of the weighted Eshelby stress $\bar{\mathbf{b}}$. Particular cases of (72) are easily discussed: (i) if the reference configuration itself is homogeneous, then the **K**-parallelism reduces to the Euclidean one and we simply have $\mathrm{div}_R \, \mathbf{b} = \mathbf{0}$. This corresponds to the absence of physical (material) inhomogeneity and of any configurational inhomogeneity (*i.e.*, artificial inhomogeneity due to the special choice of a reference configuration); (ii) if the body is materially homogeneous but the reference configuration is arbitrary, then we have the *material conservation law* $\mathrm{div}_{\mathbf{K}} \bar{\mathbf{b}} = \mathbf{0}$. We may say that an observer adapted to the crystallographic frame sees no inhomogeneity (somewhat like a geodesic observer does not feel any gravitational field in general relativity as we need two neighboring observers (notion of *geodesic separation*) to place that field in evidence); (iii) the general case is represented by (72) where not even an adapted observer can remove the inhomogeneity as the material is "*intrinsically dislocated*" (see next section). In any case, we say that the so-called *uniformity map* (in the language of Noll (1967) and Wang (1967)), $\mathbf{K}$ helps us define a local *prototype reference crystal* at each material point $\mathbf{X}$ on the material manifold.

Remark 1.1. In standard treatises on continuum mechanics it is recalled that *material symmetry* consists in studying the possible isomorphisms of a particle onto itself which leave invariant the response of the material (*cf.* Noll). This materializes in changes of the reference configuration which belong to a certain group (a *crystallographic group* in general as studied in the Appendix of Eringen and Maugin (1990), the group of *orthogonal transformations* in the case of *isotropy*). For instance, for an elastic body in large strains, we would write the invariance

$$W = \overline{W}(\mathbf{F}) = \overline{W}(\mathbf{F\,P}), \quad \mathbf{P} \in SO(3), \quad \det \mathbf{P} = +1. \tag{73}$$

The first of these looks somewhat like $(55)_2$, but for the determinant factor which is here equal to one. What occurs here (Epstein and Maugin, 1990b) is as follows. Although one often concentrates on a discrete symmetry group, here we may consider that the material is a solid with a continuous symmetry group. In a uniform body B the symmetry groups of W at different points, although generally different, are all conjugate, *via* the $\mathbf{K}$-mappings, with the symmetry group of the *reference crystal.* Let $\mathbf{G}_X(\lambda)$ be a one-parameter (λ real) subgroup of the symmetry group of the energy W at X with $\mathbf{G}_X(0) = \mathbf{I}$, the identity. From the material symmetry condition we have

$$\overline{W}(\mathbf{F};\mathbf{X}) = \overline{W}(\mathbf{F\,G_X}(\lambda), \mathbf{X}), \tag{74}$$

that is valid for all real λ's and all non-singular deformation gradients $\mathbf{F}$. The remarkable property pointed out by Epstein and Maugin (1990b) is that the Eshelby stress does not produce any work in any small change of reference which belongs, at each material point X, to the *Lie algebra* of the symmetry group, *i.e.*,

$$\operatorname{tr}\left(\mathbf{b} \cdot \left.\frac{\mathrm{d}\mathbf{G_X}(\lambda)}{\mathrm{d}\lambda}\right|_{\lambda=0}\right) = 0. \tag{75}$$

[*i.e.*, (73) is not a local rearrangement of matter that costs energy]. When applied to the special case of isotropy, this reasoning says that for this case where point-wise symmetry groups are all conjugate *via* $\mathbf{K}(\mathbf{X})$ with the proper orthogonal group $SO(3)$, the Eshelby stress is symmetric with respect to a Riemannian metric induced by the uniformity map and defined by

$$\mathbf{G}_{\mathbf{K}} := \mathbf{K}^{-T} \cdot \mathbf{K}^{-1}. \tag{76}$$

In a stress-free configuration, this reduces to ordinary Euclidean symmetry. This should be contrasted with the $\mathbf{C}$-symmetry satisfied by $\mathbf{b}$ when the Cauchy stress is symmetric (Epstein and Maugin, 1990a,b):

$$\mathbf{C} \cdot \mathbf{b} = (\mathbf{C} \cdot \mathbf{b})^{T} = \mathbf{b}^{T} \cdot \mathbf{C}. \tag{77}$$

The above reasoning involves the theory of so-called $\mathbf{G}$-structures admitting $\mathbf{G}_{\mathbf{X}}(\lambda)$ subgroups as discussed by Elzanowski *et al.* (1990).

1.5 Back to the Continuous Distribution of Dislocations

Starting with papers by Kroener (1958, 1960) and Bilby (1968), it has been proposed that the torsion of the material connection be a measure of the *dislocation density* – a special type of material inhomogeneities – in an elastic continuum presenting a continuous distribution of dislocations. The original approach of Kroener considered only small strains (see Paragraph 1.2) but now we can easily reformulate his reasoning in the finite-strain framework. To that effect, we shall note, α^{QK}, $Q, K = 1, 2, 3$, the local material components of the dislocation density tensor. The natural way of comparing the defect of closure of a Burgers' circuit given in the material framework is to write the displacement jump in the form (compare to (16))

$$\Delta \mathbf{u} = \oint_C \mathrm{d}\mathbf{X} = \oint_C (\mathbf{K}^{-1})^{\alpha}_{.K}\, \mathbf{E}_{\alpha}\, \mathrm{d}X^{K}, \tag{78}$$

and the relation to dislocation density is written as

$$\Delta u^{Q} = \int_S \alpha^{QK}\, \mathrm{d}A_{K} = \int_S \alpha^{QK}\, N_{K}\, \mathrm{d}A, \tag{79}$$

where the surface S leans on the contour C. On using Stokes' theorem, this yields

$$\Delta \mathbf{u} = \int_S \varepsilon^{KLM} (\mathbf{K}^{-1})^{\alpha}_{.K,L}\, \mathbf{E}_{\alpha}\, \mathrm{d}A_{K} = \int_S \varepsilon^{KLM} (\mathbf{K}^{-1})^{\alpha}_{.K,L}\, K^{J}_{.\alpha}\, \mathbf{G}_{J}\, \mathrm{d}A_{M} \tag{80}$$

since $\mathbf{E}_{\alpha} = K^{J}_{.\alpha}\, \mathbf{G}_{J}$, where the $\mathbf{G}$'s are a local basis on the material manifold. On comparing (79) and (81), we see that the following relation holds:

$$\alpha^{QM} = \varepsilon^{KLM}\, \Gamma^{Q}_{KL} = \varepsilon^{KLM}\, \widetilde{T}^{Q}_{KL}. \tag{81}$$

As a result, either $\boldsymbol{\alpha}$ or $\widetilde{\mathbf{T}}$ can equally characterize the dislocation density. We may conclude this paragraph by noting that the right-hand side of the first of (71) can be expanded on account of the definition of the quasi-static Eshelby stress while accounting for the skewsymmetry of the Levi–Civita permutation symbol and the symmetry of α (without loss of generality). We thus obtain the Peach–Koehler force in finite strains as

$$f_I^{(PK)} = J_{\mathbf{K}} \, F^i_{.Q} \, T^J_{.i} \, \varepsilon_{IJM} \, \alpha^{QM} = J_K \, \varepsilon_{IJM} \, M^J_{.Q} \, \alpha^{QM}, \tag{82}$$

where $\mathbf{M}$ is the so-called *Mandel (material) stress* such that

$$\mathbf{M} = \mathbf{T} \cdot \mathbf{F} = \mathbf{S} \cdot \mathbf{C} = W \, \mathbf{1}_R - \mathbf{b}, \tag{83}$$

where $\mathbf{S}$ is the second (symmetric, material, contravariant) Piola–Kirchhoff stress. The Mandel stress plays an important role in plasticity. It is the *driving force* behind volume-preserving plastic deformations (or in defining a material measure of *resolved shear stress*). In small strains this reduces to

$$f_i^{(PK)} = \varepsilon_{ijp} \, \sigma_{jk} \, \alpha_{kp}. \tag{84}$$

This coincides with the usual Peach–Koehler force (9) for a unique dislocation of Burgers' vector $\tilde{b}_i$ and unit tangent vector τ_j to the dislocation line with $\alpha_{ij} = \tilde{b}_i \, \tau_j \, \delta(\xi)$. Mura (1981) has given a derivation of this type.

1.6 Conclusion of Lecture 1

In this lecture we have established the connection between the *geometrical* description of continuum mechanics (in the tradition of Bilby, Kroener and others) for a defective body, that of *finite-strain formulation* (following modern continuum mechanics, and Mandel in particular), and that of *configurational* (and driving) *forces* (in the manner of Eshelby and Peach and Koehler). This synthesis exhibited in three lines of converging research trends in the chart 1 probably is one of the main achievements in contemporary continuum mechanics. Furthermore, the connection with the general *theory of material inhomogeneities* is shown in the next two sections. We also note that all theories involving so-called *internal strains* and *stresses* ("Eigenspannungen" in the original language of Kroener), such as those due to thermal effects, anelasticity, electro- and magneto-mechanical couplings (piezoelectricity, magnetostriction), phase transformation, can be interpreted in terms of local structural rearrangements in the thought experiment outlined in Paragraph 1.4 above (*cf.* Maugin, 2003a). By duality, an Eshelby stress and a configurational-material force are associated with all these phenomena.

Additional bibliography: Anthony, 1970; Bilby *et al.*, 1957; Elzanowski and Epstein, 1990; Epstein and Maugin, 1992, 1995b, 1997; Friedel, 1979; Forest, 2006; Indenbom, 1965; Kosevich, 1999; Kroener and Seeger, 1959; Le, 1999; Maugin, 2011; Nabarro, 1967, Mura, 1981; Zorski, 1981

Finite Deformation Line ↓	**Geometrical Line** ↓	**Configurational-force Line** ↓
1950s **Multiplicative Decomposition**	**1950s** **Riemannian Geometry**	**1950s** **Force on a singularity** PEACH–KOEHLER (1950)
	* K. KONDO (JP)	
ICTAM Brussels 1956 ←	* BILBY *et al.* (UK)	**Force on an inhomogeneity**
	STROH ↔	J.D. ESHELBY (1951)
	Attemps to relate the Einstein-Cartan tensors to density of defects	
←	**Incompatibility tensor** E. KROENER	
	Non-Riemannian geometry W. NOLL (inhomogeneity) C.C. WANG	
* E.H. LEE (1969)		**Mechanics on the Material Manifold**
	Gauge theory EDELEN, LAGOUDAS KROENER, KLEINERT **(1980)**	* G.A.M. (1969, 71) * D. ROGULA ↓ A. GOLEBIEWSKA G. HERRMANN R. KIENZLER
Elastoplasticity J. MANDEL ↓	M. EPSTEIN ↔ ↓	G.A.M. (1989) $\mathrm{div}_R\,\mathbf{b} + \mathbf{f}^{inh} = \mathbf{0}$ ↓
Mandel Stress →	**STRUCTURAL REARRANGEMENTS** *e.g.*, $\mathrm{div}_R\,\mathbf{b} = \mathbf{b} : \Gamma$	← **Eshelby Stress**

Table 1. FLOW CHART. Three converging lines of research (1950–2000).

2 Lecture 2: Canonical Momentum and Energy from a Variational Formulation (Noether's Theorem)

2.1 Elements of Field Theory

For a long time it has been thought that the notions of Eshelby stress and Eshelbian mechanics could be introduced and be meaningfully interpreted only after their introduction in the framework of a variational field-theoretical formulation. But this is blatantly erroneous as these notions should always exist as we shall show in further chapters. What is nonetheless true is that the field-theoretic formulation provides a hint to formulate dissipative cases (in the same way as Lagrange analytical mechanics suggests us how to introduce dissipative forces such as those related to friction). That is why we start with a standard variational formulation in order to introduce the celebrated Noether's theorem. In what follows it is understood that "variation" means "infinitesimal variation", and the symmetries involved are generated by infinitesimally small variations and parameters.

Here we are concerned with simple general features of field theories in a continuum with space-time parametrization $\{\mathbf{X}, t\}$, where $\mathbf{X}$ stands *for material coordinates* of classical continuum mechanics (*e.g.*, in Truesdell and Toupin, 1960), and t for a timelike scalar variable (Newton's absolute time). We consider *Hamiltonian actions* of the type

$$A(\phi; V) = \int_{V \times I} L(\phi^a, \partial_\mu \phi^a, \partial_\mu \partial_\nu \phi^a, ..; X^\mu)\, \mathrm{d}^4 X, \tag{85}$$

where ϕ^α, $\alpha = 1, 2, \ldots, N$, denotes the ordered array of fields, say the independent components of a certain geometric object, and $\mathrm{d}^4 X = \mathrm{d}V\, \mathrm{d}t$. This is a Cartesian-Newtonian notation, with

$$\left\{ \partial_\mu = \frac{\partial}{\partial X^\mu}; \mu = 1, 2, 3, 4 \right\} = \left\{ \frac{\partial}{\partial X^K}, K = 1, 2, 3; \frac{\partial}{\partial X^4} = \frac{\partial}{\partial t} \right\}. \tag{86}$$

The summation over dummy indices (Einstein convention) is enforced.

In agreement with the rather general expression (85), we say that we envisage the construction of an *n-th-order gradient theory* of the field ϕ^α when gradients of order n at most are considered in the functional dependence of the Lagrangian volume density L. Most of classical physics is based on first-order gradient theories (*cf.* Maugin, 1980). This is the case of classical elasticity which considers gradients of placement or displacement. But we note a recent attraction towards higher-order gradient theories in elasticity (a type of generalized continuum).

From expression (85) we can derive two types of equations: those relating to *each one* of the fields ϕ^α, and those which express a general *conservation law* of the system governing *all* fields simultaneously. The first group is obtained by imposing the requirement that the variation of the action A be zero when we perform a *small* variation $\delta\phi^\alpha$ of the field under well specified conditions at the boundary ∂V, of V (if V is not the whole of space), and at the end points of the time interval $I = [t_0, t_1]$ if such limitations are considered. However most field theories are developed for an infinite domain. The second group of equations are the result of the variation of the parametrization, and these results, on account of the former group, express the *invariance* or lack of invariance of the whole system under changes of this parametrization. To simplify the presentation we will assume an infinite domain V with vanishing fields at infinity, and an infinite time interval since our concern here is neither boundary conditions, nor initial conditions.

In order to perform these variations we consider ε-parametrized families of transformations of *both* coordinates (parametrization) and fields such as

$$(X^\mu, \phi^\alpha) \to (\overline{X}^\mu, \bar{\phi}^\alpha), \tag{87}$$

with

$$\overline{X}^\mu = \kappa^\mu(\mathbf{X}, \varepsilon), \quad \bar{\phi}^\alpha(\overline{\mathbf{X}}) = \Phi^\alpha(\phi^\beta(\mathbf{X}), \overline{\mathbf{X}}, \varepsilon), \tag{88}$$

where ε is an infinitesimal parameter such that for $\varepsilon = 0$ we have identically $\kappa^\mu(\mathbf{X}, 0) = X^\mu$, $\Phi^\alpha(\phi^\beta, \mathbf{X}, 0) = \phi^\alpha$. We assume that the quantity L in (85) transforms as a *scalar* quantity, *i.e.*,

$$L(\overline{\mathbf{X}}, \varepsilon) = \det\left(\frac{\partial \mathbf{X}}{\partial \overline{\mathbf{X}}}\right) L(\mathbf{X}). \tag{89}$$

We note that derivations with respect $\mathbf{X}$ and ε commute, and the same holds true of integration in $\mathbf{X}$ space and derivation with respect to ε. The variation of a field ϕ^α is then defined by

$$\delta\phi^\alpha := \left.\frac{\partial \Phi^\alpha}{\partial \varepsilon}\right|_{\varepsilon=0}. \tag{90}$$

With vanishing fields at infinity in space and vanishing variations at the ends of the time intervals, limiting ourselves to a first-order gradient theory, and applying an ε-parametrization to (85) we immediately have:

$$\delta A = \int \mathrm{d}^4 X \left(\sum_\alpha \left\{ \frac{\partial L}{\partial \phi^\alpha}\, \delta\phi^\alpha + \frac{\partial L}{\partial(\partial_\mu \phi^\alpha)}\, \delta(\partial_\mu \phi^\alpha) \right\} + \frac{\partial L}{\partial X^\mu}\, \delta X^\mu \right). \tag{91}$$

In order that δA vanish for all admissible $\delta\phi^\alpha(\mathbf{X})$, and any α, with $\mathbf{X}$ *fixed*, a classical computation yields the following *Euler–Lagrange equations*:

$$E_\alpha \equiv \frac{\delta L}{\delta \phi^\alpha} = \frac{\partial L}{\partial \phi^\alpha} - \partial_\mu \frac{\partial L}{\partial(\partial_\mu \phi^\alpha)} = 0 \tag{92}$$

for each $\alpha = 1, 2, \ldots, N$, at any space-time event $\mathbf{X}$. Equation (92) is a *strict* (divergence type) conservation law when L does not depend explicitly on ϕ^α. We note that the result (92) is left unchanged if we make the substitution

$$L \to L + \partial_\mu \Omega^\mu(\mathbf{X}),$$

i.e., if we add to L a four-divergence contribution in which Ω depends at most on the space-time coordinates, and *not* on the fields.

The second group of equations, called *conservation equations*, that can derive from (85) by variation, results from a simultaneous transformation of both the coordinates X^μ and the fields. We shall not repeat all the details of the derivation of the resulting theorem known as *Noether's theorem* (Noether, 1918; Soper, 1976; Nelson, 1979), but we shall give the main steps. That theorem states that to any *symmetry of the system there corresponds the conservation (or lack of strict conservation) of a current* (*cf.* Maugin, 1993, pp. 99–103). For L given in (85), such a current generally reads

$$J^\mu = L\,\frac{\partial \overline{X}^\mu}{\partial \varepsilon} + \sum_\alpha \frac{\partial L}{\partial(\partial_\mu \phi^\alpha)} \left[\frac{\partial \bar{\phi}^\alpha}{\partial \varepsilon} - \partial_\nu \phi^\alpha\,\frac{\partial \overline{X}^\nu}{\partial \varepsilon}\right], \tag{93}$$

where ε must be taken equal to zero. In spite of the notation J^μ is not always simply a four-vector. It all depends on the group of transformations considered.

Example 2.1. Space-Time Translation

This is the most relevant invariance for our purpose. In this case the transformations (87)–(88) reduce to a pure space-time transformation of the type

$$X^\mu \to \overline{X}^\mu(\mathbf{X}, \varepsilon) = X^\mu + \varepsilon\,\delta^\mu_{.\lambda}, \tag{94}$$

where δ^μ_λ is the Kronecker symbol, *i.e.*, it equals one only when μ takes the value granted to λ. There is no transformation of the fields themselves. On account of (93), we obtain the following current in components:

$$J^\mu \to J^\mu_{.\lambda} = T^\mu_{.\lambda} = L\,\delta^\mu_{.\lambda} - \sum_\alpha \partial_\lambda \phi^\alpha\,\frac{\partial L}{\partial(\partial_\mu \phi^\alpha)}. \tag{95}$$

This is a mixed second-order space-time tensor called the *energy-momentum tensor*. This denomination is clearly understood when we separate space and time components. If L does not depend explicitly on $\mathbf{X}$, then this tensor satisfies the following (four-dimensional) *strict conservation law*:

$$\partial_\mu T^\mu_{.\lambda} = 0, \quad \lambda = 1, 2, 3, 4. \tag{96}$$

Had we considered an explicit dependence of L on the space-time coordinates X^{μ}, instead of the strict conservation law (96) we would have a source term

$$f_{\lambda} = - \left. \frac{\partial L}{\partial X^{\lambda}} \right|_{expl} , \tag{97}$$

in the right-hand side.

With the *Cartesian notation* introduced in (86) we see that the independence of L on t, $\lambda = 4$, yields a *scalar* conservation law in the explicit form

$$\left. \frac{\partial H}{\partial t} \right|_{X} - \nabla_{R} \cdot \mathbf{Q} = 0, \tag{98}$$

while the independence of L on the spatial part of $\mathbf{X}$, $\{X^{K} = 1, 2, 3\}$, yields the *material balance law*:

$$\left. \frac{\partial \mathbf{P}}{\partial t} \right|_{X} - \mathrm{div}_{R}\, \mathbf{b} = \mathbf{0}, \tag{99}$$

where we defined the following canonical quantities:

- *energy (Hamiltonian density)*:

$$H := \sum_{\alpha} \dot{\phi}^{\alpha} \left(\frac{\partial L}{\partial \dot{\phi}^{\alpha}} \right) - L, \quad \dot{\phi}^{\alpha} \equiv \frac{\partial \phi^{\alpha}}{\partial t}; \tag{100}$$

- *energy flux vector*:

$$\mathbf{Q} = \left\{ Q^{K} := - \sum_{\alpha} \dot{\phi}^{\alpha} \frac{\partial L}{\partial(\partial_{K} \phi^{\alpha})} \right\}; \tag{101}$$

- *canonical (here material) momentum*:

$$P = \left\{ P_{K} := - \sum_{\alpha} \frac{\partial \phi^{\alpha}}{\partial X^{K}} \frac{\partial L}{\partial(\partial \phi^{\alpha} / \partial t)} \right\}; \tag{102}$$

- *canonical stress tensor*:

$$\mathbf{b} = \left\{ b^{K}_{.L} := - \left(L\, \delta^{K}_{.L} - \sum_{\alpha} \frac{\partial \phi^{\alpha}}{\partial X^{L}} \frac{\partial L}{\partial(\partial \phi^{\alpha} / \partial X^{K})} \right) \right\}. \tag{103}$$

The "explicit" independence of L on t – no right-hand side in (98) – stands for the conservation of energy (98). The "explicit" independence of L on X^{K} signifies that the material body is *materially homogeneous*. But this is not a fundamental requirement of physics, so that, in general, (99) may contain a nonzero right-hand denoted

$$\mathbf{f}^{inh} = \left\{ f_L^{inh} = \left(\frac{\partial L}{\partial X^L} \right)_{expl} \right\}, \tag{104}$$

so that (99) would be replaced by the inhomogeneous equation

$$\left.\frac{\partial \mathbf{P}}{\partial t}\right|_X - \mathrm{div}_R\, \mathbf{b} = \mathbf{f}^{inh}. \tag{105}$$

Remark 2.2. Had we considered a Lagrangian depending explicitly on time t, (98) would be replaced by

$$\left.\frac{\partial H}{\partial t}\right|_X - \nabla_R \cdot \mathbf{Q} = h, \quad h = -\left.\frac{\partial L}{\partial t}\right|_{expl}. \tag{106}$$

What must be principally gathered from the above is the essentially different nature of the Euler–Lagrange equations (92) for which one such equation is written for *each field* – or in a more mechanical jargon, *each degree of freedom* – and of the *canonical equations of energy and momentum* – *e.g.*, (105) and (106) – which pertain to the whole physical system and by necessity consider all fields simultaneously [note the summations over α in the definitions (100) through (103)]. In particular, in establishing (96) we have made use of the celebrated *Noether's identity*

$$(\partial_\mu J^\mu)\,\delta\varepsilon + \sum_\alpha (E_\alpha\, \delta\phi^\alpha) = 0, \tag{107}$$

which emphasizes the above-made remark, but naturally implies (96) whenever all *field* equations (92) are satisfied simultaneously.

Relationship to Lie-group theory: The transformations (87) and (88) can be written in the form

$$\overline{X}^\mu = X^\mu + \varepsilon \left.\frac{\partial \overline{X}^\mu}{\partial \varepsilon}\right|_{\varepsilon=0}, \quad \bar{\phi}^\alpha = \phi^\alpha + \varepsilon \left.\frac{\partial \bar{\phi}^\alpha}{\partial \varepsilon}\right|_{\varepsilon=0}. \tag{108}$$

In what follows it is understood that the partial derivatives are evaluated at $\varepsilon = 0$. Then the infinitesimal generator associated with the group (108) of transformations is defined as

$$\mathbf{V} = \frac{\partial \overline{X}^\mu}{\partial \varepsilon} \frac{\partial}{\partial X^\mu} + \frac{\partial \bar{\phi}^\alpha}{\partial \varepsilon} \frac{\partial}{\partial \phi^\alpha}. \tag{109}$$

Then the so-called *first prolongation* of $\mathbf{V}$ is given by (*cf.* Olver, 1986; Ibragimov, 1985; Kalpakides and Maugin, 2004)

$$\mathbf{V}^{(1)} = \mathbf{V} + \left(\frac{\partial}{\partial X^\mu} \left(\frac{\partial \bar{\phi}^\alpha}{\partial \varepsilon} \right) - \frac{\partial \phi^\alpha}{\partial X^\beta} \frac{\partial}{\partial X^\mu} \left(\frac{\partial \overline{X}^\beta}{\partial \varepsilon} \right) \right) \frac{\partial}{\partial (\partial \phi^\alpha / \partial X^\mu)}. \quad (110)$$

The differential operators $\mathbf{V}$ and $\mathbf{V}^{(1)}$ are to be applied to L. Indeed, the equation

$$\mathbf{V}^{(1)} L + L \frac{\partial}{\partial X^\mu} \left(\frac{\partial \overline{X}^\mu}{\partial \varepsilon} \right) = 0, \quad (111)$$

is none other that the vanishing of δL if we want to respect the invariance of the action A. The second contribution in the left-hand side of (111) accounts for the change of volume. In the same formalism Noether's identity (107) reads

$$\partial_\mu J^\mu = 0 = E_\alpha \left(\frac{\partial \phi^\alpha}{\partial \varepsilon} - \frac{\partial \phi^\alpha}{\partial X^\beta} \frac{\partial X^\beta}{\partial \varepsilon} \right), \quad (112)$$

with

$$\begin{aligned} J^\mu &= T^\mu_{.\beta} \frac{\partial \overline{X}^\beta}{\partial \varepsilon} + \frac{\partial L}{\partial (\partial \phi^\alpha / \partial X^\mu)} \frac{\partial \phi^\alpha}{\partial \varepsilon}, \\ T^\mu_{.\beta} &:= L \, \delta^\mu_\beta - \frac{\partial \phi^\alpha}{\partial X^\beta} \frac{\partial L}{\partial (\partial \phi^\alpha / \partial X^\mu)}. \end{aligned} \quad (113)$$

This is entirely equivalent to the foregoing formulation.

Remark 2.3. Had we considered a Lagrangian function L depending on higher-order space-time gradients, *e.g.*, second-order one,

$$L = L(\phi^\alpha, \partial_\mu \phi^\alpha, \partial_\mu \partial_\nu \phi^\alpha; \mathbf{X}), \quad (114)$$

we would have obtained longer expressions for E_α, J^μ, and (93); *e.g.*, in place of (95) and (97),

$$\begin{aligned} T^\mu_{.\lambda} &= L \, \delta^\mu_\lambda - \partial_\lambda \phi^\alpha \frac{\partial L}{\partial (\partial_\mu \phi^\alpha)} + 2 \, \partial_\nu \partial_\lambda \phi^\alpha \frac{\partial L}{\partial (\partial_\nu \partial_\mu \phi^\alpha)} \\ &\quad - \partial_\nu \left(\partial_\lambda \phi^\alpha \frac{\partial L}{\partial (\partial_\nu \partial_\mu \phi^\alpha)} \right), \end{aligned} \quad (115)$$

and

$$\partial_\mu T^\mu_{.\lambda} = - \left. \frac{\partial L}{\partial X^\lambda} \right|_{expl}. \quad (116)$$

But apart for some paradoxical theories with strange inertial terms (see, *e.g.*, Maugin and Christov, 1997, 2002), in classical theories L depends at most on the first-order time derivatives and the second-order gradients in (85) are purely spatial. When this is the case the formal definitions of both H and $\mathbf{P}$ are left unchanged while $\mathbf{Q}$ and $\mathbf{b}$ take on the following more general component form including second-order space-like (here material) gradients:

$$\begin{aligned} Q^K = & -\sum_{\alpha} \dot{\phi}^{\alpha} \left(\frac{\partial L}{\partial(\partial_K \phi^{\beta})} - \partial_L \left(\frac{\partial L}{\partial(\partial_L \partial_K \phi^{\alpha})} \right) \right) \\ & -\sum_{\alpha} \partial_L \dot{\phi}^{\alpha} \frac{\partial L}{\partial(\partial_L \partial_K \phi^{\alpha})} \end{aligned} \tag{117}$$

and

$$\begin{aligned} b^K_{.L} = & -\left(L\,\delta^K_L - \sum_{\alpha} \partial_L \phi^{\alpha} \frac{\partial L}{\partial(\partial_K \phi^{\alpha})} \right) - \sum_{\alpha} 2\,\partial_L \partial_M \phi^{\alpha} \frac{\partial L}{\partial(\partial_M \partial_K \phi^{\alpha})} \\ & + \sum_{\alpha} \partial_L \left(\partial_M \phi^{\alpha} \frac{\partial L}{\partial(\partial_M \partial_K \phi^{\alpha})} \right). \end{aligned} \tag{118}$$

Other symmetries, and thus other consequences of Noether's theorem can be applied – *e.g.*, rotations, dilatation (change of scale), expansion, *etc.*; *cf.* Maugin, 1993, pp. 110–112. This is not dealt with here.

2.2 Application to First-Gradient Elasticity in Finite Strains

We consider the so-called *direct-motion* formulation. In this formulation the fields ϕ^{α} are none other than the physical components of the current placement. In a general manner we can consider the following Lagrangian density per unit volume of K_R

$$L = \bar{L}(\mathbf{v}, \mathbf{F}; \mathbf{X}, \mathbf{x}, t) = \bar{K} - \bar{W} - \Phi_R, \tag{119}$$

where

$$\bar{K}(\mathbf{v}; \mathbf{X}, t) = \frac{1}{2}\,\rho_0(\mathbf{X}, t)\,\mathbf{v}^2,\ W = \bar{W}(\mathbf{F}; \mathbf{X}, t),\ \Phi_R = \rho_0(\mathbf{X}, t)\,\varphi(\mathbf{x}, t). \tag{120}$$

This is indeed quite general because it includes a potential for *physical* body forces, *via* Φ_R, where the potential φ per unit mass is necessarily dependent on the current placement, while both the reference matter density and the elasticity potential remain dependent on both the material point $\mathbf{X}$ and Newtonian time t. Dependence on $\mathbf{X}$ means *material inhomogeneity*, *i.e.*, whenever

$$\left.\frac{\partial \rho_0}{\partial \mathbf{X}}\right|_t \neq \mathbf{0}, \quad \left.\frac{\partial \bar{W}}{\partial \mathbf{X}}\right|_{expl} \neq \mathbf{0}. \tag{121}$$

Explicit dependence on time t is a much more rare occurrence in books on continuum mechanics. Indeed the condition (*cf.* Epstein and Maugin, 2000)

$$\left.\frac{\partial \rho_0}{\partial t}\right|_X \neq 0, \tag{122}$$

would mean that the reference density may evolve in time by addition or subtraction of matter by some means. This would account for the phenomenon of *growth or resorbtion* of matter. We shall return to this in due time. As to the condition (*cf.* Maugin, 2007)

$$\left.\frac{\partial \overline{W}}{\partial t}\right|_{expl} \neq 0, \tag{123}$$

it would mean a possible evolution in time of elastic properties, hence the phenomenon of *ageing*. This is seldom considered in view of the large time scales involved in the process for normal physical conditions for inert matter. From the point of view of analytical mechanics, systems in which the Lagrangian density depends explicitly on time are said to be *rheonomic*, according to a classification originally due to Boltzmann (*cf.* Lanczos, 1962). In contrast, systems in which L does not depend explicitly on time are called *scleronomic* systems. Most systems considered are scleronomic.

Remark 2.4. A typical example of rheonomic system is one in which a time-dependent spring constant yields a Mathieu differential equation with a time-dependent characteristic frequency.

A last remark concerning the Lagrangian density for the direct-motion description is that neither the kinetic energy nor the elasticity potential energy involve the field itself $\mathbf{x}$, but only its space and time derivative in the parametrization $(\mathbf{X}, t)$. The reason is that elasticity is a *gauge theory* as *Galilean invariance* (translation in physical space) eliminates this possible dependence on $\mathbf{x}$ (or displacement $\mathbf{u}$ in small strains).

In order to account for both natural boundary conditions at the boundary ∂B and initial conditions (this specifies the physical momentum, position being $\mathbf{X}$ at $t = 0$), we envisage a Hamiltonian variational principle in the following form:

$$\begin{aligned} \delta_X \int_0^t \int_{B_R} L \, \mathrm{d}V \, \mathrm{d}t + \int_0^t \int_{\partial B} \mathbf{T}^d \cdot \delta_X \chi \, \mathrm{d}a \, \mathrm{d}t \\ - \int_{B_R} (\mathbf{p}_R(t) - \mathbf{p}_R(0)) \cdot \delta_X \chi \, \mathrm{d}V = 0, \end{aligned} \tag{124}$$

where the surface term corresponding to the datum of a physical traction is expressed at the deformed (actual) boundary. The variation effected in

(124) to the price of simple calculations immediately yields the *field* equation which is none other than the Piola–Kirchhoff format of the local balance of linear (physical) momentum:

$$\left.\frac{\partial \mathbf{p}_R}{\partial t}\right|_X = \operatorname{div}_R \mathbf{T} = \rho_0\, \mathbf{f}, \quad \mathbf{f} = -\nabla\varphi \text{ in } B_R, \tag{125}$$

$$\mathbf{N}\cdot\mathbf{T} = \left(\frac{\mathrm{d}a}{\mathrm{d}A}\right)\mathbf{T}^d = \overline{\mathbf{T}}^d \text{ at } \partial B_R, \tag{126}$$

$$\mathbf{p}_R(t=0) = \mathbf{p}_R(0), \quad \mathbf{p}_R(t=t) = \mathbf{p}_R(t), \tag{127}$$

of which the last one is a mere identity.

On applying now Noether's identity, we can write down the accompanying conservation laws that follow for space-time translations $\lambda = 4$ and $\lambda = K$, $K = 1, 2, 3$,

- Conservation of energy:

$$\left.\frac{\partial H}{\partial t}\right|_X - \nabla_R\cdot\mathbf{Q} = h, \tag{128}$$

- Conservation of material momentum:

$$\left.\frac{\partial \mathbf{P}}{\partial t}\right|_X - \operatorname{div}_R \mathbf{b} = \mathbf{f}^{inh} + \mathbf{f}^{ext}, \tag{129}$$

where the have defined the following quantities:

$$\mathbf{p}_R = \rho_0\,\mathbf{v}, \quad \mathbf{T} = \frac{\partial \overline{W}}{\partial \mathbf{F}}, \quad H = K + W, \quad \mathbf{Q} = \mathbf{T}\cdot\mathbf{v}, \tag{130}$$

$$\mathbf{P} = -\rho_0\,\mathbf{v}\cdot\mathbf{F}, \quad \mathbf{b} = -(L\,\mathbf{1}_R + \mathbf{T}\cdot\mathbf{F}), \quad \mathbf{f}^{ext} = -\rho_0\,\mathbf{f}\cdot\mathbf{F}, \tag{131}$$

and

$$h = -\left(\frac{K}{\rho_0}\right)\left.\frac{\partial \rho_0}{\partial t}\right|_X + \left.\frac{\partial \overline{W}}{\partial t}\right|_{expl}, \tag{132}$$

$$\mathbf{f}^{inh} = \left(\frac{K}{\rho_0}\right)\left.\frac{\partial \rho_0}{\partial \mathbf{X}}\right|_X - \left.\frac{\partial \overline{W}}{\partial \mathbf{X}}\right|_{expl}. \tag{133}$$

In view of (129) we may say that *material momentum* is not *strictly* conserved for a materially inhomogeneous elastic material and in the presence of an external body force. As to (128), it tells us that energy is not strictly conserved for rheonomic materials, whether they suffer growth or ageing. However, the very form of the source term in the right-hand side of (128) provides some hint at a simple and naïve approach to deal with these *dissipative* effects. Indeed, we would have energy conservation if the right-hand side of (128) would vanish. Without introducing any thermal processes, this could be accomplished by adding from the outside ("out of the blue") a term that makes this right-hand side vanish. This will be a dissipation rate Φ_d. Thus we would have

$$\Phi_d = -\left(-\frac{\partial L}{\partial t}\bigg|_{expl} \right) = \Phi_{growth} + \Phi_{age}, \tag{134}$$

with (*cf.* eqn. (132))

$$\Phi_{growth} = K\,\frac{\Pi_0}{\rho_0}, \quad \Phi_{age} = -\left.\frac{\partial W}{\partial t}\right|_{expl}, \tag{135}$$

where Π_0 is the right-hand side of (122). This right-hand side will typically be of the form (see Epstein and Maugin, 2000)

$$\Pi_0 = R_0 + \nabla_R \cdot \mathbf{M}, \tag{136}$$

where R_0 is an external supply (*e.g.*, via some biophysical process, nutriments) and $\mathbf{M}$ is an influx of mass. Since both K and ρ_0 are obviously positive, so must be Π_0 in order to guarantee a positive Φ_{growth}.

As to Φ_{age} a simplistic case exemplifies the situation. We may assume that the explicit time dependence of W is only through material functions (with certain tensorial properties, a set of scalars being the simplest case) $\alpha(t)$. Then using a notation intentionally akin to that of the thermodynamics of irreversible processes with internal variables (*cf.* Maugin, 1999a), we will set and have:

$$A := -\frac{\partial W}{\partial \alpha}, \quad \Phi_{age} = A\,\dot{\alpha}, \tag{137}$$

where the appropriate inner product must be used in the last expression which is classically written as a *bilinear* form, the product of a thermodynamical force, A, and a time rate. The total expression must be non-negative according to the second law of thermodynamics. We do not pursue here these considerations and consider from here on only the case of scleronomic elastic systems.

2.3 Application to Second-Gradient Elasticity in Finite Strains

Many of the above-stated results can be generalized to the case of a field theory of elasticity that is a *second-gradient theory*, *i.e.*, whenever (119)–(120), ignoring the presence of a body force and its potential, are replaced by the more general expressions:

$$L = \bar{L}(\mathbf{v}, \mathbf{F}, \nabla_R \mathbf{F}; \mathbf{X}) = \bar{L}\left(\frac{\partial \chi}{\partial t}, \frac{\partial \chi}{\partial \mathbf{X}}, \frac{\partial^2 \chi}{\partial \mathbf{X} \otimes \partial \mathbf{X}}; \mathbf{X} \right), \tag{138}$$

and

$$L = \frac{1}{2}\,\rho_0(\mathbf{X})\,\mathbf{v}^2 - \overline{W}(\mathbf{F}, \nabla_R \mathbf{F}; \mathbf{X}). \tag{139}$$

The kinetic energy is left unchanged for there is no additional microstructure here, just a finer description of stress effects. Elastic media of the type of those governed by a Lagrangian (138) were considered by Toupin (1962, 1964) for finite strains and Mindlin and Tiersten (1962) and Mindlin and Eshel (1968) for the small-strain approximation. In the 1960s and 1970s theories of this complexity (also called second-grade or strain-gradient theories) were expected to bring some satisfactory answer to the question raised by the appearance of local singularities in some elasticity problems (*e.g.*, those involving corners, edges, *etc.*). This was not so successful. But gradient theories are also relevant to the elasticity theory of crystals, where long-range interactions and the associated wave dispersion effects cannot be neglected (*cf.* Maugin, 1999b). This appears of necessity because scale effects are present (*e.g.*, coherence length) and theories such as (139) include *internal length* scales. Recently, Lazar and Maugin (2004a to 2006b) have revisited the problem of singularities at dislocation cores for such generalized continua and others.

We are just in the framework of the variational formulation. The field (Euler–Lagrange) equations are the spatial components of the local balance of linear (physical) momentum in the Piola–Kirchhoff formulation:

$$\left.\frac{\partial \mathbf{p}}{\partial t}\right|_X - \mathrm{div}_R\, \mathbf{T}^{\mathit{eff}} = \mathbf{0}, \tag{140}$$

wherein

$$\mathbf{p} = \rho_0\,\mathbf{v}, \quad \mathbf{T}^{\mathit{eff}} = \mathbf{T} - \mathrm{div}_R\,\mathbf{M}, \tag{141}$$

$$\mathbf{T} = \frac{\partial \overline{W}}{\partial \mathbf{F}}, \quad \mathbf{M} = \frac{\partial \overline{W}}{\partial(\nabla_R \mathbf{F})}, \tag{142}$$

where the last introduced quantity has components

$$M^{LK}_{\cdot\cdot i} = \frac{\partial \overline{W}}{\partial (F^i_{\cdot K,L})} = M^{KL}_{\cdot\cdot i}. \tag{143}$$

Here $\mathbf{T}^{\mathit{eff}}$ is an *effective* first Piola–Kirchhoff stress, and $\mathbf{M}$ is a new *internal force* that may be referred to as the Piola–Kirchhoff *hyperstress*. The latter has at most eighteen independent components due to the symmetry on the two material indices. This large number of components will unfortunately be accompanied by a large number of new material coefficients to be determined from appropriate experiments or by comparison with long-wave limits of a microscopic model or a procedure of homogenization (*cf.* Forest, 2006). The application of Noether's theorem to the case of material space-time translations for scleronomic systems in the absence of body force systems yields the *conservation of energy* and that of *material momentum* in the form (Maugin and Trimarco, 1992)

$$\left.\frac{\partial \mathbf{P}}{\partial t}\right|_{\mathbf{X}} - \mathrm{div}_R\,\mathbf{b}^{eff} = \mathbf{f}^{inh}, \tag{144}$$

and

$$\left.\frac{\partial H}{\partial t}\right|_{\mathbf{X}} - \nabla_R \cdot \mathbf{G}^{eff} = 0, \tag{145}$$

where we have defined the following quantities:

$$\mathbf{P} = -\mathbf{p}\cdot\mathbf{F}, \quad \mathbf{b}^{eff} = \mathbf{b} - \mathrm{div}_R\,(\mathbf{M}\cdot\mathbf{F}), \tag{146}$$

$$\mathbf{b} = -(L\,\mathbf{1}_R + \mathbf{T}\cdot\mathbf{F} + \mathbf{M} : (\nabla_R\mathbf{F})^T), \tag{147}$$

$$\begin{gathered} H = K + W, \quad L = K - W, \\ \mathbf{G}^{eff} = \mathbf{G} - (\mathrm{div}_R\,\mathbf{M})\cdot\mathbf{v}, \quad \mathbf{G} = \mathbf{T}\cdot\mathbf{v} + \mathbf{M} : (\nabla_R\mathbf{v})^T, \end{gathered} \tag{148}$$

and

$$\mathbf{f}^{inh} = \left.\frac{\partial L}{\partial \mathbf{X}}\right|_{expl}. \tag{149}$$

This shows very little variance with the first-gradient theory. But two remarks are in order. First, we must express the rotational invariance of W. This can be obtained by performing an infinitesimal rotation in the current configuration, the result of which being a generalization of the relation $\mathbf{F}\,\mathbf{T} = (\mathbf{F}\,\mathbf{T})^T = \mathbf{T}^T\,\mathbf{F}^T$, *i.e.*,

$$skew_{[ij]}(\mathbf{F}\,\mathbf{T} + 2\,(\nabla_R\mathbf{F})^T : \mathbf{M}) = 0 \tag{150}$$

which shows that the Cauchy stress is *not* symmetric in this description since it is proportional to $\mathbf{F}\cdot\mathbf{T}$. Equation (150) is automatically satisfied if, for instance, for an anisotropic inhomogeneous material, we take

$$W = \overline{W}(\mathbf{C}, (\nabla_R\mathbf{F}) : (\nabla_R\mathbf{F})^T, (\nabla_R\mathbf{F})^T\cdot\mathbf{F}; \mathbf{X}). \tag{151}$$

The second remark concerns *boundary conditions* which must accompany (140) and (144) for a body of finite extent, as the theory clearly becomes of higher degree in space derivatives. This, however, is much better evidenced in the corresponding small-strain theory as developed by, *e.g.*, Mindlin and Eshel (1968) – obviously without the concepts of inhomogeneity, material momentum (field momentum) and Eshelby stress – to whom we refer the reader (see below). In terms of "forces", $\mathbf{N}\cdot\mathbf{M}$ must be specified in addition to $\mathbf{N}\cdot\mathbf{T}$. In terms of displacements, their **curvature** should be specified in order to match the duality of data in terms of forces.

2.4 Small Strain Theory

A. Small-strain (first-gradient) theory

In this case we have the following trivial reductions:

• Balance of linear (physical) momentum:

$$\rho_0\,\ddot{u}_i - \sigma_{ji,j} = \rho_0\,f_i \text{ in } B, \tag{152}$$

• Balance of energy:

$$\dot{W} = \sigma_{ji}\,\dot{u}_{i,j}, \quad \sigma_{ij} = \frac{\partial W}{\partial \varepsilon_{ij}}; \tag{153}$$

• Natural boundary condition:

$$\eta_j\,\sigma_{ji} = T_i^d \text{ at } \partial B_\sigma. \tag{154}$$

• Reduced form of the Eshelby stress and field momentum:

First, according to the above given introduction of this notion, we note that the Eshelby stress itself is less important than its divergence or its flux in so far as mechanics is concerned. This remark is to be taken seriously because such a tensor can then be defined up to a divergence-free tensor. This is what clearly occurs in the following consideration of the small-strain case. In the quasi-static case in the absence of external body force, the Cauchy stress has a vanishing divergence. Abandoning the distinction between upper and lower Latin indices, from the above we have for a materially inhomogeneous body

$$f_i^{inh} = -(\operatorname{div}\mathbf{b})_i = -b_{ji,j}, \quad (\operatorname{div}\sigma)_i = \sigma_{ji,j}. \tag{155}$$

On account of the definition of the deformation gradient $\mathbf{F}$ in terms of the displacement gradient, we immediately have

$$b_{ji} = W\,\delta_{ji} - \sigma_{jk}\,(\delta_{ki} + u_{k,i}). \tag{156}$$

On substituting from this into the first of (155) and accounting for the second, we can write identically

$$f_i^{inh} = -b_{ji,j} \text{ with } b_{ji} \equiv W\,\delta_{ji} - \sigma_{jk}\,u_{k,i} = (W\,\mathbf{1} - \sigma\cdot(\nabla u)^T)_{ji}, \tag{157}$$

where the last expression is that one to be considered for the Eshelby stress in small strains. This procedure is *not* a linearization of $\mathbf{b}$, a procedure which would have no meaning, because all elements of $\mathbf{b}$ should be of the energy type. Some authors, that we will not cite, have mistakenly "linearized" the second contribution in the Eshelby stress, showing by that a lack of understanding of the notion. This remark has an importance in fracture studies. What about the *dynamical case*? For small strains we will have in components

$$\begin{aligned} P_i &= -\rho_0\, \dot{u}_j\, (\delta_{ji} + u_{j,i}) = -p_i + P_i^f, \\ p_i = \rho_0\, \dot{u}_i, \quad P_i^f &:= -\rho_0\, \dot{u}_j\, u_{j,i} = -(\rho_0\, \dot{\mathbf{u}} \cdot (\nabla \mathbf{u})^T)_i, \end{aligned} \tag{158}$$

where the last equation defines the so-called *field momentum* (sometimes called *wave momentum*) as it appears in crystal physics (*cf.* Brenig, 1955). Of course, we note the remarkable identity

$$\mathbf{P}^f = \mathbf{p} + \mathbf{P}, \tag{159}$$

that is, *field momentum is the sum of linear physical momentum and material momentum.*

On account of the balance of linear physical momentum and of (158), we then check in the case of the presence of an external body force and of material inhomogeneities of both inertial and elastic origins, that the balance of material momentum takes on the following form:

$$\dot{P}_i^f - b_{ji,j} = f_i^{ext} + f_i^{inh}, \tag{160}$$

wherein

$$\begin{aligned} b_{ji} &= -(L\, \delta_{ji} + \sigma_{jk}\, u_{k,i}), \quad L = K - W, \\ K &= \frac{1}{2}\, \rho_0(\mathbf{x})\, \dot{\mathbf{u}}^2, \quad f_i^{ext} = -\rho_0\, f_{0j}\, u_{j,i}. \end{aligned} \tag{161}$$

B. Small-strain strain-gradient (or second-gradient -of displacement) theory

From the above this is trivially given by

• Balance of linear (physical) momentum:

$$\rho_0\, \ddot{u}_i - \sigma_{ji,j} = \rho_0\, f_i \text{ in } B, \tag{162}$$

• Balance of energy:

$$\dot{W} = \bar{\sigma}_{ji}\, \dot{u}_{i,j} + m_{kji}\, \dot{\varepsilon}_{ij,k}, \tag{163}$$

• Balance of field momentum:

$$\dot{P}_i^f - b_{ji,j} = f_i^{inh} - \rho_0\, f_j\, u_{j,i}, \tag{164}$$

• Constitutive equations:

$$\sigma = \frac{\delta \overline{W}}{\delta \varepsilon} = \bar{\sigma} - \operatorname{div} \mathbf{m}, \tag{165}$$

$$W = \overline{W}(\varepsilon, \nabla\varepsilon; \mathbf{x}), \quad \bar{\sigma} = \frac{\partial \overline{W}}{\partial \varepsilon}, \quad \mathbf{m} = \frac{\partial \overline{W}}{\partial(\nabla\varepsilon)}, \tag{166}$$

• Definitions:

$$\begin{aligned} P_i^f &= -\rho_0\, \dot{u}_j\, u_{j,i}, \\ b_{ji} &= -(L\,\delta_{ji} + \bar{\sigma}_{jk}\, u_{i,k} + m_{jkp}\, u_{i,kp}) - (m_{kjp}\, u_{i,p})_{,k}, \\ f_i^{inh} &= \left.\frac{\partial L}{\partial \mathbf{x}}\right|_{expl}, \end{aligned} \tag{167}$$

where σ is the Cauchy stress, the symmetric tensor $\bar{\sigma}$ is called the *intrinsic stress*, and the last object, the *hyperstress* tensor $\mathbf{m}$, has components $m_{kji} = m_{kij}$.

• Natural boundary conditions
(see, *e.g.*, Germain, 1973a; Maugin, 1980):

$$\begin{aligned} n_j\,\sigma_{ji} + (n_j\, D_p\, n_p - D_j)\,(n_k\, m_{kji}) &= T_i^d \quad \text{at } \partial B - \Gamma\uparrow \\ n_k\, m_{kji}\, n_j &= R_i \quad \text{at } \partial B - \Gamma\uparrow \\ \varepsilon_{jpq}\,\tau_p[n_k\, m_{kji}\, n_j] &= E_i \quad \text{along } \Gamma\uparrow \end{aligned} \tag{168}$$

where D_j denotes the tangential derivative at the regular parts of ∂B, $\Omega = -\frac{1}{2}\, D_j\, n_j$ is the local mean curvature at regular parts of ∂B, τ is the unit tangent along the oriented edge $\Gamma\uparrow$ (if the latter exists), R_i is a so-called *double normal traction*, and E_i is a *lineal force* along the edge. This shows the complexity of prescribing data at the irregular boundary because the *hyperstress* is a complex notion of "internal force".

On the one hand, the very expression of σ through a *functional derivative* of the Euler–Lagrange type provides the expression in terms of even-order space derivatives (that is, two orders more appear at each step of refinement in the n-th order gradient theory). On the other hand, a local experience (real or in thought only) considering a *facet* cut in the material (geometrical description at the first order of the cut *via* its unit normal only) "sees" the Cauchy stress. Only *edges* or the second-order geometrical description (tangential derivatives, curvature) can "see" $\mathbf{m}$ independently (in this regard, see the beautiful work of Dell'Isola and Seppecher (1995) on the generalization of Cauchy's lemma for the representation of stresses). This is by-passed by a weak formulation of the type of the principle of virtual power such as

$$\begin{aligned} \int_B \rho_0\, \dot{u}_i\, v_i^*\, \mathrm{d}v = &-\int_B (\bar{\sigma}_{ji}\, v_{i,j}^* + m_{kji}\, v_{(i,j),k}^*)\, \mathrm{d}v \\ &+ \int_{\partial B} (T_i^d\, v_i^* + M_{ji}^d\, v_{i,j}^*)\, \mathrm{d}a + \int_B \rho_0\, f_i\, v_i^*\, \mathrm{d}v, \end{aligned} \tag{169}$$

where $\mathbf{T}^d$ and $\mathbf{M}^d$ are an applied *traction* and an applied so-called *double force* (represented by a symmetric second-order tensor – see Maugin (1980) and Germain (1973b)). Other conservation laws and the connection to Lie groups are to be found in Lazar *et al.* (2005).

2.5 Conclusive Remarks of Lecture 2

What we have developed in this lecture is essentially the variational analytical mechanics of fields with direct application to elasticity. It happens that elasticity is paradigmatic as it appears to be an ideal illustration. The approach chosen has emphasized the role played by material and spatial (Eulerian) variations. This was acknowledged by several authors in special cases (no dynamics, no material inhomogeneities, small strains depending on the case), among them Casal (1978), Rogula (1966, 1977), Edelen (1981), and Golebiewska-Herrmann (1981). But two remarks are essential at this point.

First, the special case of elasticity is indeed more than special in the context of field theory because it leads to some misinterpretation. The reason for this is that in this case the *field* (placement) and the space-parametrization share the same nature: they are *positions* but in different spaces. It is even worst when Lagrangian coordinates are used as space parametrization because they correspond to a placement occupied in the past! This gives a feeling of strict equivalence by mere pull-back or push-forward between, or of a strict ontological status of, the balances of linear physical momentum and of material momentum. But the latter, that we better call *canonical momentum* in view of the general theory of fields, sums up contributions from all fields involved (remember the summation over α), whether mechanical or other in nature, and it will include contributions other than those of the actual placement in more complex theories of continua (such as in media with internal degrees of freedom or with coupled fields, some we examine later on in this contribution). In particular, there are field theories that admit no deformation field, but they still admit the notions of canonical momentum and "Eshelby" stress, a typical example being that of pure electromagnetic fields in a vacuum or in a rigid body.

Second, canonical equations for both momentum and energy must exist whatever the precise material behavior of the material. We shall see in the forthcoming section how this can be dealt with in a sufficiently general thermodynamical framework for deformable solids.

Additional Bibliography: Eshelby, 1975; Herrmann and Kienzler, 2000; Kienzler and Maugin, 2001; Kroener, 1993; Lazar and Maugin, 2007; Maugin, 2011.

3 Lecture 3: Canonical Momentum and Energy from Balance Laws for Dissipative and Complex Media

3.1 Introduction

Of course the variational formulation of the foregoing section has its limitations although it does yield canonical conservation laws of energy and material momentum, and provides in a direct manner the notions of Eshelby material stresses and forces. But there is no reason why these should be limited to special classes of physical processes in which the energy density is known from the start, so that, in effect, the material behavior is also specified from the start. It is only recently that a general canonical formulation of **thermo**-dynamics was formulated that accommodates large classes of dissipative phenomena and is naturally subjected to the necessary satisfaction of the second law of thermodynamics. To illustrate this general viewpoint, we consider the sufficiently general continuum physics of finitely deformable solids that may conduct heat and exhibit dissipation of different types *via* the presence of so-called internal variables of state. The uncontrollability of the internal state variables is reflected in the fact that they have no conjugated forces in the bulk and at the surface of the considered body; so that their only virtue, and essence, is to produce a dissipation. Accordingly, the general starting field equations are those of a classical thermo-deformable (scleronomic) body.

3.2 Reminder

We suppose that the following three local balance laws have been deduced from a global statement for sufficiently smooth fields. Here we directly consider the **Piola–Kirchhoff** formulation of the balance of mass, physical (linear) momentum, and energy (no external supply of energy apart from that related to the body force) at any regular material point $\mathbf{X}$ in a continuous body in the presence of a body force $\mathbf{f}_0$ per unit reference volume

$$\left.\frac{\partial \rho_0}{\partial t}\right|_X = 0, \tag{170}$$

$$\left.\frac{\partial (\rho_0\, \mathbf{v})}{\partial t}\right|_X - \mathrm{div}_R\, \mathbf{T} = \mathbf{f}_0, \tag{171}$$

$$\left.\frac{\partial (K+E)}{\partial t}\right|_X - \nabla_R \cdot (\mathbf{T}\cdot\mathbf{v} - \mathbf{Q}) = \mathbf{f}_0 \cdot \mathbf{v}, \tag{172}$$

where ρ_0 is the mass density, $\mathbf{v} = \partial\bar{\mathbf{x}}/\partial t|_X$ is the physical velocity, $\mathbf{T}$ is the first Piola–Kirchhoff stress, $K = \rho_0\, \mathbf{v}^2/2$ is the kinetic energy, E is the internal energy per unit reference volume, and $\mathbf{Q}$ is the material heat flux. This is complemented by the second law of thermodynamics written as

$$\left.\frac{\partial S}{\partial t}\right|_X + \nabla_R \cdot \mathbf{S} \geq 0, \quad \mathbf{S} = \frac{\mathbf{Q}}{\theta} + \mathbf{K}, \tag{173}$$

where S is the entropy density, θ is the absolute temperature ($\theta > 0$, $\inf \theta = 0$), and $\mathbf{S}$ is the entropy flux. The "extra entropy flux" $\mathbf{K}$ vanishes in most cases.

3.3 Canonical Balance Laws of Momentum and Energy

A. Canonical form of the energy conservation First we formulate an interesting form of the energy conservation equation. A part of the reasoning is standard. In effect, taking the scalar product of both sides of eqn. (171) by $\mathbf{v}$ and performing some elementary manipulations we obtain the so-called *theorem of the kinetic energy* as

$$\frac{\mathrm{d}K}{\mathrm{d}t} - \nabla_R \cdot (\mathbf{T} \cdot \mathbf{v}) + \operatorname{tr}(\mathbf{T} \cdot \dot{\mathbf{F}}) - \mathbf{f}_0 \cdot \mathbf{v} = 0. \tag{174}$$

Combining this with the first law of thermodynamics (172) we obtain the local statement of the so-called *theorem of internal energy*:

$$\frac{\mathrm{d}E}{\mathrm{d}t} - \operatorname{tr}(\mathbf{T} \cdot \dot{\mathbf{F}}) + \nabla_R \cdot \mathbf{Q} = 0, \tag{175}$$

(i) Case where $\mathbf{K} \equiv \mathbf{0}$

In this case, introducing the Helmholtz free energy function by $W = E - S\,\theta$, we transform the inequality $(173)_1$ into the celebrated *Clausius–Duhem inequality*

$$-\left(\frac{\mathrm{d}W}{\mathrm{d}t} + S\,\frac{\mathrm{d}\theta}{\mathrm{d}t}\right) + \operatorname{tr}(\mathbf{T} \cdot \dot{\mathbf{F}}) - \mathbf{S} \cdot \nabla_R \theta \geq 0. \tag{176}$$

As we all know, this is exploited as a constraint in the formulation of thermodynamically admissible constitutive equations, while the "conservation equation" (175) is the equation governing heat propagation in a disguise. This can be given several transformed forms. A most interesting form is obtained straightforwardly by noting that $E = W + S\,\theta$, yielding

$$\frac{\mathrm{d}(S\,\theta)}{\mathrm{d}t} + \nabla_R \cdot \mathbf{Q} = h^{int}, \quad h^{int} := \operatorname{tr}(\mathbf{T} \cdot \dot{\mathbf{F}}) - \left.\frac{\partial W}{\partial t}\right|_X. \tag{177}$$

This is of special interest because of the expression in the right-hand side which *a priori* appears as an *internal heat source*. Indeed, for a typically thermodynamically reversible behavior such as pure nonlinear elasticity (hyperelasticity), where $W = \overline{W}(\mathbf{F})$ depends only on $\mathbf{F}$, we have from the exploitation of (176),

$$\mathbf{T} = \frac{\partial \overline{W}}{\partial \mathbf{F}} \Rightarrow h^{int} \equiv 0. \tag{178}$$

Note that in the situation where (177) holds good, the inequality (176) can also be written in the following enlightening form

$$S\,\dot{\theta} + \mathbf{S} \cdot \nabla_R \theta \leq h^{int}. \tag{179}$$

We claim that $(177)_1$ in fact is the most interesting form of the energy conservation equation for our purpose (*i.e.*, establishing canonical equations). This we discover by constructing the canonical equation of momentum as follows.

B. Canonical (material) momentum conservation Guided by what is valid for pure finite-strain elasticity (Noether's identity; previous section), we apply $\mathbf{F}$ to the right of (171) and note that (T = transpose)

$$\left(\frac{\partial}{\partial t}(\rho_0\,\mathbf{v})\right) \cdot \mathbf{F} = -\left.\frac{\partial \mathbf{P}}{\partial t}\right|_X - \nabla_R\left(\frac{1}{2}\,\rho_0\,\mathbf{v}^2\right) + \left(\frac{1}{2}\,\mathbf{v}^2\right)(\nabla_R \rho_0), \tag{180}$$

and

$$(\mathrm{div}_R\,\mathbf{T}) \cdot \mathbf{F} = \mathrm{div}_R\,(\mathbf{T} \cdot \mathbf{F}) - \mathrm{tr}\big(\mathbf{T} \cdot (\nabla_R \mathbf{F})^T\big), \tag{181}$$

where we have set

$$\mathbf{P} := -\rho_0\,\mathbf{v} \cdot \mathbf{F} \tag{182}$$

the (mechanical) *material* momentum. Introducing now plus and minus the material gradient of an (unspecified) free energy density $W = \overline{W}(.,.,..,\mathbf{X})$, we then check that (171) yields the following *material balance of momentum*

$$\frac{\mathrm{d}\mathbf{P}}{\mathrm{d}t} - \mathrm{div}_R\,\mathbf{b} = \mathbf{f}^{int} + \mathbf{f}^{ext} + \mathbf{f}^{inh}, \tag{183}$$

in which we have defined the material *Eshelby stress* $\mathbf{b}$, the material *inhomogeneity force* $\mathbf{f}^{inh}$, the material *external* (or body) force $\mathbf{f}^{ext}$, and the material *internal* force $\mathbf{f}^{int}$ by

$$\mathbf{b} = -(L_W\,\mathbf{1}_R + \mathbf{T} \cdot \mathbf{F}), \quad L_W := K - W, \tag{184}$$

$$\mathbf{f}^{inh} := \left.\frac{\partial L_W}{\partial \mathbf{X}}\right|_{expl} \equiv \left.\frac{\partial L_W}{\partial \mathbf{X}}\right|_{fixed\ fields} = \frac{\mathbf{v}^2}{2}\,\nabla_R \rho_0 - \left.\frac{\partial \overline{W}}{\partial \mathbf{X}}\right|_{expl}, \tag{185}$$

$$\mathbf{f}^{ext} := -\mathbf{f}_0 \cdot \mathbf{F}, \quad \mathbf{f}^{int} := \mathrm{tr}\big(\mathbf{T} \cdot (\nabla_R \mathbf{F})^T\big) - \nabla_R W|_{impl}, \tag{186}$$

where the subscript notations *expl* and *impl* mean, respectively, the material gradient keeping the fields fixed (and thus extracting the explicit dependence on $\mathbf{X}$), and taking the material gradient only through the fields present in the function.

Equation (183) is the *canonical* balance of momentum of continuum mechanics *in the absence of specification of constitutive equations*. It is a (mathematically) strict conservation equation only when all source terms in its right-hand side vanish. Here the new notion is that of ***internal material force*** which appears in parallel and total analogy with the ***internal heat source*** $(177)_2$, the action of the material gradient replacing that of the material time derivative. We note that there is no "time-like" scalar equivalent to $\mathbf{f}^{inh}$ in equation $(177)_1$ because this inhomogeneity force which is automatically captured by that equation, has *no* dissipative nature. An explicit dependence of W on time (in a rheonomic system) would yield a nonzero term h^{inh}. Similarly, there is no equivalent to the external material force $\mathbf{f}^{ext}$ in $(177)_1$ because this equation governs essentially the internal energy. It would be easy to rewrite eqns. $(177)_1$ and (183) as a single four-dimensional space-time equation (see Maugin, 2000) but this serves no special purpose, except for an aesthetic satisfaction, in engineering applications. Still the *consistency* between the *space-like* co-vectorial equation (183) and the *time-like* equation $(177)_1$ is a fundamental requirement in the thermodynamical study of the progress of singularity sets (*e.g.*, defects) in order that the related dissipation be none other than the power expended by the material force in the material velocity of the defect.

Still, in the present approach, in order to proceed further we need to specify the full functional dependence of W. The general expressions $(177)_1$ and (183) are the most general canonical equations for momentum and energy we can write down without a postulate of the full dependency of W. However, just like for other equations in continuum mechanics, we could also write the jump relations associated with $(177)_1$ and (183) at a singular surface by using elements of the theory of hyperbolic systems or a more naïve method such as the pill-box method. But since the "conservation laws" $(177)_1$ and (183) already exhibit source terms in the bulk (*i.e.*, they are not conservation laws in a strict mathematical sense), the associated jump relations will also contain surface source terms. The latter, *a priori* unknown but responsible for the dissipation at the singularity, have to be computed with the help of the standard jump relations associated with eqns. (170)–(172).

(ii) Case where $\mathbf{K} \neq \mathbf{0}$

Without reporting the whole algebra, starting with $(173)_2$, we let the reader check that the thermodynamic inequality (176) is replaced by

$$-\left(\frac{\mathrm{d}W}{\mathrm{d}t}+S\,\frac{\mathrm{d}\theta}{\mathrm{d}t}\right)+\mathrm{tr}\big(\mathbf{T}\cdot\dot{\mathbf{F}}\big)+\nabla_R\cdot(\theta\,\mathbf{K})-\mathbf{S}\cdot\nabla_R\theta\geq 0, \tag{187}$$

where $\mathbf{S}$ is still given by the general expression $(173)_2$. Equations (177) and (183) are left unchanged:

$$\frac{\mathrm{d}(S\,\theta)}{\mathrm{d}t}+\nabla_R\cdot\mathbf{Q}=h^{int},\quad h^{int}:=\mathrm{tr}\big(\mathbf{T}\cdot\dot{\mathbf{F}}\big)-\left.\frac{\partial W}{\partial t}\right|_X, \tag{188}$$

$$\frac{\mathrm{d}\mathbf{P}}{\mathrm{d}t}-\mathrm{div}_R\,\mathbf{b}=\mathbf{f}^{int}+\mathbf{f}^{ext}+\mathbf{f}^{inh}; \tag{189}$$

On account of (187), (179) is now replaced by

$$S\,\dot{\theta}+\mathbf{S}\cdot\nabla_R\theta\leq h^{int}+\nabla_R\cdot(\theta\,\mathbf{K}). \tag{190}$$

Now let us illustrate these general equations by specific cases, some of them trivial, and some others nontrivial.

3.4 Examples Without Body Force

A. Pure homogeneous elasticity In that case $\rho_0=$ const., and $W=\overline{W}(\mathbf{F})$ only. We have $h^{int}\equiv 0$, $\mathbf{f}^{int}\equiv\mathbf{0}$ since (178) holds good, and also $\mathbf{f}^{inh}=\mathbf{0}$, $\mathbf{Q}\equiv 0$ since the body is homogeneous and non conducting. Equations (177) and (183) reduce to the following [in fact Hamiltonian for a $(3+1)$-dimensional canonical momentum $(\mathbf{P},\theta_0\,S)$] system ($\theta_0=$ const.):

$$\frac{\mathrm{d}\mathbf{P}}{\mathrm{d}t}-\mathrm{div}_R\,\mathbf{b}=\mathbf{0},\quad \theta_0\,\frac{\mathrm{d}S}{\mathrm{d}t}=0. \tag{191}$$

In four-dimensional form this is the formulation of Kijowski and Magli (1998), in which the second of (191) is trivial and 4-momentum is given by $\mathbf{P}_{(4)}=(\mathbf{P},\theta_0\,S)$.

B. Inhomogeneous thermoelasticity of conductors In that case $\rho_0=\bar{\rho}_0(\mathbf{X})$, and $W=\overline{W}(\mathbf{F},\theta;\mathbf{X})$. We have the constitutive equations

$$\mathbf{T}=\frac{\partial\overline{W}}{\partial\mathbf{F}},\quad S=-\frac{\partial\overline{W}}{\partial\theta}, \tag{192}$$

that follow from a standard exploitation of the Clausius–Duhem inequality. Accordingly, we obtain that

$$\mathbf{f}^{int}\equiv\mathbf{f}^{th},\quad h^{int}\equiv h^{th}:=S\,\dot{\theta}, \tag{193}$$

where

$$\mathbf{f}^{th} := S\,\nabla_R \theta, \tag{194}$$

is the *material thermal force* first introduced by Bui (1978) in small strains – in fact in a little known conference paper of 1977 – and independently by Epstein and Maugin (1995a) in their geometrical considerations, so that (177) and (183) are replaced by the following canonical (*non*-Hamiltonian) system of balance of momentum and energy:

$$\frac{\mathrm{d}\mathbf{P}}{\mathrm{d}t} - \mathrm{div}_R\,\mathbf{b} = \mathbf{f}^{inh} + \mathbf{f}^{th}, \qquad \frac{\mathrm{d}(S\,\theta)}{\mathrm{d}t} + \nabla_R \cdot \mathbf{Q} = h^{th}, \tag{195}$$

as first found in Maugin (2000, 2006a).

C. Homogeneous dissipative solid material described by means of a diffusive internal variable Let α the internal variable of state whose tensorial nature is not specified. This may relate to damage, or anelasticity of some sort with a possible diffusion of the said variable so that its material gradient must be taken into account (*e.g.*, in strain-gradient plasticity). This is in the spirit of the thermodynamics developed at length in a book; Maugin (1999a). Then W is specified as the general sufficiently regular function

$$W = \overline{W}(\mathbf{F}, \theta, \alpha, \nabla_R \alpha). \tag{196}$$

(i) Case where $\mathbf{K} \equiv \mathbf{0}$

First we assume that $\mathbf{K}$ vanishes. The *equations of state* (in a sense mere definition of the partial derivatives of the free energy) are given by *Gibbs' equation* as

$$\mathbf{T} = \frac{\partial \overline{W}}{\partial \mathbf{F}}, \quad S = -\frac{\partial \overline{W}}{\partial \theta}, \quad A := -\frac{\partial \overline{W}}{\partial \alpha}, \quad \mathbf{B} := -\frac{\partial \overline{W}}{\partial (\nabla_R \alpha)}. \tag{197}$$

Accordingly, we find that

$$\mathbf{f}^{int} = \mathbf{f}^{th} + \mathbf{f}^{intr}, \quad h^{int} = h^{th} + h^{intr}, \tag{198}$$

where the thermal sources have already been defined and the "intrinsic" sources are given by

$$\mathbf{f}^{intr} := A\,(\nabla_R \alpha)^T + \mathbf{B} \cdot \nabla_R (\nabla_R \alpha)^T, \quad h^{intr} := A\,\dot{\alpha} + \mathbf{B} \cdot (\nabla_R \dot{\alpha})^T, \tag{199}$$

so that we have the following consistent (non-Hamiltonian) system of canonical balance laws:

$$\frac{\mathrm{d}\mathbf{P}}{\mathrm{d}t} - \mathrm{div}_R\,\mathbf{b} = \mathbf{f}^{th} + \mathbf{f}^{intr}, \qquad \frac{\mathrm{d}(S\,\theta)}{\mathrm{d}t} + \nabla_R \cdot \mathbf{Q} = h^{th} + h^{intr}, \tag{200}$$

while the dissipation reads

$$\Phi = h^{intr} - \mathbf{S} \cdot \nabla_R \theta \geq 0, \quad \mathbf{K} \equiv \mathbf{0}. \tag{201}$$

Here the thermodynamical forces A and $\mathbf{B}$ are purely dissipative by virtue of the "internal" character of the state variable α. This approach with $\mathbf{K} = \mathbf{0}$ favors the *continuum-mechanics* (Coleman–Noll) *standard viewpoint* (*cf.* Truesdell, 1969, 1984) by accepting the classical relationship between heat and entropy flux, and assuming that α and its material gradient are essentially independent. A more *field-theoretic* viewpoint is to envisage the set (187) through (190) as holding true and selecting the non-zero $\mathbf{K}$ such that the divergence term in (187) would not be present in that equation.

(ii) Case where $\mathbf{K} \neq \mathbf{0}$

In order to make the divergence term in (187) to vanish identically, we set

$$\mathbf{K} = -\frac{\mathbf{B}}{\theta}\,\dot{\alpha}. \tag{202}$$

This follows the scheme originally developed in Maugin (1990) for materials with *diffusive* dissipative processes described by means of internal variables of state. We let the reader check that eqns. (200) and (201) are then replaced by the following equations:

$$\frac{\mathrm{d}\mathbf{P}}{\mathrm{d}t} - \mathrm{div}_R\,\tilde{\mathbf{b}} = \mathbf{f}^{th} + \tilde{\mathbf{f}}^{intr}, \quad \frac{\mathrm{d}(S\,\theta)}{\mathrm{d}t} + \nabla_R \cdot \tilde{\mathbf{Q}} = h^{th} + \tilde{h}^{intr}, \tag{203}$$

and

$$\Phi = \tilde{h}^{intr} - \tilde{\mathbf{S}} \cdot \nabla_R \theta \geq 0, \quad \tilde{h}^{intr} := \tilde{A}\,\dot{\alpha} \tag{204}$$

where we have introduced the new definitions

$$\begin{gathered}\tilde{A} \equiv -\frac{\delta \overline{W}}{\delta \alpha} := -\left(\frac{\partial \overline{W}}{\partial \alpha} - \nabla_R \cdot \frac{\partial \overline{W}}{\partial(\nabla_R \alpha)}\right) = A - \nabla_R \cdot \mathbf{B}, \\ \tilde{\mathbf{S}} := \frac{\tilde{\mathbf{Q}}}{\theta}, \quad \tilde{\mathbf{Q}} = \mathbf{Q} - \mathbf{B}\,\dot{\alpha}\end{gathered} \tag{205}$$

and

$$\tilde{\mathbf{b}} = -(L\,\mathbf{1}_R + \mathbf{T} \cdot \mathbf{F} - \mathbf{B} \cdot (\nabla_R \alpha)^T), \quad \tilde{\mathbf{f}}^{intr} := \tilde{A}\,\nabla_R \alpha. \tag{206}$$

The two thermodynamical approaches just illustrated are to be compared to the constructive comments of Ireman and Nguyen Quoc Son (2004). Here we additionally show that alteration in the entropy flux definition goes along with a parallel alteration in the expression of the Eshelby stress tensor, thus reinforcing the space-like complementarity of equations (203). Alteration of the definition of the entropy flux is classical in the theory of irreversible processes in the presence of diffusion (*cf.* de Groot and Mazur, 1962). More on this with the possible interpretation of α as an additional degree of freedom when it is equipped with its own inertia in a work by Maugin, 2006b (also next paragraph).

3.5 Variable α as an Additional Degree of Freedom

Had we considered α as an observable field endowed with an inertia and a flux, where the latter is not necessarily purely dissipative (on the contrary, it could be purely **non**-dissipative), we would have started with:

- a global statement of the *principle of virtual power* (**PVP**) following Germain (1973a,b) and Maugin (1980), and
- a global statement of first and second laws of thermodynamics.

The first statement (**PVP**) would read for a body B occupying the regular material region B_R with regular boundary surface ∂B_R (of outward pointing unit normal $\mathbf{N}$) in its reference configuration K_R

$$P^*_{(inertia)} = P^*_{(i)} + P^*_{(data)} \tag{207}$$

with

$$P^*_{(inertia)} = \int_{B_R} \left(\rho_0 \, \mathbf{v} \cdot \mathbf{v}^* + \rho_0 \, (\mathbf{I} \cdot \dot{\alpha}) \cdot (\dot{\alpha})^*\right) \mathrm{d}V, \tag{208}$$

$$P^*_{(data)} = \int_{B_R} \left(\hat{\mathbf{f}} \cdot \mathbf{v}^* + \widehat{A} \cdot (\dot{\alpha})^*\right) \mathrm{d}V_0 + \int_{\partial B_R} \left(\widehat{\mathbf{T}}_{\mathbf{n}} \cdot \mathbf{v}^* + \widehat{\mathbf{B}}_{\mathbf{n}} \cdot (\dot{\mathbf{v}})^*\right) \mathrm{d}S_0, \tag{209}$$

and

$$P^*_{(i)} = -\int_{B_R} \left(\mathrm{tr}\left(\mathbf{T} \cdot (\nabla_R \mathbf{v})^*\right) - A \cdot (\dot{\alpha})^* - \mathbf{B} \cdot \nabla_R (\dot{\alpha})^*\right) \mathrm{d}A, \tag{210}$$

where

$$\hat{\mathbf{f}}, \quad \widehat{A}, \quad \widehat{\mathbf{T}}_{\mathbf{n}}, \quad \widehat{\mathbf{B}}_{\mathbf{n}} \tag{211}$$

are prescribed – hence the subscript "*data*" in the corresponding power –, and

$$A, \quad \mathbf{T}, \quad \mathbf{B} \tag{212}$$

are *generalized internal forces* to be given (nondissipative or dissipative, or both) *constitutive equations.*

Remarks.

(*i*) In equations (207) to (210) an asterisk means that the expression or the field to which it is attached is a virtual field at our choice disposal.

Deleting the asterisk means that the expression or the field in question takes its actual value in the initial-value boundary-value problem to be solved. Prescriptions to formulate the various virtual powers, once the basic fields such as the motion of the body and the variable α are chosen, have been clearly enunciated in Maugin (1980). In particular, the virtual power of internal forces $P^*_{(i)}$ must be written as a linear continuous form on a set of **objective** generalized velocity fields in order that the internal forces introduced by the inherent duality as the set (212), be objective fields (*i.e.*, invariant under superimposed rigid body motions in the current configuration K_t). If α simply is a scalar under such transformations, there are no problems in writing directly the linear form (210). For instance this is done by Frémond and B. Nedjar (1993, 1996) when α is a scalar damage variable. Then the "inertia" $\boldsymbol{I}$ in the expression of the virtual power of inertia force is also a scalar, if it exists at all. The objectivity requirement does not apply to the other virtual powers since neither inertial forces nor externally applied forces are objective. Whenever α is a field of higher tensorial order (*e.g.*, a vector field such as in the cases of liquid crystals, internally electrically polarized materials – *cf.* Maugin, 1974b, 1976 –, or internally magnetized material – *cf.* Maugin, 1974a), one has to be more thoughtful in writing down $P_{(i)}$; in particular, objective (convected or Jaumann) time derivatives of α and its gradient must be introduced. This remark obviously holds true if $\boldsymbol{\alpha}$ is a tensor of second order (*e.g.*, representative of a rigid or deformable microstructure). The tensorial order of the "inertia" $\mathbf{I}$ is accordingly formulated. The dots between symbols in eqns. (207) through (210) take their full meaning of contractors of indices or inner products in the appropriate space. These technicalities are more or less obvious. The formulation given in eqns. (207)–(210) is of the type of a *first-gradient theory* in so far as both the classical continuum motion *and* the variable α are concerned. Higher-order gradient theories, *i.e.*, "stronger non-local theories" may be constructed following the same generalized pattern and rules (*cf.* Maugin, 1980). "First order" is enough to compare with the foregoing "diffusive internal variable theory. Note that there is no obligation that the theory be of the same order for the classical motion and for the additional variable α. This choice is at our disposal and depends essentially on our apprehending of the spatial range of interactions.

(*ii*) Here we call α an observable field. We could as well say "*controllable*" field, meaning by this that, contrary to internal variables of state, their values can be adjusted by a proper action in the bulk and at the surface of the body by the introduced generalized forces present in $P_{(data)}$.

The two laws of thermodynamics are now set forth in the following global form:

First law of thermodynamics

$$\frac{\mathrm{d}}{\mathrm{d}t}\int_{B_R}(K+E)\,\mathrm{d}V = P_{(data)} + \dot{Q}, \quad \dot{Q} = -\int_{\partial B_R}\widehat{Q}_{\mathbf{n}}\,\mathrm{d}A \tag{213}$$

and

Second law of thermodynamics

$$\frac{\mathrm{d}}{\mathrm{d}t}\int_{B_R} S\,\mathrm{d}V + \int_{\partial B_R}\mathbf{N}\cdot\widetilde{\mathbf{S}}\,\mathrm{d}A \geq 0, \quad \widetilde{\mathbf{S}} = \frac{\mathbf{Q}}{\theta} + \mathbf{K} \tag{214}$$

In these two equations, K, E, and S are the kinetic energy, internal energy, and entropy density per unit **reference** volume. $\dot{Q}$ is the energy rate supply to the body through its boundary. This occurs through heat. We do not introduce any energy supply per unit volume. In writing the second law, we have somewhat anticipated by introducing an entropy flux with the more general relationship to heat and dissipative processes since, while θ is the standard thermodynamic temperature, we have admitted the possible presence of a nonzero extra entropy flux $\mathbf{K}$. The kinetic energy here is given by

$$K = \frac{1}{2}\rho_0\,\mathbf{v}^2 + \frac{1}{2}\rho_0\,\dot{\alpha}\cdot\mathbf{I}\cdot\dot{\alpha}. \tag{215}$$

As a first outcome from the application of (207) for any virtual fields $(\mathbf{v}^*, (\dot{\alpha})^*)$ in any volume and surface element, we obtain the following balance laws and natural boundary conditions:

$$\left.\frac{\partial(\rho_0\,\mathbf{v})}{\partial t}\right|_X - \mathrm{div}_R\,\mathbf{T} = \hat{\mathbf{f}} \text{ in } B_R, \quad \widehat{\mathbf{T}}_{\mathbf{n}} = \mathbf{N}\cdot\mathbf{T} \text{ at } \partial B_R, \tag{216}$$

$$\left.\frac{\partial(\rho_0\,\mathbf{I}\cdot\dot{\alpha})}{\partial t}\right|_X = \widehat{A} + A - \nabla_R\cdot\mathbf{B} \text{ in } B_R, \quad \widehat{\mathbf{B}}_{\mathbf{n}} = \mathbf{N}\cdot\mathbf{B} \text{ at } \partial B_R. \tag{217}$$

A second result is obtained for real virtual velocity fields (no asterisks), on account of (215). It is the so-called *equation of kinetic energy* in global form as

$$\frac{\mathrm{d}}{\mathrm{d}t}\int_{B_R} K\,\mathrm{d}V = P_{(i)} + P_{(data)} \tag{218}$$

On combining this with the first law (213), we obtain the so-called *equation of internal energy* in global form:

$$\frac{\mathrm{d}}{\mathrm{d}t}\int_{B_R} E\,\mathrm{d}V + P_{(i)} = \dot{Q}, \tag{219}$$

the localization of which yields

$$\begin{gathered}\left.\frac{\partial E}{\partial t}\right|_{X} = \mathrm{tr}\big(\mathbf{T}\cdot(\nabla_R\mathbf{v})^T\big) - A\cdot\dot{\alpha} - \mathbf{B}\cdot\nabla_R\dot{\alpha} - \nabla_R\cdot\mathbf{Q} \text{ in } B_R,\\ \widehat{Q}_{\mathbf{n}} = \mathbf{N}\cdot\mathbf{Q} \text{ at } \partial B_R.\end{gathered} \tag{220}$$

Introducing now the Helmholtz free energy per unit reference volume W by

$$W = E - S\,\theta, \tag{221}$$

and combining with the local form of the inequality $(214)_1$, we arrive at the following *Clausius–Duhem inequality*:

$$-\left(\frac{\mathrm{d}W}{\mathrm{d}t} + S\,\frac{\mathrm{d}\theta}{\mathrm{d}t}\right) + p_{(i)} + \nabla_R\cdot(\theta\,\mathbf{K}) - \widetilde{\mathbf{S}}\cdot\nabla_R\theta \geq 0; \tag{222}$$

while evaluating $\theta\,\mathrm{d}S/\mathrm{d}t$ from $(220)_1$ on account of (221) we obtain the "entropy equation" which will ultimately provide the heat-propagation equation, in the form

$$\theta\,\frac{\mathrm{d}S}{\mathrm{d}t} = -\left(\frac{\mathrm{d}W}{\mathrm{d}t} + S\,\frac{\mathrm{d}\theta}{\mathrm{d}t}\right) + p_{(i)} - \nabla_R\cdot\mathbf{Q}. \tag{223}$$

We have introduced the following notation

$$p_{(i)} := \mathrm{tr}\big(\mathbf{T}\cdot\dot{\mathbf{F}}\big) - A\,\dot{\alpha} - \mathbf{B}\cdot\nabla_R\dot{\alpha}, \tag{224}$$

and accounted for the fact that

$$(\nabla_R\mathbf{v})^T \equiv \dot{\mathbf{F}}. \tag{225}$$

Now we consider three cases of exploitation of the above-developed scheme.

A. The only dissipative process is heat conduction We have the following obvious reduction:

$$\mathbf{K} \equiv \mathbf{0}, \quad \widetilde{\mathbf{S}} \equiv \frac{\mathbf{Q}}{\theta}, \tag{226}$$

$$-\left(\frac{\mathrm{d}W}{\mathrm{d}t}+S\,\frac{\mathrm{d}\theta}{\mathrm{d}t}\right)+p_{(i)}=0, \quad \Phi^{th}:=-\widetilde{\mathbf{S}}\cdot\nabla_R\theta\geq 0, \tag{227}$$

$$\theta\,\frac{\mathrm{d}S}{\mathrm{d}t}+\nabla_R\cdot\mathbf{Q}=0. \tag{228}$$

The last of these can also be written as (since $\theta>0$ always)

$$\frac{\mathrm{d}S}{\mathrm{d}t}+\nabla_R\cdot\widetilde{\mathbf{S}}=\sigma^{th}, \quad \sigma^{th}=\frac{\Phi^{th}}{\theta}. \tag{229}$$

With a functional dependence $W=\overline{W}(\mathbf{F},\theta,\alpha,\nabla_R\alpha)$, a classical reasoning applied to the first of (227) yields the constitutive equations:

$$S=-\frac{\partial W}{\partial\theta}, \quad \mathbf{T}=\frac{\partial W}{\partial\mathbf{F}}, \quad A=-\frac{\partial W}{\partial\alpha}, \quad \mathbf{B}=-\frac{\partial W}{\partial(\nabla_R\alpha)}. \tag{230}$$

It remains to give an expression to the heat flux $\mathbf{Q}$ in order to arrive at a fully explicit expression for (220).

B. The Coleman–Noll continuum thermodynamics viewpoint In this case eqns.(226) are valid *a priori* independently of any deformation and time internal evolution of α, *i.e.*,

$$\mathbf{K}\equiv\mathbf{0} \quad\Rightarrow\quad \widetilde{\mathbf{S}}\equiv\frac{\mathbf{Q}}{\theta}. \tag{231}$$

The generalized internal forces (212) are each the sum of a thermodynamically reversible part and a *dissipative* thermodynamically irreversible part, *i.e.*,

$$A=A_{rev}+A_{irrev}, \quad \mathbf{T}=\mathbf{T}_{rev}+\mathbf{T}_{irrev}, \quad \mathbf{B}=\mathbf{B}_{rev}+\mathbf{B}_{irrev}. \tag{232}$$

The dependent functions $\{W,S,A_{rev},\mathbf{T}_{rev},\mathbf{B}_{rev}\}$ depend all on the same set as

$$W=\overline{W}(\mathbf{F},\theta,\alpha,\nabla_R\alpha), \tag{233}$$

while the remaining dissipative fields $\{\widetilde{\mathbf{S}},A_{irrev},\mathbf{T}_{irrev},\mathbf{B}_{irrev}\}$ depend on the same set as **also** on the set $\{\nabla_R\theta,\dot{\mathbf{F}},\dot{\alpha},\nabla_R\dot{\alpha}\}$. Here also a classical reasoning yields constitutive equations similar to those in eqns. (230) for the nondissipative contributions, *i.e.*,

$$\begin{gathered}\mathbf{T}_{rev}=\frac{\partial\overline{W}}{\partial\mathbf{F}}, \quad A_{rev}=-\frac{\partial\overline{W}}{\partial\alpha}, \quad \mathbf{B}_{rev}=-\frac{\partial\overline{W}}{\partial(\nabla_R\alpha)}, \\ S=-\frac{\partial\overline{W}}{\partial\theta}, \quad W=\overline{W}(\mathbf{F},\theta,\alpha,\nabla_R\alpha)\end{gathered} \tag{234}$$

while there remains the following **residual dissipation inequality**:

$$\Phi := \mathbf{T}_{irrev} \cdot \dot{\mathbf{F}} - A_{irrev} \cdot \dot{\alpha} - \mathbf{B}_{irrev} \cdot \nabla_R \cdot \dot{\alpha} - \widetilde{\mathbf{S}} \cdot \nabla_R \theta \geq 0. \tag{235}$$

In exploiting this inequality, one does as if the gradient of $\dot{\alpha}$ and $\dot{\alpha}$ itself were independent variables. This is a standard method reported and advised in continuum mechanics books. Note that here the entropy flux satisfies the classical formula $(231)_2$. For instance, this is applied to liquid crystals (where α is a unit "director") by Leslie (1968) or to elastic dielectrics by Maugin (1974b, 1977) and elastic ferromagnets in Maugin (1974a) – also Maugin (1972) where this was first applied to a sufficiently sophisticated case. We shall not pursue this exploitation of eqn. (235) here. Suffice it to notice that if the time parity of α as a tensor allows for it, there might be a direct linear coupling between the irreversible effect associated with α and heat conduction. On another occasion such a direct coupling could arise between heat conduction and the dissipative *process* associated with the gradient of α.

C. The field-theoretic viewpoint This is quite different in the sense that a scientist trained in analytical mechanics will not *a priori* assume that $\dot{\alpha}$ and $\nabla_R \dot{\alpha}$ are independent. He may prefer to keep the general expressions $(214)_2$ and (222) so that (235) will be replaced by the following expression in which $\widetilde{\mathbf{S}}$ is given by $(214)_2$ while eqns. (234) still hold true:

$$\Phi = \mathbf{T}_{irrev} \cdot \dot{\mathbf{F}} - A_{irrev} \cdot \dot{\alpha} - \mathbf{B}_{irrev} \cdot \nabla_R \cdot \dot{\alpha} + \nabla_R \cdot (\theta \mathbf{K}) - \widetilde{\mathbf{S}} \cdot \nabla_R \theta \geq 0. \tag{236}$$

But we note that the very form of eqn. $(205)_1$ suggests us to consider the grouping

$$\widetilde{A} := A - \nabla_R \cdot \mathbf{B}. \tag{237}$$

According to eqns. (237) the reversible part of this is given by

$$\widetilde{A}_{rev} \equiv -\frac{\delta \overline{W}}{\delta \alpha} := -\left(\frac{\partial \overline{W}}{\partial \alpha} - \nabla_R \cdot \frac{\partial \overline{W}}{\partial (\nabla_R \alpha)} \right). \tag{238}$$

On the basis of (237) we also set

$$\widetilde{A}_{irrev} := A_{irrev} - \nabla_R \cdot \mathbf{B}_{irrev}. \tag{239}$$

On selecting the extra entropy flux as

$$\mathbf{K} = \frac{\mathbf{B}_{irrev}}{\theta} \dot{\alpha}, \tag{240}$$

it is easily shown that the inequality (236) reduces to the following inequality

$$\Phi = \mathbf{T}_{irrev} \cdot \nabla_R \mathbf{v} - \widetilde{A}_{irrev} \cdot \dot{\alpha} - \widetilde{\mathbf{S}} \cdot \nabla_R \theta \geq 0, \tag{241}$$

where $\widetilde{\mathbf{S}}$ is given by eqns. $(214)_2$ and (240). The exploitation of this inequality would follow the usual formalism of irreversible thermodynamics.

Simultaneously, the heat-propagation equation (220) is shown to take on the following form (compare to eqn. (223))

$$\theta \frac{\mathrm{d}S}{\mathrm{d}t} + \nabla_R \cdot \widetilde{\mathbf{Q}} = \mathbf{T}_{irrev} : \dot{\mathbf{F}} - \widetilde{A}_{irrev}\, \dot{\alpha}, \quad \widetilde{\mathbf{Q}} := \mathbf{Q} + \mathbf{B}_{irrev}\, \dot{\alpha}, \tag{242}$$

providing thus a completely consistent theory.

Comment. Equation (240) is a sufficient condition for the non-existence of a divergence term in the residual dissipation inequality.

3.6 Example: Homogeneous Dissipative Solid Material Described by Means of a Diffusive Internal Variable

In general α is an internal variable of state whose tensorial nature is not specified. This may relate to damage, or anelasticity of some sort with a possible diffusion of the said variable so that its material gradient must be taken into account (*e.g.*, in strain-gradient plasticity). Then W is specified as the general sufficiently regular function $W = \overline{W}(\mathbf{F}, \theta, \alpha, \nabla_R \alpha)$. Here we specify that α is a scalar variable c akin to a concentration. We keep the possibility that $\mathbf{K}$ be not zero. The equations of state are given by Gibbs' equation as

$$\begin{aligned} \mathbf{T} &= \frac{\partial \overline{W}}{\partial \mathbf{F}}, \quad S = -\frac{\partial \overline{W}}{\partial \theta}, \\ \mu &:= \frac{\partial \overline{W}}{\partial c}, \quad \mathbf{M} := \frac{\partial \overline{W}}{\partial (\nabla_R c)}. \end{aligned} \tag{243}$$

so that μ is a chemical potential. We find that

$$\mathbf{f}^{int} = \mathbf{f}^{th} + \mathbf{f}^{intr}, \quad h^{int} = h^{th} + h^{intr}, \tag{244}$$

where the thermal sources have already been defined and the "intrinsic" sources are given by

$$\mathbf{f}^{intr} := -\mu\, (\nabla_R c)^T - \mathbf{M} \cdot \nabla_R (\nabla_R c)^T, \quad h^{intr} := -\mu\, \dot{c} - \mathbf{M} \cdot (\nabla_R \dot{c})^T, \tag{245}$$

so that we have the following consistent (non-Hamiltonian) system of canonical balance laws:

$$\frac{\mathrm{d}\mathbf{P}}{\mathrm{d}t} - \mathrm{div}_R\, \mathbf{b} = \mathbf{f}^{th} + \mathbf{f}^{intr}, \quad \frac{\mathrm{d}(S\,\theta)}{\mathrm{d}t} + \nabla_R \cdot \mathbf{Q} = h^{th} + h^{intr}, \tag{246}$$

while the dissipation reads

$$\Phi = h^{intr} - \mathbf{S} \cdot \nabla_R \theta \geq 0, \quad \mathbf{K} \equiv \mathbf{0}. \tag{247}$$

This approach favors the *continuum mechanics* (Coleman–Noll) *standard viewpoint* by accepting the classical relationship between heat and entropy flux, and assuming that c and its material gradient are essentially independent.

A more *field-theoretic* viewpoint is to envisage a non-zero $\mathbf{K} = \theta^{-1}\mathbf{M}\,\dot{c}$. We let the reader check that eqns. (177) and (183) are then replaced by the following equations:

$$\frac{\mathrm{d}\mathbf{P}}{\mathrm{d}t} - \mathrm{div}_R\,\tilde{\mathbf{b}} = \mathbf{f}^{th} + \tilde{\mathbf{f}}^{intr}, \quad \frac{\mathrm{d}(S\,\theta)}{\mathrm{d}t} + \nabla_R \cdot \widetilde{\mathbf{Q}} = h^{th} + \tilde{h}^{intr}, \tag{248}$$

and

$$\Phi = \tilde{h}^{intr} - \widetilde{\mathbf{S}} \cdot \nabla_R \theta \geq 0, \quad \tilde{h}^{intr} = -\tilde{\mu}\,\dot{c}, \tag{249}$$

where we have introduced the new definitions

$$\tilde{\mu} \equiv \frac{\delta \overline{W}}{\delta c} := \left(\frac{\partial \overline{W}}{\partial c} - \nabla_R \cdot \frac{\partial \overline{W}}{\partial(\nabla_R c)} \right) = \mu - \nabla_R \cdot \mathbf{M},$$
$$\widetilde{\mathbf{S}} := \frac{\widetilde{\mathbf{Q}}}{\theta}, \quad \widetilde{\mathbf{Q}} = \mathbf{Q} + \mathbf{M}\,\dot{c} \tag{250}$$

and

$$\tilde{\mathbf{b}} = -(L\,\mathbf{1}_R + \mathbf{T} \cdot \mathbf{F} + \mathbf{M} \otimes (\nabla_R c)), \quad \tilde{\mathbf{f}}^{intr} := -\tilde{\mu}\,\nabla_R c. \tag{251}$$

This is in the spirit of the approach that we advocated before (Maugin, 1990).

The closure of the thermodynamical system requires the setting of relationships between $(\mu, \mathbf{M})$ and $(\dot{c}, \nabla_R \dot{c})$ on the one hand, or $\tilde{\mu}$ and $\dot{c}$ on the other, and a relationship *à la* Fourier between $\widetilde{\mathbf{S}}$ and $\nabla_R \theta$. We pursue the second line (the field-theoretic one), considering as an example a free energy function W of the type

$$\overline{W}(\mathbf{F}, \theta, c, \nabla_R c) = \frac{1}{2}\gamma\,(\nabla_R c) \cdot (\nabla_R c) + \widehat{W}(\mathbf{F}, \theta, c), \tag{252}$$

where the positive scalar coefficient γ depends at most on the temperature θ. In agreement with (249) we select (sufficient conditions) the evolution equation and heat transport equation as

$$\dot{c} = -\frac{\tilde{\mu}}{\tau}, \quad \widetilde{\mathbf{S}} = -\chi\,\nabla_R \theta, \tag{253}$$

with positive coefficients τ and χ (they could be temperature dependent). On account of these and eqns. (252) and (250), we obtain a *nonlinear evolution-diffusion equation* for c in the following form:

$$\tau\,\dot{c} + f(c; \mathbf{F}, \theta) = \nabla_R \cdot (\gamma\,\nabla_R c), \tag{254}$$

where

$$f(c; \mathbf{F}, \theta) \equiv \frac{\partial \widehat{W}}{\partial c} \equiv \mu. \tag{255}$$

One could assume that $\widehat{W}$ behaves like c^2 for small c's, but more generally it may be of a higher degree in c, or may even be non convex in c, remaining nonetheless positive. To the same degree of approximation, the nonlinear evolution-diffusion equation for temperature is given by

$$\theta\,\dot{S} + \big(f - \nabla_R \cdot (\gamma\,\nabla_R c)\big)\,\dot{c} + \nabla_R \cdot (\theta\,\chi\,\nabla_R \theta) = 0, \tag{256}$$

where $S = -\partial W / \partial \theta$ is usually linear in θ and such that W is *concave* in this variable.

Finally, eqn. (248), considered with inertia neglected, yields the equation

$$\mathrm{div}_R\,\big(W\,\mathbf{1}_R - \mathbf{T} \cdot \mathbf{F} - \lambda\,(\nabla_R c) \otimes (\nabla_R c)\big) + S\,\nabla_R \theta + \tilde{\mu}\,\nabla_R c = 0, \tag{257}$$

or equivalently

$$\mathrm{div}_R\,\big(\widetilde{G}\,\mathbf{1}_R - \mathbf{T} \cdot \mathbf{F} - \lambda\,(\nabla_R c) \otimes (\nabla_R c)\big) - \theta\,\nabla_R S - c\,\nabla_R \tilde{\mu} = 0, \tag{258}$$

where $\widetilde{G} = W + S\,\theta + \tilde{\mu}\,c = E + \tilde{\mu}\,c$ is Gibbs' energy density.

In the absence of temperature effects and for a nondiffusive variable c, this reduces to the simple equation

$$\mathrm{div}_R\,\big(G\,\mathbf{1}_R - \mathbf{T} \cdot \mathbf{F}\big) - c\,\nabla_R \mu = 0, \tag{259}$$

where $G = W + \mu\,c$. In this case (254) reduces to a (generally) *nonlinear relaxation equation*:

$$\tau\,\dot{c} + f(c; \mathbf{F}) = 0, \tag{260}$$

Remarks.

(*i*) First, one may think that we have given a peculiar status to the free energy W in expressing the Eshelby stress in terms of W. What is important is that it be the same potential that appears in both canonical equations of energy and momentum. It could have been the *internal energy*, and then the situation would be more prepared to treat adiabatic situations. This critical dependence of both equations of energy and material momentum on the same thermodynamic potential was especially noticed by the author (Maugin, 2002a) and Abeyaratne and Knowles (2000) when dealing with the jump relations associated with these canonical equations.

(*ii*) Second but nonetheless important is the fact that the expression for the Eshelby stress contains the gradient of α. This means that the conjugate force **B** will directly play a role in the evaluation of the critical driving forces if we follow the scheme given in Maugin (2000), while the associated expression of $\tilde{h}^{int}$ will directly yield an *evolution-diffusion equation* for α in a simple application of the thermodynamics of irreversible processes.

3.7 Conclusion and Comments

The above given developments show that there is no unique thermomechanical description of continuous media exhibiting diffusion of some property. A first choice is between considering the additional variable required to describe this property either as a true additional degree of freedom (with an inertia and then safely following an application of the principle of virtual power) or as or an internal variable of state. If the second possibility is selected (essentially when the new property is fully dissipative) then the main question is whether this variable is diffusive and therefore controllable to some extent *via* a flux at a material boundary. Finally, it is shown that the field-theoretic viewpoint certainly prevails and is more economical from the point of view of thought even though both notions of entropy and Eshelby's material stress tensor must then be consistently revisited. The resulting formulation offers new perspectives of research in both biomechanics and the mechanics of complex materials exhibiting a microstructure of some kind. These may be inert materials or materials met in biological situations. The example of micromorphic materials is treated in greater detail in the last lecture.

Two viewpoints on the equation of material momentum Here we shall emphasize that there exist two opposite viewpoints concerning the status of the equation of material (or canonical) momentum in continuum mechanics. The viewpoint of the author (Maugin, 1993, 1995) exposed in this chapter is that this equation is never independent of the classical (physical) equation of linear momentum, in Cauchy or Piola–Kirchhoff form, being essentially deduced from the latter by a complete pull-back to the reference configuration, even when constitutive equations are not known to start with (*i.e.*, there exists no variational formulation). It is, therefore, an identity at all regular material points – but it still is extremely useful on any singular manifold. This viewpoint agrees with the application of Noether's identity when one considers a variational formulation for a nondissipative material, a point of view shared by J. D. Eshelby in his original works, *e.g.*, Eshelby, 1975.

The second viewpoint is that of M. E. Gurtin (several works but particularly Gurtin, 1999) who repeatedly claims that the equation of material momentum equation (bulk equation for "configurational forces") is an *a priori* statement independent of the classical balance laws although in the end it is, for sure (our opinion), always shown to be related to the physical balance of momentum so that Gurtin's statement is somewhat inappropriate. Hereabove we expanded the view that the balance of canonical or material momentum, albeit following from the balance of physical momentum, can be formulated independently of any constitutive behavior. Moreover, accounting for the fact that this equation is the space-like equation associated with a particular form of the energy equation, it was shown that the former and the latter can be used in parallel to build a consistent thermomechanics of many behaviors. As a matter of fact, the two canonical equations of momentum and energy must be consistent if, for instance, dissipation due to the irreversible movement of a singularity set is none other than the power expanded by the driving force acting on this set, and this in *all* cases (Maugin, 2000).

Additional Bibliography: Bowen, 1967, 1976; Epstein and Maugin, 1995a; Eringen and Maugin, 1990; Forest, 2006; Forest and Sievert, 2003; Maugin, 1979, 1994, 2007, 2011.

4 Lecture 4: Exploiting the Canonical Conservation Laws in Defect Mechanics

4.1 Introduction

From the contents of the second and third lectures it is clear that we are now equipped to treat many problems where the Eshelby stress and its related thermodynamics play a prominent role. There are two types of applications depending on whether we work with sufficiently smooth fields or with field singularities.

In the first case, the equation of canonical momentum appears as a strict identity deduced from the classical field equations either in the variational framework *via* Noether's identity, or in a general postulational framework of balance equations as was the case in the Third lecture. This was done for elastic, anelastic, second-gradient and microstructured (*e.g.*, micropolar or micromorphic) materials opening up wide horizons in materials sciences). Applications then rely either on the meaning of the Eshelby material stress as the *driving force behind volume matter rearrangements* such as in finite strain plasticity, damage, growth (of, *e.g.*, biomaterials), and bulk phase transformations, or one can use the interplay between two equipresent formulations of a similar concept, momentum balance, to exploit one (the physical momentum equation or its equibilibrium degenerate form) in the solution of a boundary-value problem (analytically or numerically) and use the second (the conservation of canonical momentum or its quasi-static version) as a criterion of accuracy of the first, since satisfaction of the first does not necessarily entail that of the second (which is of higher order in the fields than the first) is automatically satisfied (but only theoretically). This is the case in numerical methods such as *finite-difference ones* used in dynamical studies of complex systems (works by C. I. Christov and G. A. Maugin) or *finite-element ones* as many recent works have shown since 2001 (works mainly by M. Braun, R. Mueller, G. A. Maugin, D. Gross, P. Steinmann and others). This type of applications is not dealt with here, but we shall deal at length with these in a future book (Thermomechanics of configurational forces – Adventures of GAMM in the Wonderland of material forces published as Maugin (2011)). Here we are satisfied with noting that configurational-material forces, just like energy, being of higher order in the fields than the classical entities are fundamentally *indicators of the presence of singularities* over sets of various dimensions, points, lines, surfaces, and provide their *driving forces in a* phenomenological *thermodynamic setting.* Phase-transition fronts and shock waves are examples of the surface type; Strongly localized dynamical solutions such as solitonic solutions also belong in this framework. Dislocations and cracks are examples belonging to the line type (in two dimensions). We are therefore back to our initial considerations in Lecture 1, except that we are now equipped with a powerful apparatus to treat so many material cases. This fourth lecture pinpoints some of these applications.

4.2 The Case of Cracks in Pure Homogeneous Elasticity

Here like in many cases, pure elasticity (whether linear or nonlinear, isotropic or anisotropic, but homogeneous) serves as a standard towards which all other cases must be compared. The reason for this is two-fold. In order to relate the driving force acting on the tip of the crack and the accompanying released energy as the crack propagates, one must know the degree of singularity of the (here pure) mechanical fields as we approach the tip of the crack. This is exactly and relatively simply obtained in 2D (plane strain or plane stress) linear elasticity, where the singularity of the displacement field is essentially that exhibited by the Laplacian operator. Second, in the framework of this ideal but paradigmatic case we can establish without approximation the remarkable result that we called the *analytical theory of brittle fracture* ("brittle" relates to the fact that fracture occurs in the elastic regime of the mechanical behavior; *cf.* Maugin, 1992 for the standard theory of fracture). The following is rigorously mathematically proven Dascalu and Maugin (1993): *the driving force (DRF) is deduced from the equation of conservation of material (canonical) momentum and, simultaneously the corresponding energy-release rate (ERR) is deduced from the canonical energy equation so that the latter (ERR) is none other than the inner product of the first (DRF) in material space by the "material" velocity of progress of the crack tip. This product has to be nonnegative (second law of thermodynamics).* This reflects the dissipation of topological origin (not of intrinsic origin like in plasticity) related to the irreversible progress of the crack: the faces of the crack cannot solder back unless some additional energy is imputed in the system. This is expressed mathematically by the following formulas that we recall without derivation (for this see Dascalu and Maugin, 1993):

- *Balance of configurational forces* (here "non"-conservation of material momentum) for the steadily fracturing body:

$$\frac{\partial}{\partial t}\int_B \mathbf{P}\,\mathrm{d}V + \mathbf{F}^{crack} = \int_{\partial B} \mathbf{N}\cdot\mathbf{b}\,\mathrm{d}A, \tag{261}$$

- *Balance of energy* (here "non"-conservation of energy) in the same conditions:

$$\frac{\partial}{\partial t}\int_B H\,\mathrm{d}V + G^{crack} = \int_{\partial B} \mathbf{N}\cdot\mathbf{T}\cdot\mathbf{v}\,\mathrm{d}A, \tag{262}$$

In these equations the two source terms are given by (axis X_1 in the direction of the crack)

$$F_1^{crack} = \int_\Gamma \left((\mathbf{N} \cdot \mathbf{b})_1 + P_1 \left(\overline{\mathbf{V}} \cdot \mathbf{N} \right) \right) \mathrm{d}A - \frac{\partial}{\partial t} \int_G P_1 \, \mathrm{d}V, \tag{263}$$

and

$$G^{crack} = \int_\Gamma \left(H \left(\overline{\mathbf{V}} \cdot \mathbf{N} \right) + \mathbf{N} \cdot \mathbf{T} \cdot \mathbf{v} \right) \mathrm{d}A - \frac{\partial}{\partial t} \int_G H \, \mathrm{d}V, \tag{264}$$

and they are such that

$$G^{crack} = \mathbf{F}^{crack} \cdot \overline{\mathbf{V}} \geq 0. \tag{265}$$

In the above set equations, the regular material region of volume extension

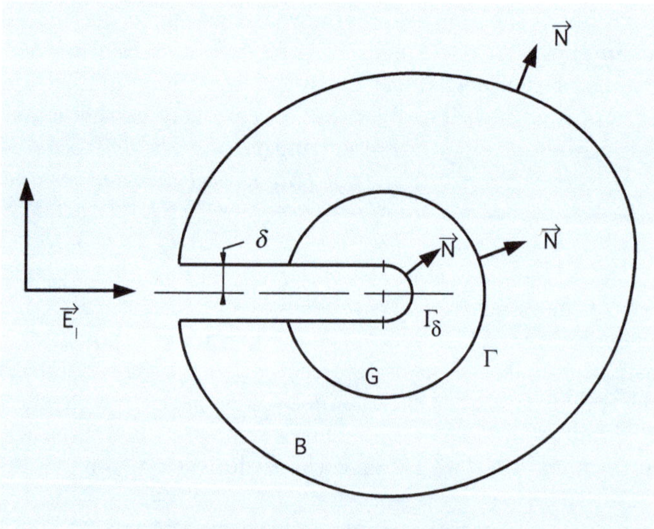

Figure 1. Notch δ in a nonlinear elastic body B

Figure 3. Notch δ in nonlinear elastic body B (Dascalu and Maugin, 1993)

G is taken around the crack tip; it has external (cylindrical) boundary Γ equipped with unit outward normal $\mathbf{N}$ and the crack is moving inward with uniform material velocity $\overline{\mathbf{V}}$. As a matter of fact, we obtained for the driving force on the crack and the corresponding energy-release rate the following

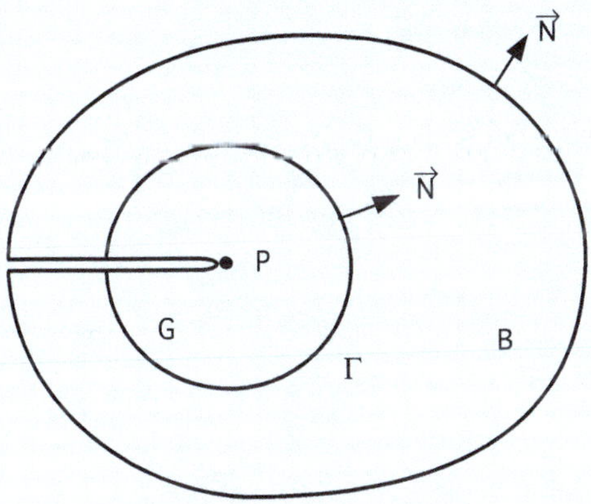

Figure 2. The straight-through crack as a linear of a notch

Figure 4. The straight-through crack as a linear of a notch (Dascalu and Maugin, 1993)

illuminating formulas valid for finite-strain and dynamical elasticity:

$$\mathbf{F}^{crack} = -\lim_{\delta\to 0} \mathbf{E}_1 \int_{\Gamma_\delta} \left(L\,N_1 - P_1\,(\overline{\mathbf{V}}\cdot\mathbf{N})\right) \mathrm{d}A, \qquad (266)$$

$$G^{crack} = \lim_{\delta\to 0} \int_{\Gamma_\delta} H\,(\overline{\mathbf{V}}\cdot\mathbf{N})\,\mathrm{d}A, \qquad (267)$$

where the limit corresponds to the procedure of considering the crack of vanishing thickness as the uniform limit of a sequence of rounded-end notches (of radius δ), following the original argument of Rice (1968) – see Figures 3 and 4. We have referred to these two basic formulas as those of the *analytical theory of brittle fracture*, because they involve the elasticity Lagrangian and Hamiltonian densities, respectively. The meaning of (267) is clear: it represents the energy consumed ("swallowed") per unit time by the inward crack motion. The tip of the crack acts as an *energy sink*. The proof of (265) involves considerations on the singularity of the velocity field (following arguments of Gurtin, 1979; and Nguyen Quoc Son, 1980) and a Legendre transformation between L and H. It is readily checked that (263) reduces to the celebrated path-independent J-integral of brittle fracture in quasi-statics and small strains. Indeed, in these conditions, the material momentum and kinetic-energy contributions are discarded, and

$$\mathbf{F} = \mathbf{1} + (\nabla_R \mathbf{u})^T, \tag{268}$$

where $\mathbf{u}(\mathbf{X}, t)$ is the elastic displacement. On expanding the expression of $\mathbf{T} \cdot \mathbf{F}$ in $\mathbf{b}$, we find that the first contribution in the stress contributes nothing because of its order of singularity and there remains the well-known expression

$$F^{crack} = J = \int_\Gamma \left(W\, N_1 - \mathbf{t} \cdot \frac{\partial \mathbf{u}}{\partial X_1} \right) \mathrm{d}L \tag{269}$$

per unit thickness of the body, where $\mathbf{t}$ is the traction at Γ. Expression (269) has the good taste to be *contour independent*, hence a great advantage in computations. But this is a peculiarity of the pure homogeneous elastic case when no tractions are acting on the faces of the crack.

Since we are interested in singularities and distributions (generalized functions in the sense of Schwartz and Sobolev) were especially introduced to deal with such situations, the following remark is in order (*cf.* Dascalu and Maugin, 1994a): from the very structure of pure elasticity theory and the singularity order of elastic solutions there results (i) the expression of the *energy balance law* in presence of cracks (or another defect) and (ii) that a *Griffith-like surface energy criterion* follows of necessity. The argument applies to cracks, but also to other defects such as cavities in expansion or progressing dislocation lines. For a straight-through crack in quasi-statics, we obtain the energy equation in distributional form including both bulk and surface terms (no body force) as

$$D_t\, W + G^{crack}\, \delta(\ell, 0) = \mathrm{div}\,(\mathbf{T} \cdot \mathbf{v}) + \overline{\mathbf{T}}^d \cdot \mathbf{v}\, \delta(\partial B), \tag{270}$$

where we have set (compare (264) with $c = \overline{\mathbf{V}} \cdot \mathbf{E}_1$; δ's are Dirac's distributions)

$$G^{crack} = \int_\Gamma (c W\, N_1 + \mathbf{N} \cdot \mathbf{T} \cdot \mathbf{v})\, \mathrm{d}s - \frac{\mathrm{d}}{\mathrm{d}t} \int_G W\, \mathrm{d}a, \tag{271}$$

and the time and divergence operators are meant in the sense of distributions.

Because all terms in (270) do not depend on the domain G, the same is true for G^{crack}. The integral driving force F_1, which is an energy released during a unit length extension, is usually involved in a quasi-static fracture criterion $F = F_{cr}$. If this holds at every time t, then the energy relation (270) takes on the form

$$D_t \left(W + G_{cr}\, H(\ell - x_1)\, \delta(x_2) \right) = \mathrm{div}\,(\mathbf{T} \cdot \mathbf{v}) + \overline{\mathbf{T}}^d \cdot \mathbf{v}\, \delta(\partial B), \tag{272}$$

which is a *distributional* conservation law. We call the term

$$W_D := G_{cr}\, H(\ell - x_1) \cdot \delta(x_2), \tag{273}$$

the *energy of the crack*. Here $H(\ell - x_1)$ is the Heaviside function of the interval $[0, \ell]$, and the dot in the right-hand side denotes the tensor product of distributions. Equation (272) shows that "the material-plus-defect system is a *conservative* one". The above-given derivation, in contrast to that of Cherepanov (1967), shows that a quasi-static propagation criterion and the elasticity equations suffice to deduce the existence of the defect energy in (272). We also note that although the above analysis was carried on linear materials, known crack tip singularities in nonlinear elastic materials (see, for example, Knowles and Sternberg, 1972, Stephenson, 1982) conduce to the same result. Although it was not mentioned since we focused on energy considerations, there also exists a distributional balance of material momentum which includes the driving force in the form

$$\operatorname{div} \mathbf{b} + F_1^{crack}\, \mathbf{E}_1\, \delta(S) + (\mathbf{N} \cdot \mathbf{b})\, \delta(\partial B) = \mathbf{0}. \tag{274}$$

4.3 The Case of Cracks in More Complex Situations

The complication may have two origins. It may be that the medium itself is more complex by the adjunction of a microstructure described by non dissipative additional degrees of freedom, for instance in the case of micropolar materials (rigidly rotating microstructure) or micromorphic materials (deformable microstructure). According to the contents of Lecture 3 the relative complication simply is the presence of an additional term in the kinetic energy, and a contribution due to the gradient of the new variable in the Eshelby stress. If inertia is neglected, then there will simply remain the contribution of the quasi-static generalized Eshelby stress. Thus the expression of the driving force on a crack has easily been established in linear micropolar homogeneous elastic solids (*cf.* Atkinson and Leppington, 1974) or nonlinear elastic solids Maugin (1998c), or in linear piezoelectric materials Parton (1976) or in finite-strain electroelasticity (Pack and Herrmann, 1986a,b; Maugin and Epstein, 1991, Dascalu and Maugin, 1994b) and the evaluation of the corresponding generalized J-integral is straightforward once we know the field solution at a certain instant by any means (*e.g.*, by finite-elements in piezoelectricity; see, *e.g.*, Benkaci and Maugin, 2001). The path independence may still hold in these conditions. The equivalence and energy approaches hold on account of the degree of singularity of the additional field. The situation is altogether different in the case when the material considered is *inhomogeneous* or there are present pseudo-inhomogeneity or pseudo-plasticity effects (in the sense of Maugin,

2003b), in which case there are at first unremovable bulk source terms in the local balance of material momentum and the energy balance. These are due to the material inhomogeneity force, the material thermal force, and the intrinsic material forces introduced in Section 3. For instance, considering only true material inhomogeneities, we shall have to modify the expression (263) accordingly. Then it is shown that for materially inhomogeneous elastic bodies (263) is replaced by the dynamical expression

$$F_1^{crack} = \int_\Gamma ((\mathbf{N}\cdot\mathbf{b})_1 + P_1\,(\overline{\mathbf{V}}\cdot\mathbf{N}))\,\mathrm{d}A - \left(\frac{\partial}{\partial t}\int_G P_1\,\mathrm{d}V - \int_G f_1^{inh}\,\mathrm{d}V\right), \quad (275)$$

Then there is no question *a priori* to obtain path-independent integrals since equations such as (263) and (264) will necessarily contain volume integrals of these material forces. Nice examples of this type of numerical treatment for inhomogeneous pure elasticity are illustrated in Haddi and Weichert (1995) and in more recent works by Mueller and Maugin (2002), Mueller *et al.* (2004). The results are drastically altered by inhomogeneities (*e.g.*, inclusions or voids) close to the crack tip. In particular, such localized "inhomogeneities" (of compact support) may "attract" or "repulse" a crack.

We let the reader show that in fracture (neglecting thermal effects, and considering a materially homogeneous material with internal dissipative (but nondiffusive) internal variable) we can obtain the expression of the *energy-release rate* as:

$$\begin{aligned} G^{crack} &= \int_\Gamma (H\,(\overline{\mathbf{V}}\cdot\mathbf{N}) + \mathbf{N}\cdot\mathbf{T}\cdot\mathbf{v})\,\mathrm{d}A \\ &\quad -\left(\frac{\partial}{\partial t}\int_G H\,\mathrm{d}V + \int_G A\,\left.\frac{\partial\alpha}{\partial t}\right|_X \mathrm{d}V\right), \end{aligned} \quad (276)$$

while the canonical momentum equation in global form yields, componentwise, a *global material force*, the driving force acting on the crack tip,

$$F_1^{crack} = \int_\Gamma ((\mathbf{N}\cdot\mathbf{b})_1 + P_1\,(\overline{\mathbf{V}}\cdot\mathbf{N}))\,\mathrm{d}S - \frac{\partial}{\partial t}\int_G P_1\,\mathrm{d}V + \int_G f_1^{intr}\,\mathrm{d}V \quad (277)$$

with, in the limit as G shrinks to the crack tip, the *global dissipative inequality*

$$G^{crack} = F_1^{crack}\,\overline{V}_1 \geq 0 \quad (278)$$

in which we have accounted for the fact that in these limits we have the following local behavior: $\mathbf{v} \approx -\overline{\mathbf{V}}\cdot\mathbf{F}^{-T}$, $\dot{\alpha} \approx -\overline{\mathbf{V}}\cdot\nabla_R\alpha$, $A\,\dot{\alpha} \approx -\overline{\mathbf{V}}\cdot\mathbf{f}^{intr}$. The bilinear form (278) governs the irreversible thermodynamics of the extension of C independently of the precise physical *significance* of α.

4.4 The Case of Dislocations

From the developments in previous sections it is clear that the model of medium and the approach chosen have direct consequences on the evaluation, first of the singularities associated with a crack line, and next with the expression of the driving force acting on such a line. That is why M. Lazar and the author have been engaged in a somewhat systematic evaluation of singularities of different modellings of elastic crystals (micropolar, micromorphic, microstretched, strongly nonlocal or weakly non local (gradient theories) models) and the construction of the associated canonical conservation laws and path-independent or non-independent integrals. This is a field of study by itself with a lot of technicalities. We shall mention only a few results. Perhaps more interesting is the treatment of dislocation singularities by means of distributions (generalized functions) as achieved by Dascalu and Maugin (1994a). Indeed, the dislocation motion was one reason that led Eshelby to introduce the notion of force on an elastic singularity. It is thus salient to revisit this problem using distribution theory and obtaining the "energy of a dislocation" in that formalism. To that purpose we consider a straight edge or *screw* dislocation in a linear homogeneous elastic body. We choose a coordinate system so that the dislocation line is parallel to the x_3 axis and pierces the $x_1 x_2$-plane at the point A of plane coordinates $(\ell, 0)$. The dislocation is supposed to glide with the speed $c(t) = \dot{\ell}(t)$ along the x_1- axis. The situation is the same as in Figure 4, but now C is the intersection of the dislocation slip plane with the $x_1 x_2$- plane. If $\tilde{\mathbf{b}}$ is the Burgers vector, then

$$[\mathbf{u}] = \tilde{\mathbf{b}} \text{ on } C. \tag{279}$$

The structure of the solution is the same as that given in standard texts with (*cf.* Teodosiu, 1982; Mura, 1981)

$$\mathbf{T}^s = O(r^{-1}), \quad (\nabla \mathbf{u})^s = O(r^{-1}), \tag{280}$$

and $\mathbf{u}^r$ and $\mathbf{T}^r$ finite on B. The singular terms $\mathbf{u}^s$ and $\mathbf{T}^s$ do not depend explicitly on time, so that the order of singularity of their time derivatives is the same as that of their gradient. Because W and $\mathbf{T} \cdot \mathbf{v}$ are not integrable we need to construct for them the regularizations $W|_G$ and $\mathbf{T} \cdot \mathbf{v}|_G$. These distributions depend on G and, restricted to $R^2 - \{A\}$, coincide with W and $\mathbf{T} \cdot \mathbf{v}$, respectively. These are technicalities that we bypass here. Without giving the tedious mathematical details (see Dascalu and Maugin, 1994a, pp. 33–36 for these), we mention that we have to evaluate $D_t(W|_G)$ and $\operatorname{div}(\mathbf{T} \cdot \mathbf{v}|_G)$. Then writing the energy conservation law for these two terms, after some lengthy computations we arrive at a distributional energy conservation in the form

$$\begin{aligned} D_t(W|_G) &+ \left(W_1 - \frac{\mathrm{d}}{\mathrm{d}t} \int_G (x_1 - \ell)\, W \,\mathrm{d}A \right) \delta_{,1}(\ell, 0) \\ &+ \left(W_2 - \frac{\mathrm{d}}{\mathrm{d}t} \int_G x_2\, W \,\mathrm{d}A \right) \delta_{,2}(\ell, 0) \\ &+ \left(\int_\Gamma (c\, W\, N_1 + \mathbf{N} \cdot \mathbf{T} \cdot \mathbf{v}) \,\mathrm{d}\Gamma \right) \delta(\ell, 0) \\ &= \operatorname{div}(\mathbf{T} \cdot \mathbf{v}|_G) + \overline{\mathbf{T}}^d \cdot \mathbf{v}\, \delta(\partial B), \end{aligned} \tag{281}$$

wherein

$$\begin{aligned} W_1 &:= \int_\Gamma (x_1 - \ell)\, (W\, N_1 - \mathbf{N} \cdot \mathbf{T} \cdot \mathbf{u}_{,1}) \,\mathrm{d}\Gamma, \\ W_2 &:= \int_\Gamma x_2\, (W\, N_1 - \mathbf{N} \cdot \mathbf{T} \cdot \mathbf{u}_{,1}) \,\mathrm{d}\Gamma, \end{aligned} \tag{282}$$

Equation (281) is an energy equation depending on G, which provides the energy balance on $B - G$. To obtain the dislocation energy we have to consider the limit as $G \to 0$. The limit

$$\lim \langle D_t(W|_G), \varphi \rangle \text{ as } G \to 0 \tag{283}$$

exists (here the function φ is sufficiently regular and the brackets denote the inner product in the sense of distribution theory). This is essentially given by the principal value of W in A. On the other hand the limit as $G \to 0$ of the last contribution in the left-hand side in (281) is none other that $c\,J$, where J is the Cherepanov–Eshelby–Rice path-independent integral. Furthermore, it can be shown that the implicit dependence on time of the singular parts of $\mathbf{u}$ and $\mathbf{T}$ are such that

$$\lim \frac{\mathrm{d}}{\mathrm{d}t} \int_G (x_1 - \ell)\, W \,\mathrm{d}A = \lim \frac{\mathrm{d}}{\mathrm{d}t} \int_G x_2\, W \,\mathrm{d}A \text{ as } G \to 0, \tag{284}$$

so that the other distributions concentrated in A are also convergent, their limit depending on the singular parts of the fields. The integrals (282) are calculated only for these singular parts. In the end, in the limit, (281) will take the following reduced form

$$D_t\, W + c\, f^{PK}\, \delta(\ell, 0) = \operatorname{div}(\mathbf{T} \cdot \mathbf{v}) + \overline{\mathbf{T}}^d \cdot \mathbf{v}\, \delta(\partial B), \tag{285}$$

where the integral f^{PK} gives the Peach–Koehler force on a dislocation (Eshelby, 1951). Then the energy-like criterion is obtained from (285) in the form $f^{PK} = f_{cr}$. Such a *criterion of activation* is usually encountered in the dislocation theory. On account of this, we can state the following energy conservation

$$D_t\left(W + f_{cr}\, H(\ell - x_1)\,\delta(x_2)\right) = \operatorname{div}(\mathbf{T}\cdot\mathbf{v}) + \bar{\mathbf{T}}^d \cdot \mathbf{v}\,\delta(\partial B). \tag{286}$$

The second contribution in the time derivative represents the energy of the dislocation which is uniformly distributed on its slip plane. The material "force" f_{cr} represents the energy expended in the dislocation glide with a unit length. A similar calculation can be carried out for edge dislocations.

4.5 Microcontinuum Field Theories, Incompatibility, Eshelby Stresses, and Material Forces

A. Compatible case We consider a sufficiently general case but in small strains and quasi-statics in order to alleviate the notation. This is the theory of *micromorphic media, i.e.*, media whose material point itself presents a deforming microstructure, so that a second-order tensor ϕ_{ij} has to be added to the usual three components u_i of the displacements to fully describe the kinematics of the material (*cf.* Eringen, 1999), which kinematics becomes nine-dimensional. Accounting for gradients of these yields a large number of generalized deformation measures and by duality a large number of generalized "internal" forces. Such a theory was developed essentially by Eringen and Suhubi in 1964, and independently by Mindlin in the same years. The principle of virtual power is the safest tool to construct such a theory without any ambiguity in the concepts, the duality between deformation fields and generalized forces, and the formulation of natural boundary conditions (*cf.* Germain, 1973b, Maugin, 1980, Forest, 2006). We note $u_{i,j}$ the displacement gradient, ϕ_{ij} the microdeformation tensor and $\kappa_{ijk} = \phi_{ij,k}$ its gradient. For compatible fields, the relevant infinitesimal deformation measures are the relative deformation (nine components) γ_{ij}, the microstrain (6 components) e_{ij}, and the micro-wryness tensor (27 components) κ_{ijk} such that

$$\gamma_{ij} = u_{i,j} - \phi_{ij}, \quad e_{ij} = e_{ji} = \phi_{(ij)}, \quad \kappa_{ijk} = \phi_{ij,k}. \tag{287}$$

These admit themselves canonical decompositions such as

$$\gamma_{ij} = \gamma_{(ij)} + \gamma_{[ij]}, \quad \gamma_{(ij)} = \gamma\,\delta_{ij} + \bar{\gamma}_{ij}, \tag{288}$$

$$e_{(ij)} = e\,\delta_{ij} + \bar{e}_{ij}, \tag{289}$$

$$\kappa_{ijk} = \delta_{ij}\,\kappa_k + \kappa_{[ij]k} + \bar{\kappa}_{(ij)k}, \quad \kappa_k := \kappa_{iik}/3. \tag{290}$$

Each of the measures (287) admits a dual in internal-force space. For an elastic solid with volume energy $W = \overline{W}(\gamma_{ij}, e_{ij}, \kappa_{ijk})$, these are the force stress (Cauchy stress) σ_{ji}, the micro stress s_{ji} and the micro-hyperstress μ_{kji} such that

$$\sigma_{ji} = \frac{\partial \overline{W}}{\partial u_{i,j}}, \quad s_{ji} = \frac{\partial \overline{W}}{\partial e_{ij}} = s_{ij}, \quad \mu_{kij} = \frac{\partial \overline{W}}{\partial \kappa_{ijk}}, \tag{291}$$

which satisfy jointly the following equilibrium equations at each regular material point (*cf.* Eringen and Suhubi, 1964; Eringen, 1999 in a different notation):

- Balance of linear (physical) momentum:

$$\sigma_{ji,j} + f_i = 0, \tag{292}$$

- Balance of micro-motion (here equilibrium form):

$$\mu_{kij,k} + \sigma_{ji} - s_{ji} + \ell_{ji} = 0, \tag{293}$$

where bulk data are the body force f_i and the body moment tensor ℓ_{ji}. The micro-hyperstress μ_{kij} naturally admits a decomposition similar to that of κ_{ijk}, *i.e.*, in antisymmetric part, its trace, and its symmetric traceless part:

$$\mu_{kij} = \mu_{k[ij]} + \frac{1}{3}\mu_k\,\delta_{ij} + \bar{\mu}_{k(ij)}. \tag{294}$$

B. Incompatible case In this case the "elastic" deformation measures γ_{ij}, e_{ij} and κ_{ijk} (where we avoid putting a superscript "e"), are *incompatible*, which means they are not gradients, and we can introduce the *total* same measures – which, them, are compatible, by (we follow Forest and Sievert, 2003; Lazar and Maugin, 2007)

$$\gamma_{ij}^T := \gamma_{ij} + \gamma_{ij}^p = u_{i,j} - \phi_{ij}, \tag{295}$$
$$e_{ij}^T = \phi_{(ij)} = e_{ij} + e_{ij}^p, \tag{296}$$
$$\kappa_{kij}^T = \phi_{ij,k} = \kappa_{kij} + \kappa_{kij}^p, \tag{297}$$

Accordingly, the "plastic" parts denoted by a superscript "p" are not compatible. The total quantities are compatible and, therefore, satisfy the following compatibility conditions (*cf.* Eringen, 1999)

$$\varepsilon_{jkl}\,(\gamma_{il,k}^T + \kappa_{kil}) = 0, \quad \varepsilon_{jkl}\,\kappa_{lin}^T = 0, \quad 2\,e_{ij,k}^T - \kappa_{kij}^T - \kappa_{kji}^T = 0. \tag{298}$$

The "plastic" parts – or the "elastic" parts – will have to fulfill incompatibility conditions. These will read (*e.g.*, for the "elastic" parts):

$$\begin{aligned} \alpha_{ij} &= \varepsilon_{jkl}\,(\gamma_{il,k} + \kappa_{ilk}), \quad \Theta_{inj} = \varepsilon_{jkl}\,\kappa_{lin,k}, \\ Q_{ijk} &= -(2\,e_{ijk} - \kappa_{kji} - \kappa_{kij}), \end{aligned} \tag{299}$$

and the nonvanishing objects introduced in the left-hand side of these equations must be given a geometrical interpretation if we want to have at hand a generalization of Kroener's vision of incompatibility. The first of these, just like in Kroener's theory, is the *dislocation density tensor* which is identified with the linearized version of the *Cartan torsion* tensor $\widetilde{T}_{ijk}$ according to (9 independent components)

$$\alpha_{ij} = \frac{1}{2}\,\varepsilon_{jkl}\,\widetilde{T}_{ikl}. \tag{300}$$

The second object introduced in (299) can be identified to the (linearized) *Riemann–Cartan curvature* R_{ijkl} by (27 independent components)

$$\Theta_{ijk} = \frac{1}{2}\,\varepsilon_{kmn}\,R_{ijmn},$$

hence it is a curvature tensor. This can also be decomposed in such a way that

$$\begin{aligned} \Theta_{ijk} &= \Theta_{(ij)k} + \Theta_{[ij]k} = \Theta_{(ij)k} - \varepsilon_{ijl}\,\Theta_{lk}, \\ \Theta_{ij} &= -\frac{1}{2}\,\varepsilon_{ikl}\,\Theta_{klj} = -\frac{1}{4}\,\varepsilon_{ikl}\,\varepsilon_{jmn}\,R_{klmn}. \end{aligned} \tag{301}$$

The last defined quantity is the *disclination density* tensor. The quantity $\Theta_k = \Theta_{jjk}/3$ is called the Weyl "distance curvature".

Finally, the third object introduced in (299) is the *tensor of nonmetricity.* The last of equations (299) therefore means that the micro-strain e_{ij} is *nonmetric.* The quantity $Q_k = Q_{jjk}/3$ is called the *Weyl covector* and will describe defects of the "*dilatation*" type. In all we can say from a differential-geometrical viewpoint that γ_{ij}, κ_{kij} and e_{ij} are, respectively, a coframe, a connection, and a metric. The allied torsion, curvature and nonmetricity fullfil the following *Bianchi identities* (*cf.* Schouten, 1954):

- First identity:

$$\alpha_{ij,j} = \Theta_{ij,j}; \tag{302}$$

- Second identity:

$$\Theta_{inj,j} = 0; \tag{303}$$

- Third identity:

$$\varepsilon_{klm}\,Q_{ijm,l} = 2\,\Theta_{(ij)k}. \tag{304}$$

The first of these means that dislocations can interact with disclinations and point defects. The second is just a divergence-free condition. And the last shows that one part of the curvature is given in terms of the nonmetricity tensor which describes point defects. Of course this greatly simplifies if there are only dislocations present in the material (classical Kroener's theory).

What about the Eshelby stress? We can write it down at once since its formula is canonical, capturing all gradients of the theory. That is, from the energy $W = \overline{W}(\gamma_{ij}, e_{ij}, \kappa_{ijk}; x_i)$ where we envisage a possibility of standard material inhomogeneity:

$$b_{ji} = \overline{W}\,\delta_{ji} - \sigma_{jk}\,(u_{k,i} - \gamma^{p}_{ik}) - \mu_{jkl}\,\phi_{lk,i}. \tag{305}$$

In order to evaluate the corresponding configurational-material forces, we must compute the divergence of this Eshelby stress. The results of this lengthy computation is as follows:

$$b_{ij,i} + f^{(m)}_j = 0, \tag{306}$$

where the *material force source* is given by

$$\begin{aligned} f^{(m)}_j = -f^{(PK)}_j - f^{(M)}_j + \frac{1}{2}\, s_{mn}\, Q_{mnj} + \sigma_{lk}\,\kappa^{P}_{kjl} \\ -f_i\,(u_{i,j} - \gamma^{P}_{ij}) - \ell_{kl}\,\kappa_{jkl} + f^{inh}_j, \end{aligned} \tag{307}$$

where the various contributions are, respectively the negative of the Peach–Koehler force in the theory of continuously distributed dislocations (compare to eqn. (1.83)), the negative of the so-called Mathisson–Papapetrou force introduced by Maugin in defect theory by analogy with gravitational theories with spin (*cf.* Maugin, 1993). The contribution of nonmetricity is original to micromorphic theory, then there is the convected of the body force changed of sign with a convection effected with an effective distortion, an equivalent term for the body couple, and finally the standard material inhomogeneity force given by the negative of the explicit gradient of $\overline{W}$. Here,

$$f^{(PK)}_j = \varepsilon_{jkl}\,\sigma_{pk}\,\alpha_{pl}, \quad f^{(M)}_j = \varepsilon_{jkl}\,\mu_{kmn}\Theta_{mnl}, \quad f^{inh}_j = -\left.\frac{\partial \overline{W}}{\partial x_j}\right|_{expl}. \tag{308}$$

C. Special cases Several cases of great interest can be extracted from the general micromorphic scheme. These are (in the classification of Eringen, 1999):

(*i*) *microstretch elasticity*: only dilatational degrees of freedom are kept in addition to the rotational and translational ones. Therefore, this is both a generalization of micropolar elasticity (Cosserat continua) and a special case of the micromorphic elasticity. The relevant equations can be deduced from the micromorphic ones by setting to zero the shear parts of the micro-fields. This can be represented by directors with stretch and rotation only, and no micro-shears of them.

(*ii*) *micropolar (Cosserat) elasticity*: only the rotational degrees of freedom in addition to the translational ones are kept. This is the most popular case. This will obviously raise some interest when dealing with defects such as disclinations.

(*iii*) *dilatation elasticity* (also called elasticity with voids; Cowin and Nunziato, 1983): only the dilatation of the microstructure is kept in addition to the macro-displacement. We can easily imagine the interest for this scheme for defects such as voids or vacancies.

We briefly examine the last two cases.

Case of micropolar (Cosserat) elasticity (see Lazar and Maugin, 2007)

In this case the equilibrium equations read as follows at any regular material point:

$$\sigma_{ji,j} + f_i = 0, \quad \mu_{ji,j} - \varepsilon_{jik}\,\sigma_{ki} + \ell_i = 0, \tag{309}$$

where μ_{ji} is the moment-stress or couple-stress tensor and ℓ_i is the body couple. The standard elastic constitutive equations are given by

$$\sigma_{ji} = \frac{\partial \overline{W}}{\partial \gamma_{ij}}, \quad \mu_{ji} = \frac{\partial \overline{W}}{\partial \kappa_{ij}}, \quad W = \overline{W}(\gamma_{ij}, \kappa_{ij}; x_i). \tag{310}$$

The total fields are considered to be the sum of elastic and plastic parts (elastic parts without superscript e), *i.e.*,

$$\gamma_{ij}^T = \gamma_{ij} + \gamma_{ij}^p = u_{i,j} + \varepsilon_{ijk}\,\phi_k, \quad \kappa_{ij}^T = \kappa_{ij} + \kappa_{ij}^p = \phi_{i,j}. \tag{311}$$

The total fields satisfy compatibility conditions (Eringen, 1999) such as

$$\varepsilon_{jkl}\,(\gamma_{il,k}^T + \varepsilon_{ikm}\,\kappa_{ml}^T) = 0, \quad \varepsilon_{jkl}\,\kappa_{il,k}^T = 0. \tag{312}$$

The incompatibility conditions then read (we write them for the elastic parts)

$$\alpha_{ji} = \varepsilon_{jkl}\,(\gamma_{il,k} + \varepsilon_{ikm}\,\kappa_{ml}), \quad \Theta_{ji} = \varepsilon_{jkl}\,\kappa_{il,k}. \tag{313}$$

The Bianchi identities for such micropolar media are obtained by differentiating these equations, yielding the result of Claus jr and Eringen (1969)

$$\alpha_{ji,j} = \varepsilon_{imk}\,\Theta_{km}. \tag{314}$$

The canonical writing of the Eshelby stress produces the following formula in the incompatible case

$$b_{ji} = \overline{W}\,\delta_{ji} - \sigma_{jk}\,\bar{\gamma}_{ki} - \mu_{jk}\,\kappa_{ki} \tag{315}$$

and

$$b_{ji} = \overline{W}\,\delta_{ji} - \sigma_{jk}\,u_{k,i} - \mu_{jk}\,\phi_{ki} \tag{316}$$

in the compatible case. The last formula corresponds to the result know from repeated studies (Jaric, 1978; Atkinson and Leppington, 1974; Lubarda and Markenscoff, 2000; and Maugin, 1998c, for finite strains and rotations). Computing the divergence of the Eshelby stress in the incompatible case, we find the equations (306) but with

$$f_j^{(m)} = -f_j^{(PK)} - f_j^{(discl)} + \varepsilon_{jki}\,\sigma_{lk}\,\kappa_{il}^{p} - f_i\,\bar{\gamma}_{ij} - \ell_i\,\kappa_{ij} + f_j^{inh}, \tag{317}$$

with

$$f_j^{(PK)} = \varepsilon_{jkl}\,\sigma_{ki}\,\alpha_{il}, \quad f_j^{(discl)} = \varepsilon_{jkl}\,\mu_{ki}\,\Theta_{il}, \quad f_j^{inh} = -\left.\frac{\partial \overline{W}}{\partial x_j}\right|_{expl}, \tag{318}$$

in which we recognize the Peach–Koehler force, an analogous force acting on the disclination density, and the pure inhomogeneity force. In the compatible case (307) reduces to

$$f_j^{(m)} = -f_i\,u_{i,j} - \ell_i\,\phi_{i,j} + f_j^{inh}. \tag{319}$$

Case of dilatation elasticity (see Lazar and Maugin, 2007)

In this case the equilibrium equations at each regular material point are reduced to

$$\sigma_{ji,j} + f_i = 0, \quad \mu_{k,k} + \sigma - s + \ell = 0, \tag{320}$$

along with the constitutive equations

$$\sigma_{ji} = \frac{\partial \overline{W}}{\partial \gamma_{ij}}, \quad s - \sigma = \frac{\partial \overline{W}}{\partial \phi}, \quad \mu_k = \frac{\partial \overline{W}}{\partial \kappa_k}, \quad W = \overline{W}(\gamma_{ij}, \phi, \kappa_k; x_i). \tag{321}$$

In the presence of incompatibilities, the total fields are given by

$$\gamma_{ij}^{T} = u_{i,j} - \phi\,\delta_{ij} = \gamma_{ij} + \gamma_{ij}^{p}, \quad e^{T} = e + e^{p} = \phi, \quad \kappa_k^{T} = \kappa_k + \kappa_k^{p} = \phi_{,k}. \tag{322}$$

The compatibility conditions for these total fields read

$$\varepsilon_{jkl}\,(\gamma_{il,k}^{T} - \delta_{ik}\,\kappa_l^{T}) = 0, \quad \varepsilon_{jkl}\,\kappa_{l,k}^{T} = 0, \quad -e_{,k}^{T} + \kappa_k^{T} = 0. \tag{323}$$

The incompatibility conditions (here written for the elastic parts) read

$$\alpha_{ji} = \varepsilon_{jkl}\,(\gamma_{il,k} - \delta_{ik}\,\kappa_l), \quad \Theta_j = \varepsilon_{jkl}\,\kappa_{l,k}, \quad Q_k = 2\,(e_{,k} - \kappa_k^{p}), \tag{324}$$

and the Bianchi identities are given by

$$\alpha_{ji,j} = \Theta_i, \quad \Theta_{k,k} = 0, \quad \varepsilon_{klm} \, Q_{m,l} = 2\,\Theta_k. \tag{325}$$

Finally, in presence of incompatibilities, the Eshelby stress is given by

$$b_{ji} = \overline{W}\,\delta_{ji} - \sigma_{jk}\,\bar{\gamma}_{ki} - \mu_j\,\kappa_i, \tag{326}$$

and the material force source in the equilibrium version of the conservation of material momentum is given by

$$f_j^{(m)} = -f_j^{(PK)} - f_j^{(PD)} + \frac{1}{2}\,s\,Q_j + \sigma_{jl}\,\kappa_l^p - f_i\,\bar{\gamma}_{ij} - \ell\,\kappa_j + f_j^{inh}, \tag{327}$$

where

$$f_j^{(PK)} = \varepsilon_{jkl}\,\sigma_{ki}\,\alpha_{il}, \quad f_j^{(PD)} = \varepsilon_{jkl}\,\mu_k\,\Theta_l, \quad f_j^{inh} = -\left.\frac{\partial \overline{W}}{\partial x_i}\right|_{expl}, \tag{328}$$

are, respectively, the Peach–Koehler force, the material force acting on *point defects/microvoids*, and the material force of pure inhomogeneity. In the compatible case, (326) reduces to

$$b_{ji} = \overline{W}\,\delta_{ji} - \sigma_{jk}\,u_{k,i} - \mu_j\,\phi_{,i}. \tag{329}$$

This concludes our excursion in the world of material and configurational forces, originally discovered by J. D. Eshelby, in relation to the theory of defects and the generalization of Kroener's geometrical incompatibility theory to modern thermomechanical theories of complex continua.

Additional Bibliography: Capriz, 1989; Dell'Isola and Seppecher, 1995; Lazar and others (2004–2007); Maugin, 2011.

(all references are not cited in the text)

Note: The above given lecture notes are those that were distributed to students attending the CISM lectures in 2007. Some references have been completed or corrected. A rather thick book on the subject has been published by the author in 2011 as Maugin (2011). The present lecture notes of 2007 have been essential in preparing that book.

Bibliography

R. Abeyaratne and J. K. Knowles, (2000). A note on the driving traction on a propagating interface: adiabatic and non-adiabatic processes of a continuum, *Trans. ASME. J. Appl. Mech.*, **67**, 829–831.

K. H. Anthony, (1970), Die Theorie der Disklinationen, Arch. Rat. Mech. Anal., **39**, 43–88.

C. Atkinson and F. G. Leppington, (1974), Some Calculations of the Energy-release rate G for Cracks in Micropolar and Couple-stress Elastic Media, *Int. J. Fracture*, **10**, 599–602.

N. Benkaci and G. A. Maugin, (2001), J-integral Computations for Piezoelectric Ceramics, *Revue Européenne des Eléments Finis*, **10**, 99–128.

B. A. Bilby, (1968), Geometry and continuum mechanics, in: Mechanics of generalized Continua (IUTAM Symp. Freudenstadt, 1967), ed. E. Kroener, Springer-Verlag, Berlin.

B. A. Bilby, L. R. T. Lardner and A. N. Stroh, (1957), Continuum Theory of Dislocations and the Theory of Plasticity, in: *Proc. Xth ICTAM (Brussels, 1956)*, Vol. **8**, pp. 35–44, Presses de l'Université de Bruxelles.

R. M. Bowen, (1967), Towards a Thermodynamics and Mechanics of Mixtures, Arch. Rat. Mech. Anal., **24**, 370–403.

R. M. Bowen, (1976), Theory of Mixtures, in: Continuum Physics, Ed. Eringen A. C., Vol. **III**, pp. 1–127, Academic Press, New York.

W. Brenig, (1955), Bessitzen Schallwellen einen Impulz, *Zeit. Phys.*, **143**, 168–172.

H. D. Bui, (1978), Mécanique de la rupture fragile, Masson, Paris.

G. Capriz, (1989), Continua with microstructure, Springer, New York.

E. Cartan, (1931), Le parallélisme absolu et la théorie unitaire du champ, *Revue de métaphysique et de morale* (Paris), pp. 13–28.

P. Casal, (1978), Interpretation of the Rice Integral in Continuum Mechanics, *Lett. Appl. Engng. Sci.*, **16**, 335–348.

G. P. Cherepanov, (1967), Crack Propagation in Continuous Media, PMM (*Appl. Math. Mech.*, Trans. From the Russian), **31**, 467–488.

Y. Choquet–Bruhat, (1968), Géométrie différentielle et systèmes extérieurs, Dunod, Paris.

W. D. Claus jr and A. C. Eringen, (1969), Three dislocation concepts and micromorphic mechanics, in: Developments in Mechanics, Vol. **6**, *Proc. 12th Midwestern Mechanics Conference*, pp. 125–147.

S. W. Cowin and J. W. Nunziato, (1983), Linear elastic materials with voids, J. Elasticity, **13**, 125–147.

C. Dascalu and G. A. Maugin, (1993), Forces matérielles et taux de restitution de l'énergie dans les corps élastiques homogènes avec défauts, C. R. Acad. Sci. Paris, **II-317**.

C. Dascalu and G. A. Maugin, (1994a), The Energy of Elastic Defects: A Distributional Approach, *Proc. Roy. Soc. Lond.*, A445, 23–37.

C. Dascalu and G. A. Maugin, (1994b), Energy-release Rates and Path-independent Integrals in Electroelastic Crack Propagation, *Int. J. Engng. Sci.*, **32**, 755–765.

F. Dell'Isola and P. Seppecher, (1995), The relationship between edge contact forces, double forces and intersticial working allowed by the principle of virtual power, C. R. Acad. Sci. Paris IIb, **321**, 303–308.

D. G. B. Edelen, (1981), Aspects of variational arguments in the theory of elasticity: facts and folklore, *Int. J. Solids Structures*, **17**, 729–740.

M. Elzanowski and M. Epstein, (1990), On the symmetry group of second-grade materials, Report No. 452, Dept. of Mech. Engineering, Univ. of Calgary.

M. Elzanowski, M. Epstein and J. Sniatycki, (1990), G-structures and material homogeneity, J. Elasticity, **23**, 167–180.

M. Epstein and G. A. Maugin, (1990a), Sur le tenseur de moment matériel d'Eshelby en élasticité non linéaire, C. R. Acad. Sci. Paris, **II-310**, 675–678.

M. Epstein and G. A. Maugin, (1990b), The Energy-momentum Tensor and Material Uniformity in Finite Elasticity, Acta Mechanica, **83**, 127-133.

M. Epstein and G. A. Maugin, (1992), Somc geometrical aspects of inhomogeneous elasticity, in: Analysis, Manifolds and Physics, ed. R. Kerner, pp. 331–335, Kluwer, Amsterdam.

M. Epstein and G. A. Maugin, (1995a), Thermoelastic Material Forces: Definition and Geometric Aspects, C. R. Acad. Sci. Paris, **II-320**, 63–68.

M. Epstein and G. A. Maugin, (1995b), On the Geometrical Material Structure of Anelasticity, Acta Mechanica, **115**, 119–131.

M. Epstein and G. A. Maugin, (1997), Notions of Material Uniformity and Homogeneity, in: *Theoretical and Applied Mechanics*, (*Proc. ICTAM'96, Kyoto*), Ed. T. Tatsumi, pp. 201–215, Elsevier, Amsterdam.

M. Epstein and G. A. Maugin, (2000), Thermomechanics of Volumetric Growth in Uniform Bodies, *Int. J. Plasticity*, **16**, 951–978.

A. C. Eringen, (1999), Microcontinuum field theories I: Foundations and solids, Springer, New York.

A. C. Eringen and G. A. Maugin, (1990), *Electrodynamics of continua*, Vol. **I**, Springer-Verlag, New York.

A. C. Eringen and E. S. Suhubi, (1964), Nonlinear theory of simple microelastic solids, *Int. J. Engng. Sci.*, **2**, 189–203, 389–404.

J. D. Eshelby, (1951), Force on an Elastic Singularity, *Phil. Tran. Roy. Soc. Lond.*, **A244**, 87–112.

J. D. Eshelby, (1975), Elastic energy-momentum tensor, J. Elasticity, **5**, 321–335.

J. D. Eshelby, (1982), Aspects of Dislocation Theory, in: *Mechanics of Solids* (The Rodney Hill 60th Anniversary Volume), eds. H. G. Hopkins and M. S. Sewell, pp. 185–255, Pergamon Press, Oxford.

S. Forest, (2006), Milieux continus généralisés et matériaux hétérogènes, Presses de l'Ecole des Mines, Paris.

S. Forest and J. M. Cordona, (2000), Thermoelasticity of Second-grade Materials, in: Continuum Thermoechanics: The Art and Science of Modeling Materials' Behaviour, Eds. G. A. Maugin, R. Drouot and F. Sidoroff, pp. 163–176, Kluwer, Dordrecht.

S. Forest and R. Sievert, (2003), Elastoviscoplastic Constitutive Frameworks for Generalized Continua, Acta Mechanica, **160**, 71–111.

M. Frémond and B. Nedjar, (1993), Endommagement et principe des puissances virtuelles, C. R. Acad. Sci. Paris, **II-317**, 857–864.

M. Frémond and B. Nedjar, (1996), Damage, Gradient of Damage and Principle of Virtual Power, *Int. J. Solids Structures*, **33**, 1083–1103.

J. Friedel, (1979), Introduction, in: *Dislocations in solids*, Vol. **I**, ed. F. R. N. Nabarro, pp. 3–32, North-Holland, Amsterdam.

P. Germain, (1973a), La méthode des puissances virtuelles en mécanique des milieux continus - I: Théorie du second gradient, *J. Mécanique* (Paris), **12**, 135–274.

P. Germain, (1973b), The Method of Virtual Power in Continuum Mechanics - II: Microstructure, *SIAM J. Appl. Math.*, **25**, 556–575.

A. Golebiewska-Herrmann, (1981), On Conservation Laws of Continuum Mechanics, *Int. J. Solids Structures*, **17**, 1–9.

S. de Groot and P. Mazur, (1962), Non-equilibrium thermodynamics, North-Holland, Amsterdam.

M. E. Gurtin, (1979), Energy-release Rate in Quasi-static Crack Propagation, J. Elasticity, **9**, 187–195.

M. E. Gurtin, (1994), The Characterization of Configurational Forces, *Arch. Rat. Mech. Anal.*, **126**, 387–394.

M. E. Gurtin, (1995), On the Nature of Configurational Forces, *Arch. Rat. Mech. Anal.*, **131**, 67–100.

M. E. Gurtin, (1999), Configurational forces as basic concepts of continuum physics, Springer, New York.

M. E. Gurtin and P. Podio-Guidugli, (1996), On configurational inertial forces at a phase interface, *J. Elasticity*, **44**, 255–269.

A. Haddi and D. Weichert, (1995), On the Computation of the J-integral for Three-dimensional Geometries in Inhomogeneous Materials, Comput. Mat. Sci., **5**, 143–150.

G. Herrmann and R. Kienzler, (2000), Mechanics in Material Space, Springer-Verlag, Berlin.

N. H. Ibragimov, (1985), Transformation group applied to mathematical physics, Reidel, Dordrecht, The Netherlands.

V. L. Indenbom, (1965), Internal Stress in Crystals, in: Theory of Crystal Defects (*Proc. Summer School, Hrazany, Czec. Sept. 1964*), ed. B. Gruber pp. 257–274, Acad. Publ. House, Prague and Academic Pres, New York.

P. Ireman and Nguyen Quoc Son, (2004), Using the gradients of the temperature and internal parameters in continuum thermodynamics, C. R. Mécanique (Acad. Sci., Paris), **333**, 249–255.

J. Jaric, (1978), Conservation Laws of J-integral Type in Micropolar Elastostatics, *Int. J. Engng. Sci.*, **16**, 967–984.

V. K. Kalpakides and G. A. Maugin, (2004), Canonical Formulation and Conservation Laws of Thermoelasticity "without dissipation", Reports in Math. Physics., **53**, 371–391.

R. Kienzler and G. A. Maugin, (Eds. 2001), Configurational Mechanics of Materials, Springer-Verlag, Vienna.

J. Kijowski and G. Magli, (1998), Unconstrained Hamiltonian Formulation of General Relativity with Thermo-elastic Sources, *Classical Quantum Grav.*, **15**, 3891–3916.

J. K. Knowles and E. Sternberg, (1972), Class of Conservation Laws in Linearized and Finite Elastostatics, *Arch. Rat. Mech. Anal.*, **44**, 187–211.

K. Kondo, (1952), On the Geometrical and Physical Foundations of the Theory of Yielding, in: *Proc. 2nd Japanese National Congress of Applied Mechanics*, pp. 41–47.

K. Kondo, (1955), Non-Riemannian Geometry of Imperfect Crystals from a Macroscopic Viewpoint, in: RAAG Memoirs of the Unifying Study of Basic Problems in Engineering and Physical Sciences by Means of geometry, Vol. **1**, pp. 459–480, ed. K. Kondo, Gakujutsu Bunken Fukyukai, Tokyo.

A. M. Kosevich, (1962), Zh. Eksper. teor. Fiz., **43**, 637) (in Russian), (*Sov. Phys. JETP*, **16**, 455, 1963).

A. M. Kosevich, (1964), Usp. Fiz. Nauk., **84**, 579 (in Russian), (*Sov. Phys. Usp.*, **7**, 837, 1965).

A. M. Kosevich, (1979), Crystal Dislocations and the Theory of Elasticity, in: *Dislocations in Solids*, Ed. F. R. N. Nabarro, Vol. **I**, pp. 33–141, North-Holland, Amsterdam.

A. M. Kosevich, (1988), Theory of Crystal Lattices (in Russian), Vishtsha Shkola, Kiev.

A. M. Kosevich, (1999), The Crystal lattice: Phonons, Solitons, Dislocations, Wiley-VCH, Berlin.

E. Kroener, (1955), Inneren Spannungen und der Inkompatibilitätstensor in der Elastizitätstheorie, *Zeit. angew. Physik*, **7**, 249–257.

E. Kroener, (1958), Kontinuumstheorie der Versetzungen und Eigenspannungen, Springer-Verlag, Berlin.

E. Kroener, (1960), Allgemeine Kontinuumstheorie der Versetzungen und Eigenspannungen, *Arch. Rat. Mech. Anal.*, **4**, 273–344.

E. Kroener, (1993), A variational principle in nonlinear dislocation theory, in: *Proc. 2nd International Conference on Nonlinear mechanics*, ed. Chien Wei-zang, pp. 59–64, Peking University Press.

E. Kroener and A. Seeger, (1959), Nicht-lineare Elastiztätstheorie und Eigenspannungen, *Arch. Rat. Mech. Anal.*, **3**, 97–119.

C. Lanczos, (1962), Variational principles, Toronto University Press, Toronto. (1962)

R. W. Lardner, (1974), Mathematical theory of dislocations and fracture, University of Toronto Press, Toronto.

M. Lazar, (2007), On conservation and balance laws in micromorphic elastodynamics, *J. Elasticity*, **88**, 63–78.

M. Lazar and C. Anastassiadis, (2006), Lie point symmetries and conservation laws in microstretch and micromorphic elasticity (preprint 2006; unpublished).

M. Lazar and G. A. Maugin, (2004a), Defects in Gradient Micropolar Elasticity- I- Screw dislocation, J. Mech. Phys. Solids, **52**, 2263–2284.

M. Lazar and G. A. Maugin, (2004b), Defects in Gradient Micropolar Elasticity- II- Edge dislocation and disclinations, *J. Mech. and Phys. Solids*, **52**, 2285–2307 (2004).

M. Lazar and G. A. Maugin, (2005), Nonsingular Stress and Strain Fields of Dislocations and Disclinations in First Strain Gradient Elasticity, *Int. J. Engng. Sci.*, **43**, 1157–1184.

M. Lazar and G. A. Maugin, (2006a), A Note on Line Forces in Gradient Elasticity, *Mechanics Research Communications*, **33/5**, 674–680.

M. Lazar and G. A. Maugin, (2006b), Dislocations in Gradient Elasticity Revisited, *Proc. Roy. Soc. Lond.*, **A, 462**, 3465–3480 (2006).

M. Lazar and G. A. Maugin, (2007), On microcontinuum field theories: The Eshelby stress tensor and incompatibility conditions, *Philosophical Magazine*, **87**, 3853-3870.

M. Lazar, G. A. Maugin and E. C. Aifantis, (2005), On Dislocations in a Special Class of Generalized Elasticity, physica status solidi, **242**, No. 12, 2365–2390.

M. Lazar, G. A. Maugin and E. C. Aifantis, (2006a), On a Theory of Nonlocal Elasticity of Bi- Helmholtz Type and Some Applications, *Int. J. Solids and Structures*, **43**, 1404–1421.

M. Lazar, G. A. Maugin and E. C. Aifantis, (2006b), Dislocations in Second Strain Gradient Elasticity, *Int. J. Solids and Structures*, **43**, 1787–1817.

K. Ch. Le, (1999), Thermodynamically Based Constitutive Equations For Single Crystals, in: Geometry, Continua and Microstructure (1st International Seminar on), ed. G. A. Maugin, pp. 87–97, Hermann, Paris.

E. H. Lee, (1969), Elastic-plastic Deformation at Finite Strain, *ASME. Trans. J. Appl. Mech.*, **36**, 1 6.

F. M. Leslie, (1968), Constitutive Equations for Liquid Crystals, *Arch. Rat. Mech. Anal.*, **28**, 265–283.

A. Lichnerowicz, (1976), Global theory of connections and holonomy groups, Noordhoff, Leyden.

H. A. Lorentz, (1952), Theory of Electrons, Dover Reprint, New York.

E. Lorentz and S. Andrieux, (1999), A Variational Formulation of Nonlocal Damage Models, *Intern. J. Plasticity*, **15**, 119–198.

V. A. Lubarda and X. Markenscoff, (2000), Conservation Integrals in Couple Stress Elasticity, *J. Mech. Phys. Solids*, **48**, 553–564.

G. A. Maugin, (1972), Remarks on Dissipative Processes in the Continuum Theory of Micromagnetics, J. Phys. (UK), **A5**, 1550–1562.

G. A. Maugin, (1974a), Sur la dynamique des milieux déformables avec spin magnétique- Théorie classique, *J. Mécanique* (Paris), **13**, 75–96.

G. A. Maugin, (1974b), Quasi-electrostatics of Electrically Polarized Continua, *Lett. Appl. Engng. Sci.*, **2**, 293–306.

G. A. Maugin, (1976), On the Foundations of the Electrodynamics of Deformable Media with Interactions, *Lett. Appl. Engng. Sci.*, **4**, 3–17.

G. A. Maugin, (1977), Deformable Dielectrics II- Voigt's Intramolecular Force Balance in Elastic Dielectrics, *Arch. Mechanics* (PL), **29**, 143–159.

G. A. Maugin, (1979), Nonlocal Theories or Gradient-type Theories: A Matter of Convenience?, Arch. Mechanics (PL, *Proc. Euromech Coll. on Nonlocal Theories, Warsaw, 1977*), **31**, 1–26.

G. A. Maugin, (1980), The Method of Virtual Power in Continuum Mechanics: Application to Coupled Fields, *Acta Mechanica*, **35**, 1–80.

G. A. Maugin, (1988), Continuum Mechanics of Electromagnetic Solids, North-Holland, Amsterdam.

G. A. Maugin, (1990), Internal Variables and Dissipative Structures, *J. Non-Equilibrium Thermodyn.*, **15**, 173–192.

G. A. Maugin, (1992), Thermomechanics of plasticity and fracture, Cambridge University Press, UK.

G. A. Maugin, (1993), Material Inhomogeneities in Elasticity, Chapman and Hall, London.

G. A. Maugin, (1994), Eshelby Stress in Plasticity and Fracture, *Int. J. Plasticity*, **10**, 393–408.

G. A. Maugin, (1995), Material Forces: Concepts and Applications, *ASME. Appl. Mech. Rev.*, **48**, 213–245.

G. A. Maugin, (1997), Thermomechanics of inhomogeneous-heterogeneous systems: application to the irreversible progress of two- and three-dimensional defects, ARI (Springer-Verlag), **50**, 41–56.

G. A. Maugin, (1998a), On Shock waves and Phase-transition Fronts in Continua, ARI, **50**, 141–150.

G. A. Maugin, (1998b), Thermomechanics of Forces Driving Singular point Sets (Anniversary Volume of H. Zorski), *Arch. Mech.*, (Poland), **50**, 477–487.

G. A. Maugin, (1998c), On the Structure of the Theory of Polar Elasticity, *Phil. Trans. Roy. Soc. Lond.*, **A356**, 1367–1395.

G. A. Maugin, (1999a), Thermomechanics of Nonlinear Dissipative Behaviors, World Scientific, Singapore, and River Edge, NJ.

G. A. Maugin, (1999b), Nonlinear Waves in Elastic Crystals, Oxford University Press, UK.

G. A. Maugin, (2000), On the universality of the thermomechanics of forces driving singular Sets, *Arch. Appl. Mech.*, **70**, 31–45.

G. A. Maugin, (2002a), Remarks on the Eshelbian thermomechanics of materials, *Mech. Res. Commun.*, **29**, 537–542.

G. A. Maugin, (2002b), Kroener-Eshelby Approach to Continuum Mechanics with Dislocations, Material Inhomogeneities and Pseudo-inhomogeneities, in: *Structured Media* (TRECOP'01 in Memory of E. Kroener), Poznan, 2001; ed. B. T. Maruszewski, pp. 182–195, Publ House Poznan Univ. Press, Poznan.

G. A. Maugin, (2003a), Geometry and Thermomechanics of Structural Rearrangements: Ekkehart Kroener's Legacy (GAMM'2002, Kroener's Lecture, Augsburg, 2002), *Zeit. angew. Math. Mech.*, **83**, 75–83.

G. A. Maugin, (2003b), Pseudo-plasticity and pseudo-inhomogeneity effects in materials mechanics, *J. Elasticity*, **71**, 81–103.

G. A. Maugin, (2006a), On Canonical Equations of Continuum Thermomechanics, *Mech. Res. Com.*, **33**, 705–710 (2006).

G. A. Maugin, (2006b), On the thermomechanics of continuous media with diffusion and /or weak nonlocality, *Arch. Appl. Mechanics* (Anniversary Issue) **75**, 723–738 (2006).

G. A. Maugin, (2007), On inhomogeneity, growth and ageing, Intern. Conference on configurational mechanics and defects, Aussois (March 2007) [published in *J. of the Mechanics of Materials and Structures*, Special issue (in memory of G. Herrmann), 4/4, 731–741 (2009)].

G. A. Maugin, (2011), Configurational forces: thermomechanics, physics, mathematics and numerics, CRC/Chapman & Hall/Taylor and Francis, Boca Raton, FL.

G. A. Maugin and C. I. Christov, (1997), Nonlinear Duality Between Elastic Waves and Quasi-particles in Microstructured Solids, *Proc. Est. Acad. Sci., Math. Phys.*, **46**, 78–84.

G. A. Maugin and C. I. Christov, (2002), Nonlinear Waves and Conservation Laws (Nonlinear Duality Between Elastic waves and Quasi-particles), in: *Selected Topics in Nonlinear Wave Mechanics*, eds. C. I. Christov and A. Guran, pp. 117–160, Birkhauser, Boston.

G. A. Maugin and M. Epstein, (1991), The Electroelastic Energy-momentum tensor, *Proc. Roy. Soc. Lond.*, **A433**, 299–312.

G. A. Maugin and C. Trimarco, (1992), Pseudo-momentum and Material Forces in Nonlinear Elasticity: Variational Formulation and Application to Fracture, *Acta Mechanica*, **94**, 1–28.

R. D. Mindlin and N. N. Eshel, (1968), On the first strain gradient theories in linear elasticity, *Int. J. Solids and Structures*, **4**, 109–124.

R. D. Mindlin and H. F. Tiersten, (1962), Effects of couple stresses in linear elasticity, *Arch. Rat. Mech. Anal.*, **11**, 415–448.

R. Mueller and G. A. Maugin, (2002), On Material Forces and Finite Element Discretizations, *Computational Mechanics*, **29**, no. 1, 52–60.

R. Mueller, D. Gross and G. A. Maugin, (2004), Use of material forces in adaptive Finite Element Method, *Computational Mechanics*, **33**, 421–434.

T. Mura, (1981), Energy release rate and the J-integral, in: *Three dimensional constitutive relations and ductile fracture*, ed. S. Nemat-Nasser, pp. 147–153, North-Holland, Amsterdam.

T. Mura, (1987), Micromechanics of defects in Solids, Martinus Nijhoff, The Hague.

F. R. N. Nabarro, (1967), Theory of Crystal Dislocations, Oxford Univ. Press, UK.

D. F. Nelson, (1979), Electric, Optic and Acoustic Interactions in Dielectrics, John Wiley, New York.

Nguyen Quoc Son, (1980), Méthodes énergétiques en mécanique de la rupture, *J. Mécanique*, **19**, 363–386.

E. Noether, (1918), Invariante Variationsproblem, Klg-Ges. Wiss. Nach. Göttingen. Math. Phys., **Kl.2**, 235–257.

W. Noll, (1967), Materially Uniform Simple Bodies with Inhomogeneities, *Arch. Rat. Mech. Anal.*, **27**, 1–32.

P. J. Olver, (1986), Applications of Lie Groups to Differential Equations, Springer, New-York.

Y. E. Pack and G. Herrmann, (1986a), Conservation Laws and the Material Momentum Tensor for the Elastic Dielectric, *Int. J. Engng. Sci.*, **24**, 1365–1374.

Y. E. Pack and G. Herrmann, (1986b), Crack Extension Force in Elastic Dielectrics, *Int. J. Engng. Sci.*, **24**, 1375–1388.

V. Z. Parton, (1976), Fracture Mechanics of Piezoelectric Crystals, ACTA ASTRONAUT., **3**, 671–683.

M. O. Peach and J. S. Koehler, (1950), Force Exerted on Dislocations and the Stress Produced by Them, *Phys. Rev.*, **II-80**, 436-439.

G. Piola, (1848), Intorno alle equazioni fondametali del movimento di corpi qualsivoglioni considerati la naturale loro forma e costituva, Mem. Mat. Soc. Ital. Modena, **24**(1), 1–186.

P. Podio-Guidugli, (2002), Configurational Forces: Are They Needed?, *Mech. Res. Commun.*, **29**, 513–519.

J. R. Rice, (1968), Path-independent Integral and the Approximate Analysis of Strain Concentrations by Notches and Cracks, *Trans. ASME. J. Appl. Mech.*, **33**, 379–385.

D. Rogula, (1965), Dynamical balance of Forces on Dislocation Line, *Bull. Acad. Pol. Sci.*, Ser. Sci. Techn., **13**, 337–343.

D. Rogula, (1966), Noether Theorem for a Continuous Medium Interacting with External Fields, *Proc. Vibration Problems (Warsaw)*, **7**, 337–344.

D. Rogula, (1977), Forces in Material Space, *Arch. Mechanics* (Poland), **29**, 705–715.

J. A. Schouten, (1954), Ricci calculus, 2nd edition, Springer-Verlag, Berlin.

D. E. Soper, (1976), Classical Field Theory, J. Wiley, New York.

R. A. Stephenson, (1982), The equilibrium field near the tip of a crack for finite plane strain of incompressible elastic materials, *J. Elasticity*, **12**, 65–99.

C. Teodosiu, (1982), Elastic models of crystal defects, Springer-Verlag, Berlin.

R. A. Toupin, (1962), Elastic materials with couple stress, *Arch. Rat. Mech. Anal.*, **11**, 395–414.

R. A. Toupin, (1964), Theories of elasticity with couple-stress, *Arch. Rat. Mech. Anal.*, **17**, 85-112.

C. A. Truesdell, (1969), Rational Thermodynamics, McGraw Hill, New York.

C. A. Truesdell, (1984), Rational Thermodynamics, 2nd Enlarged Edition, Springer-Verlag, New York.

C. A. Truesdell and W. Noll, (1965), Nonlinear Field Theories of Mechanics, in: *Handbuch der Physik*, **Bd.III/3**, ed. S. Flügge, Springer-Verlag, Berlin.

C. A. Truesdell and R. A. Toupin, (1960), The classical field theories, in: *Handbuch der Physik*, ed. S. Flügge, **Bd.III/1**, Springer-Verlag, Berlin.

L. M. Truskinovskii, (1983), The Chemical Tensor, *Geokhimiya*, **12**, 1730–1744.

C. C. Wang, (1967), On the Geometric Structure of Simple Bodies, or Mathematical Foundations for the Theory of Continuous Distributions of Dislocations, *Arch. Rat. Mech. Anal.*, **27**, 33–94.

H. Zorski, (1981), Force on a Defect in Nonlinear Elastic Medium, *Int. J. Engng. Sci.*, **19**, 1573–1579.

Approaches to Generalized Continua

Carlo Sansour * and Sebastian Skatulla †

* Department of Civil Engineering, INSA Rennes, France; on leave from The University of Nottingham, United Kingdom

† CERECAM, Department of Civil Engineering, University of Cape Town, South Africa

1 Introduction

Decades ago, it has been recognized that for some materials the kinematics on meso- and micro-structural scale needs to be considered, if the material's resistance to deformation exhibits a finite radius of interaction on atomic or molecule level, e.g. Kröner (1963); Mindlin (1964) outlined that this is the case if the deformation wave length approaches micro-structural length scale. Differently said, if the external loading corresponds material entities smaller than the *representative volume element* (RVE), then the statistical average of the macro-scopical material behaviour does not hold anymore. In this sense the fluctuation of deformation on micro-structural level as well as relative motion of micro-structural constituens, such as granule, crystalline or other heterogeneous aggregates, influence the material response on macro-structural level. Consequently, field equations based on the assumption of micro-scopically homogeneous material have to be supplemented and enriched to also include non-local and higher-order contributions.

In particular, generalized continua aim to describe material behaviour based on a deeper understanding of the kinematics at smaller scales rather than by pure phenomenological approximation of experimental data obtained at macro-scopical level. The meso- or micro-structural kinematics and its nonlocal nature is then treated either by incorporating higher-order gradients or by introducing extra degrees of freedom. For the latter, the small-scale kinematics at each material point can be thought to be equipped with a set of directors which specify the orientation and deformation of a surrounding a micro-space. The resulting theories can be categorized into three main groups: (1) rigid body motion where the directors only rotate but do not deform or change their angles between each other referring to the so-called *Cosserat* (Cosserat and Cosserat, 1909) or *micropolar continuum* (Eringen, 1967); (2) directors undergoing additional to rotation also stretch referring to the so-called *micro-stretch continuum* (Eringen, 1969); (3) rotation, stretch and change of angles between directors referring to the so-called *micromorphic continuum* (Ericksen and Truesdell, 1957; Eringen, 1972). Accordingly, the number of extra degrees of freedom depends on the dimension of macro- and small-scale spaces but also what kind of meso- or micro-deformation modes are considered and the order of the coresponding ansatz formulations. Thus, the number of additional degrees of freedom can vary between 1 and 18 or more (Eringen, 1999; Sansour and Skatulla, 2010) which in turn leads to large diversity of derived strain and stress measures describing deformation phenomena originated on meso- or micro-structural level.

In solid mechanics generalized continuum formulations initially only considered elastic material behaviour aiming at introducing an internal length which can describe scale effects. These kinds of physical phenomena were reported in literature for a variety of materials such as epoxy polymers, foam, mild steel, copper, marble, e.g. (Richards, 1958; Yang and Lakes, 1981; Vardoulakis et al., 1998; Fleck et al., 1994; Zhu et al., 1995; Lam et al., 2003; Al-Rub and Voyiadjis, 2004). Also for multi-walled carbon nanotubes it was shown that the buckling behaviour and more specifically the critical axial strain depended on the choice for an intrinsic length scale parameter (Sudak, 2003). The incorporation of first as well as second strain gradients proved to be helpful to deal with strain and stress singularities (see e.g. (Lazar et al., 2006)), or address scale effects (Aifantis, 1999; Akarapu and Zbib, 2006; Ohashi et al., 2007). It was discovered that such formulations can be utilized to overcome loss of ellipticity of the governing equations in material instability problems such as shear band formations (Knowles and Sternberg, 1978; Triantafyllidis and Aifantis, 1986; R. De Borst, 1992). Furthermore, direction and magnitude of localized stress could be controlled by an additional material parameter. Higher gradients were also shown to be physically meaningful with regards to wave dispersion (Bennett et al., 2007). Recently, Eshelby's inclusion problem was solved accounting for strain gradients and boundary effects (Gao and Ma, 2010). Examples of linear micromorphic continuum theories were reported in (Manzaria and Regueiro, 2005; Kumar and McDowell, 2004; Dillard et al., 2006). More specifically, questions related to heat conduction were treated in (Iesan, 2006) and some aspects of work and energy in micromorphic continua were discussed in (Tian-min, 2001).

The transition of non-local meso- or micro-structural to the macroscopical constitutive response can be achieved by employing homogenization schemes integrating over the microspace and in that way averaging or "smearing" of the behaviour at small scales. The accuracy of theses homogenization schemes largely depends on the choices made on the meso-micro-space, as to the order of the ansatz of the small-scale fields, boundary conditions, detail of representation of the actual meso- or micro-structure. In this sense generalized continua can be related to first, second- and higher-order homogenization schemes that allow for the formulation of higher gradient continuum theories. In (Kouznetsova et al., 2002; Larsson and Diebels, 2006) an interesting aspect was the incorporation of micro-space boundary conditions. The homogenization can also be based on micropolar formulations, e.g. modelling elastic size-scale together with interface effects of composites where the material nonlocality was addressed by the micropolar length scale parameter whereas critical surface-to-volume ratio of matrix and the reinforced phase by intrinsic length scale of theinterface model (Chen et al., 2007). Also, a cohesive zone model was shown to be formulated on the micro-level within a micropolar homogenization framework addressing size-dependent interface degradation of layered micro-mechanical systems (Zhang and Larsson, 2009). In (Jänicke et al., 2009) different homogenization schemes and projection rules based on a micromorphic continuum were discussed and applied to FE^2 to investigate discrete micro-structure.

Geometrically nonlinear formulations are sparsely found in literature, e.g. strain gradient theories applied to granular geo-materials modelling their cohesive-frictional deformation behaviour (Chambon et al., 2004), or to heterogeneous material considering periodic micro-boundary conditions (Kouznetsova et al., 2002). McVeigha and Liu (2008) proposed a multi-resolution approach accounting for multiple scaling levels and featuring a higher-order variational principle as well as corresponding governing equations. In the context of a micromorphic continuum approach, issues related to material forces of in the hyperelastic case were discussed in (Hirschberger et al., 2007). The main attention of geometrically exact generalized continuum theories, however, was focused on micropolar or *Cosserat* theories, e.g. (Steinmann, 1994; Sansour, 1998a; Forest et al., 1997; Epstein and de Leon, 1998).

Inelastic *Cosserat*-type formulations first considered only the geometrically linear settings ((Mühlhaus, 1989; Mühlhaus and Aifantis, 1991; de Borst, 1993)). In small strain micromorphic elasto-plasticity the combined kinematic and isotropic hardening case was addressed by Zhang et al. (2011). An additionally evolution equation of micro-couple-backstress was formulated and related to an internal length scale parameter which allowed to control the magnitude of scale effects in the Bauschinger effect under cyclic loading. Ductile fracture toughness of multiphase materials were dealt with a micromorphic multi-scale approach by Vernerey et al. (2009). The incorporation of spatial gradient of the micromorphic degrees of freedom complemented with a corresponding balance of angular momentum equation (Forest, 2009) was applied to nonlinear kinematic hardening. Plastic strain gradients also have a long history in the field of dislocation theories, e.g. (Fleck et al., 1994; Zhu et al., 1997; Han et al., 2005; Gurtin and Ohno, 2011; Yalcinkaya et al., 2011), where they were linked to statistically stored plus geometrically necessary dislocation densities. When dealing with rate-independent and rate-dependent strain gradient inelasticity elastic and plastic parts can be associated with separate energetic and dissipative constitutive length parameters, respectively (Gudmundson, 2004; Fleck and Willis, 2009; Niordson and Legarth, 2010). Additionally, interface conditions for plastic strain and moment traction were considered. In realm of finite strain plasticity micropolar continuum-based isotropic and kinematic hardening was considered to describe size effects exhibited for torsion of thin wires (Grammenoudis and Tsakmakis, 2009). Also in dislocation theory-based crystal plasticity the micropolar approach was found to be useful (Forest et al., 2000; J.R. Mayeur, 2011). Micromorphic plasticity two-scale models have been proposed by (Grammenoudis et al., 2009; Regueiro, 2009) addressing micro-structural damage as well as granular material behaviour.

So far formulations of generalized continua are faced with two major problems. The first one relates to the fully non-linear and inelastic material behaviour. Classical inelastic formulations are based on decompositions of strain measures. Since generalized continua exhibit more than one strain or deformation measure the question arises as to how these can be decomposed into elastic and inelastic parts. Few suggestions were made in (Steinmann, 1994; Forest et al., 1997; Sansour, 1998b; Forest and Sievert, 2006; Chambon et al., 2004). These formulations remained, however, less satisfactory since the decomposition of the two deformation measures were, strictly speaking, independent of each other, which raises many questions regarding the adequate formulation of evolution laws for the inelastic parts. The second problem relates to the observation of scale effects also in a homogenously deforming specimen. *Cosserat* and higher gradient theories cannot predict such scale effects, because the extra strain measures are identically zero for homogenous deformation. Furthermore, it is desirable to set out from a general and unified formulation of continua with meso- and micro-structure. We propose a framework based on the mathematical concept of fibre bundles embedded into a generalized continuum formulation. More specifically, we want to consider the Cartesian product of the macroscopic and further meso- or microscopic spaces and, accordingly, the generalized deformation is composed of a macro-, meso- and/or micro-components. In principle, every point of the macroscopic space would have an infinite number of degrees of freedom and dimensions. In practice, the number of degrees of freedom is finite corresponding to the chosen level of accuracy. In this sense, strain gradient, micropolar, micro-stretch and micromorphic continua appear just as special approximations of the general case (Sansour, 1998a). From the generalized deformation description nonlinear strain measures are derived and corresponding stress measures are defined which allow for the formulation of generalized variational principles and corresponding field equations as well as Dirichlet boundary conditions. Here, it is assumed that the generalized deformation field can only be varied within the macro-space so that the balance equations and the Dirichlet boundary conditions are established for the macro-space. In this sense the field equations as well as the boundary conditions represent the meso- or microscopic average at every point of the macro-space. The main features of the framework are summarized as follows:

- The transition from meso- or micro- to the macro-quantities is carried out in a mathematically sound way using projection maps which are natural operation for fibre bundles.
- Size scale effects can be dealt with as the approach provides an internal length relating to the size of the meso- or micro-space.
- The issue of nonlinear constitutive laws, a delicate matter in generalized continua, is solved in an elegant manner. Since the constitutive law is defined pointwise at the meso- or micro-level, well established nonlinear constitutive laws can be utilized directly. In the nonlinear hyperelastic case we avoid cumbersome applications of representation theorems which introduce a large number of obscure material parameters. In the inelastic case classical well established decompositions of strain measures into elastic and inelastic parts can be utilized directly.
- The method is numerically appealing, since it is based on numerical integration over the meso- and micro-spaces. The dimensions of the same being the only extra material parameters introduced into the formulation.

Now, before the generalized continuum framework is explored in detail in Sec. 7 to derive from it the special cases of the micromorphic, micropolar and strain gradient formulations, we briefly introduce in Sec. 2 basic relations of transformation groups and their corresponding Lie algebras which are of relevance to us. In Sec. 4, classical stretch tensor-based continuum formulations are investigated as they provide the first step in introducing rotational degrees of freedom. In Sec. 5 the conventional *Cosserat* continuum theory is outlined and the update of the rotational field in numerical computations is tackled. A short discussion of the conventional micromorphic approach follows in Sec. 6.

2 Elements of Linear Transformation Groups (Lie Groups)

2.1 Definition of transformation groups and the exponential map

A group is a set M with an internal operation $\circ$ (usually called multiplication), such that the following axioms hold:

- $(a \circ b) \in M$ for all $a, b \in M$
- $(a \circ b) \circ c = a \circ (b \circ c)$ for all $a, b, c \in M$ (associativ law)
- their exists a neutral element $i \in M$, such that $i \circ a = a \circ i = a$ for all $a \in M$
- for every $a \in M$ an element $a^{-1} \in M$ exists such that $a \circ a^{-1} = a^{-1} \circ a = i$.

With $Mat(3, \mathbb{R})$ we refer to the set of all 3×3 matrices defined over the field of real numbers $\mathbb{R}$. The subset defined by all invertible matrices is denoted by $GL(3, \mathbb{R})$ (General Linear Group):

$$GL(3, \mathbb{R}) := \{\mathbf{A} \in Mat(3, \mathbb{R}) | \; det\mathbf{A} \neq 0\}, \tag{1}$$

where det denotes the determinant of a matrix. Under the operation of usual matrix multiplication, $GL(3, \mathbb{R})$ is a group. The group consists of two connected but otherwise disjunct components. The elements of the one component have positive determinants, while those of the other have negative determinants. The first component, which we denote by $GL^+(3, \mathbb{R})$ is of special interest to us since the deformation gradient $\mathbf{F}$ is an element of $GL^+(3, \mathbb{R})$. $GL^+(3, \mathbb{R})$ is itself a subgroup of $GL(3, \mathbb{R})$. Two subgroups of $GL^+(3, \mathbb{R})$ are of special importance: the special linear group $SL(3, \mathbb{R})$,

$$SL(3, \mathbb{R}) := \{\mathbf{A} | \; \mathbf{A} \in GL^+(3, \mathbb{R}), det\mathbf{A} = 1\}, \tag{2}$$

and the special orthogonal group $SO(3, \mathbb{R})$,

$$SO(3, \mathbb{R}) := \{\mathbf{A} | \; \mathbf{A} \in GL^+(3, \mathbb{R}), det\mathbf{A} = 1, \mathbf{A}\mathbf{A}^T = \mathbf{1}\}. \tag{3}$$

Here the identity tensor is denoted by $\mathbf{1}$. In the mathematical text books on algebra it is shown that only $SO(3, \mathbb{R})$ carries the special feature of being a compact group.

The exponential map plays within the theory of Lie groups a fundamental role. By understanding that all groups are defined over $\mathbb{R}$, from now on we omit the explicit reference to it. If we restrict ourselves to linear Lie groups, the exponential map

$$\exp : Mat(3) \to GL^+(3), \quad \exp \mathbf{Z} = \mathbf{A}, \quad \mathbf{Z} \in Mat(3), \mathbf{A} \in GL^+(3), \tag{4}$$

is defined as

$$\mathbf{A} = \exp \mathbf{Z} = \mathbf{1} + \mathbf{Z} + \frac{\mathbf{Z}^2}{2!} + \frac{\mathbf{Z}^3}{3!} + \cdots \tag{5}$$

The restriction of this map to skew-symmetric matrices defined in $Mat(3)$ gives us elements defined in the group $SO(3)$. That is

$$\mathbf{A} = \exp \mathbf{Z} |\ \mathbf{Z} = -\mathbf{Z}^T \Longrightarrow \mathbf{A} \in SO(3). \tag{6}$$

On the other hand the restriction of the exponential map to traceless elements in $Mat(3)$ results in elements which are defined in $SL(3)$. One has

$$\mathbf{A} = \exp \mathbf{Z} |\ tr\mathbf{Z} = 0 \Longrightarrow \mathbf{A} \in SL(3), \tag{7}$$

where tr defines the trace of a tensor. This statement is a direct result of the important relation

$$det(\exp \mathbf{Z}) = \exp(tr\mathbf{Z}). \tag{8}$$

The exponential map restricted to skew-symmetric tensors is surjective in $SO(3)$. But it is not surjective in $GL^+(3)$ or $SL(3)$. That is, while every element of $SO(3)$ can be written in form of an exponential map, this is not true for $GL^+(3)$ or $SL(3)$. Also the image of the exponential map in $GL^+(3)$ does not constitute a subgroup. That is the product of two elements of $GL^+(3)$, which both are images of exponential maps is not necessary an image of an exponential map.

The above statements are of global nature. Of great importance is the following local statement. Near the identity any element of a linear Lie group can be written down as the image of an adequate exponential map. Hence, the product of two images of exponential maps (near the identity) can be formulated as an exponential map of an adequate element of the groups algebras (to be specified later).

As an example let us consider the case of simple shear described by

$$\mathbf{F} = \begin{pmatrix} 1 & \alpha \\ 0 & 1 \end{pmatrix},$$

where α is a shearing parameter. In this special case $\mathbf{F}$ can globally be written down as an exponential map:

$$\mathbf{F} = \exp \begin{pmatrix} 0 & a \\ 0 & 0 \end{pmatrix}.$$

On the other hand a combination of simple shear and compression in the form, say

$$\mathbf{F} = \begin{pmatrix} -1 & \alpha \\ 0 & -1 \end{pmatrix},$$

cannot be written down as an exponential map (for a proof see Varadarajan (1984)).

2.2 Tangent spaces and Lie algebras

An important aspects is understanding transformation groups, e.g. $GL^+(3)$, as Lie groups. Lie groups are equipped with the structure of differentiable manifolds, such that group operations of multiplication and inversion are smooth. Hence it makes sense to talk about differentiation. Rates can be naturally understood as tangent vectors along a curve defined on the manifold.

To illustrate the concepts of Lie groups we exemplify them on the group $GL^+(3)$. First we describe a curve C on $GL^+(3)$ as a map from $\mathbb{R}$ to $GL^+(3)$:

$$C : \mathbb{R} \to GL^+(3),\ C(t) = \{\mathbf{A}|\ t \in \mathbb{R}, \mathbf{A} \in GL^+(3)\}. \tag{9}$$

Differentiation along this curve with respect to the parameter t gives us the tangent vectors on $GL^+(3)$. The set of all tangent vectors at any point in $GL^+(3)$ constitutes the tangent space at that point. Accordingly, different points have different tangent spaces.

Let us have a look at the neutral element of the group which is then identity $\mathbf{1}$. The tangent space at the identity is denoted by $gl^+(3)$ and is called the Lie algebra of the group. Elements of the Lie algebra of a group fulfill the following two conditions. First, the operation of the Lie bracket (the commutator) is closed in the Lie algebra. That is the relation holds:

$$[\mathbf{a}, \mathbf{b}] = (\mathbf{ab} - \mathbf{ba}) \in gl^+(3), \qquad \mathbf{a}, \mathbf{b} \in gl^+(3), \tag{10}$$

and, second, the *Jacobi Identity* is fulfilled

$$[\mathbf{a}, [\mathbf{b}, \mathbf{c}]] + [\mathbf{c}, [\mathbf{a}, \mathbf{b}]] + [\mathbf{b}, [\mathbf{c}, \mathbf{a}]] = \mathbf{0}. \tag{11}$$

Near the identity the group elements themselves are approximated by their algebra. Accordingly, an element of the algebra defines an infinitesimal transformation. Transformation groups are characterised by the fact that any element of the group can be connected to the identity through a curve adequately parameterised by a scalar.

Consider a curve in $GL^+(3)$, which goes through the identity. Any tangent vector at the identity can be transformed through left or right transformation using an element of the group to a tangent vector at that point of the group. For $\mathbf{A} \in GL^+(3)$ and $\mathbf{X} \in gl^+(3)$ the product $\mathbf{AX}$ is called a left invariant tangent vector at $\mathbf{A}$. We also make use of the notation

$L_\mathbf{A}\mathbf{X} = \mathbf{AX}$. The adjective left invariant is motivated by the relation

$$\mathbf{B}L_\mathbf{A}\mathbf{X} = \mathbf{BAX} = L_\mathbf{BA}\mathbf{X}, \qquad \mathbf{B}, \mathbf{A} \in GL^+(3). \tag{12}$$

Now, the parameter-dependent exponential map $\exp t\boldsymbol{\beta}$, with $\boldsymbol{\beta} \subset gl^+(3)$, constitutes a one parameter subgroup. It has already been mentioned that, in general, the product of two exponential maps need not be given by an exponential map. The multiplication of exponential maps is not a closed operation, which means that exponential maps do not define a subgroup. However, a condition for the validity of the relation $\exp\boldsymbol{\alpha}\exp\boldsymbol{\beta} = \exp(\boldsymbol{\alpha}+\boldsymbol{\beta})$ is that $[\boldsymbol{\alpha}, \boldsymbol{\beta}] = \mathbf{0}$. Since this condition is valid for the Lie bracket of a tensor with itself, the set of elements $\exp t\boldsymbol{\beta}$ (defined by different values of t) is in fact a subgroup of $GL^+(3)$.

Let us consider a curve in $GL^+(3)$ parameterised by t. We assume that for $t = t_0$ the curve passes through the point $\mathbf{F}$. The neighbourhood of $\mathbf{F}$ can be reached through a left transformation according to $\mathbf{FF}^*$. $\mathbf{F}^*$ is an element of $GL^+(3)$ defined near the identity. In the vicinity of $\mathbf{F}$ the curve through $\mathbf{F}$ can be described as a one-parameter subgroup with the exponential map:

$$\mathbf{FF}^* = \mathbf{F}\exp\{(t - t_0)\mathbf{L}\}, \mathbf{L} \in gl^+(3). \tag{13}$$

Tangent vectors along the curve at $t = t_0$ are given by a differentiation with respect to t:

$$\frac{D}{Dt}\mathbf{F}\exp((t - t_0)\mathbf{L})|_{t=t_0} = \mathbf{FL}. \tag{14}$$

The quantity $\mathbf{FL}$ defines the tangent vector at $\mathbf{F}$. Accordingly, tangent vectors are nothing but left invariant vector fields. Alternatively, $\mathbf{FL}$ can be written down as $L_\mathbf{F}\mathbf{L}$. Since $det(\exp\mathbf{M}) = \exp(tr\mathbf{M}) > 0$ for all $\mathbf{M} \in Mat(3)$, we conclude that $gl^+(3)$ can be identified with $Mat(3)$.

In the same way we can construct right invariant tangent vectors instead of left invariant ones. For $\mathbf{A} \in GL^+(3)$ and for $\mathbf{X} \in gl^+(3)$ we call the product $\mathbf{XA}$ a right invariant tangent vector in the group $GL^+(3)$ at the point $\mathbf{A}$. Similarly we make use of the notation $R_\mathbf{A}\mathbf{X} = \mathbf{XA}$. The manifestation of the right invariance is documented through

$$(R_\mathbf{A}\mathbf{X})\mathbf{B} = \mathbf{XAB} = R_\mathbf{AB}\mathbf{X}, \quad \mathbf{B}, \mathbf{A} \in GL^+(3). \tag{15}$$

In analogy to 13 we can define the right action of $\mathbf{F}$ on $\mathbf{F}^*$. One has

$$\frac{D}{Dt}\exp((t - t_0)\mathbf{l})\mathbf{F}|_{t=t_0} = \mathbf{lF}, \quad \mathbf{l} \in gl^+(3). \tag{16}$$

A relation between both tangent vectors $\mathbf{lF}$ and $\mathbf{FL}$ is provided by the adjunct action of the group:

$$\mathbf{l} = \mathbf{FLF}^{-1}. \tag{17}$$

The algebra of the group $SO(3)$ will be denoted by $so(3)$ and that of $SL(3)$ by $sl(3)$. The elements of $so(3)$ and $sl(3)$ are defined by

$$so(3) := \{\mathbf{X}|\ \mathbf{X} \in gl^+(3), \mathbf{X}^T = -\mathbf{X}\}, \tag{18}$$

$$sl(3) := \{\mathbf{X}|\ \mathbf{X} \in gl^+(3), tr\mathbf{X} = 0\}. \tag{19}$$

From the above it becomes clear that rates can be understood as elements of tangent spaces. Within solid mechanics we are confronted with the deformation gradient $\mathbf{F}$, the rotation tensor $\mathbf{R}$ or the plastic part of the deformation gradient $\mathbf{F}^p$. The latter has the constraint $det\mathbf{F}^p = 1$ and is, hence, an element of $SL(3)$. The configuration spaces are non-linear manifolds and the rates in these spaces are best understood as elements of the algebra of an appropriately defined group. We already can see that the well known spatial tensor $\mathbf{l} = grad\dot{\mathbf{x}}$ with $\mathbf{x}$ being the placement, also called the velocity gradient, is nothing but a right invariant tangent vector. However, we are now motivated to introduce a left invariant rate $\mathbf{L}$, which in fact proves to be a suitable rate within purely material descriptions.

2.3 Time derivatives and variations

In the last section, tangent vectors have been introduced through a differentiation process along a curve parameterised with a scalar t. This parameter can now be understood as representing the time scale. The tangent vectors then would correspond to time derivatives or velocities. Alternatively, the parameter t can simply be used to describe the neighbourhood of a point in the group of interest. The tangent vectors then correspond to linearised expressions. Variations belong to the latter type of expressions.

In any physical process, the time is already present as a parameter. Such processes are often described using functionals which depend on physical quantities defined as groups and their velocities. Also spatial derivatives of these quantities must be considered. The variation of these functionals means that expressions must be derived for the variation of velocities, variations of spatial derivatives, velocities of variations or spatial derivatives, etc. The following expressions provide the necessary relations. However, we have to bear in mind that what we so far called rate can be now understood as a real time derivative, as a spatial derivative, or as a variation. According to the preceding sections there is always the possibility of deriving left and right time rates, variations, or spatial derivatives.

Let $\mathbf{F} \in GL(3), \mathbf{l}, \mathbf{L}, \mathbf{a}, \mathbf{A} \in gl(3)$, such that we have

$$\dot{\mathbf{F}} = \mathbf{lF}, \quad (20)$$

$$\dot{\mathbf{F}} = \mathbf{FL}, \quad (21)$$

$$\delta\mathbf{F} = \mathbf{aF}, \quad (22)$$

$$\delta\mathbf{F} = \mathbf{FA}. \quad (23)$$

In the above, a dot denotes the time derivative and δ is the variation symbol.

The expressions for the time derivative of a variation and vice versa, are derived by making use of the equality

$$\delta\dot{\mathbf{F}} = \frac{D\delta\mathbf{F}}{Dt}, \quad (24)$$

which is nothing but the statement that variations and time derivatives can be interchanged in their order. The exploitation of this equality leads to

$$\delta(\mathbf{lF}) = \frac{D\mathbf{aF}}{Dt}, \quad (25)$$

resulting in

$$\delta\mathbf{lF} + \mathbf{laF} = \dot{\mathbf{a}}\mathbf{F} + \mathbf{alF}. \quad (26)$$

As a general expression one has

$$\delta\mathbf{l} = \dot{\mathbf{a}} + [\mathbf{a}, \mathbf{l}]. \quad (27)$$

This expression is of general nature and holds for all groups together with their corresponding algebras.

Analogously, one has for the variation of the rate the relation

$$\delta\mathbf{L} = \dot{\mathbf{A}} + [\mathbf{L}, \mathbf{A}]. \quad (28)$$

Similarly one can derive the alternative expressions

$$\delta\mathbf{l} = \mathbf{F}\dot{\mathbf{A}}\mathbf{F}^{-1}, \quad (29)$$

$$\delta\mathbf{L} = \mathbf{F}^{-1}\dot{\mathbf{a}}\mathbf{F}. \quad (30)$$

In contrast to (27) and (28) in (29) and (30) it is the group element itself which is involved.

Along the same ideas one can consider time derivatives of variations, spatial derivatives of variations, spatial derivatives of rates, as well as time derivatives of spatial derivatives. This will be done explicitly for the special rotation group in the next subsection.

3 The rotation group SO(3)

In this section we consider in some detail the group $SO(3)$. In addition to the parameterisation with the help of a scaler parameter, we want to consider also fields of elements of $SO(3)$ given as functions of some co-ordinates ϑ^i, where $i = 1, 2, 3$. Derivatives with respect to these co-ordinates will be denoted by a comma.

First we recall that any skew-symmetric tensor $\mathbf{W} \in so(3)$ can be related to a vector $\mathbf{w} \in \mathbb{R}^3$ called its axial vector and vice versa. The map which relates every *Euclidean* vector to a skew-symmetric tensor is often called the adjoint representation of the *Euclidean* space and is defined by the following relation

$$
\begin{aligned}
ad : \mathbf{w} \in \mathbb{R}^3 &\longrightarrow \mathbf{W} \in so(3) \\
\mathbf{W}\mathbf{a} &= \mathbf{w} \times \mathbf{a},
\end{aligned}
\tag{31}
$$

where $\mathbf{a}$ is an arbitrary vector and $\times$ denotes the ordinary cross or vector product.

Let $\mathbf{R} \in SO(3)$, from the previous elaborations we have immediately the general rates (left invariant)

$$
\dot{\mathbf{R}} = \mathbf{R}\hat{\boldsymbol{\Omega}}, \tag{32}
$$
$$
\delta\mathbf{R} = \mathbf{R}\hat{\mathbf{W}}, \tag{33}
$$
$$
\mathbf{R}_{,i} = \mathbf{R}\hat{\mathbf{K}}_i. \tag{34}
$$

Likewise, the right invariant rates are given as

$$
\dot{\mathbf{R}} = \boldsymbol{\Omega}\mathbf{R}, \tag{35}
$$
$$
\delta\mathbf{R} = \mathbf{W}\mathbf{R}, \tag{36}
$$
$$
\mathbf{R}_{,i} = \mathbf{K}_i\mathbf{R}. \tag{37}
$$

All the quantities $\hat{\boldsymbol{\Omega}}, \boldsymbol{\Omega}, \hat{\mathbf{W}}, \mathbf{W}, \hat{\mathbf{K}}_i, \mathbf{K}_i \in so(3)$. That is, all these quantities are skew-symmetric tensors. In fact, the skew-symmetry is a direct result of the orthogonality condition

$$
\mathbf{R}^T\mathbf{R} = 1.
$$

Then, upon taking any derivative of this expression (say with respect to time)

$$
\frac{D}{Dt}(\mathbf{R}^T\mathbf{R}) = \mathbf{0}, \tag{38}
$$

we conclude that

$$
\mathbf{R}^T\dot{\mathbf{R}} = -\left(\mathbf{R}^T\dot{\mathbf{R}}\right)^T, \tag{39}
$$

Inserting (32) or (35) results immediately in the skew-symmetry of $\mathbf{\Omega}$ or $\hat{\mathbf{\Omega}}$. The same can be shown for all other quantities using related derivatives. Now, comparison of (32) with (35), (33) with (36), and (34) with (37) reveals

$$\begin{aligned} \mathbf{\Omega} &= \mathbf{R}\hat{\mathbf{\Omega}}\mathbf{R}^T, && (40)\\ \mathbf{W} &= \mathbf{R}\hat{\mathbf{W}}\mathbf{R}^T, && (41)\\ \mathbf{K}_i &= \mathbf{R}\hat{\mathbf{K}}_i\mathbf{R}^T,. && (42) \end{aligned}$$

Moreover, following the results (27)-(30) we immediately infer that

$$\begin{aligned} \delta\hat{\mathbf{\Omega}} &= \dot{\hat{\mathbf{W}}} + \mathbf{\Omega}\hat{\mathbf{W}} - \hat{\mathbf{W}}\mathbf{\Omega} = \dot{\hat{\mathbf{W}}} + [\mathbf{\Omega}, \hat{\mathbf{W}}], && (43)\\ \dot{\hat{\mathbf{K}}}_i &= \hat{\mathbf{\Omega}}_{,i} + \mathbf{K}_i\hat{\mathbf{\Omega}} - \hat{\mathbf{\Omega}}\mathbf{K}_i = \hat{\mathbf{\Omega}}_{,i} + [\mathbf{K}_i, \hat{\mathbf{\Omega}}], && (44)\\ \delta\hat{\mathbf{K}}_i &= \hat{\mathbf{W}}_{,i} + \hat{\mathbf{K}}_i\hat{\mathbf{W}} - \hat{\mathbf{W}}\hat{\mathbf{K}}_i = \hat{\mathbf{W}}_{,i} + [\hat{\mathbf{K}}_i, \hat{\mathbf{W}}], && (45) \end{aligned}$$

as well as

$$\begin{aligned} \delta\mathbf{\Omega} &= \dot{\mathbf{W}} + [\mathbf{W}, \mathbf{\Omega}], && (46)\\ \dot{\mathbf{K}}_i &= \mathbf{\Omega}_{,i} + [\mathbf{\Omega}, \mathbf{K}_i], && (47)\\ \delta\mathbf{K}_i &= \mathbf{W}_{,i} + [\mathbf{W}, \mathbf{K}_i]. && (48) \end{aligned}$$

Also the relation holds

$$\begin{aligned} \delta\mathbf{\Omega} &= \mathbf{R}\dot{\hat{\mathbf{W}}}\mathbf{R}^T, \quad \text{or} \quad \dot{\hat{\mathbf{W}}} = \mathbf{R}^T\delta\mathbf{\Omega}\mathbf{R} && (49)\\ \mathbf{\Omega}_{,i} &= \mathbf{R}\dot{\hat{\mathbf{K}}}_i\mathbf{R}^T, \quad \text{or} \quad \dot{\hat{\mathbf{K}}}_i = \mathbf{R}^T\mathbf{\Omega}_{,i}\mathbf{R}^T && (50)\\ \mathbf{W}_{,i} &= \mathbf{R}\delta\hat{\mathbf{K}}_i\mathbf{R}^T \quad \text{or} \quad \delta\hat{\mathbf{K}}_i = \mathbf{R}^T\mathbf{W}_{,i}\mathbf{R} && (51) \end{aligned}$$

Finally,

$$\begin{aligned} \delta\hat{\mathbf{\Omega}} &= \mathbf{R}^T\dot{\mathbf{W}}\mathbf{R} \quad \text{or} \quad \dot{\mathbf{W}} = \mathbf{R}\delta\hat{\mathbf{\Omega}}\mathbf{R}^T && (52)\\ \hat{\mathbf{\Omega}}_{,i} &= \mathbf{R}^T\dot{\mathbf{K}}_i\mathbf{R}, \quad \text{or} \quad \dot{\mathbf{K}}_i = \mathbf{R}\hat{\mathbf{\Omega}}_{,i}\mathbf{R}^T && (53)\\ \hat{\mathbf{W}}_{,i} &= \mathbf{R}^T\delta\mathbf{K}_i\mathbf{R}, \quad \text{or} \quad \delta\mathbf{K}_i = \mathbf{R}\hat{\mathbf{W}}_{,i}\mathbf{R}^T. && (54) \end{aligned}$$

All the above relations can be reformulated in terms of the axial vectors of the skew-symmetric quantities. We select some of the relevant relations, which often prove useful. If we denote the axial vectors by lower case letters, then the following holds:

$$\begin{aligned} \delta\hat{\boldsymbol{\omega}} &= \dot{\hat{\mathbf{w}}} + \boldsymbol{\omega} \times \hat{\mathbf{w}} && (55)\\ \dot{\hat{\mathbf{k}}}_i &= \hat{\boldsymbol{\omega}}_{,i} + \mathbf{k}_i \times \hat{\boldsymbol{\omega}}, && (56)\\ \delta\hat{\mathbf{k}}_i &= \hat{\mathbf{w}}_{,i} + \hat{\mathbf{k}}_i \times \hat{\mathbf{w}}, && (57) \end{aligned}$$

as well as

$$\delta\boldsymbol{\omega} = \dot{\mathbf{w}} + \mathbf{w} \times \boldsymbol{\omega}, \quad (58)$$
$$\dot{\mathbf{k}}_i = \boldsymbol{\omega}_{,i} + \boldsymbol{\omega} \times \mathbf{k}_i, \quad (59)$$
$$\delta\mathbf{k}_i = \mathbf{w}_{,i} + \mathbf{w} \times \mathbf{k}_i. \quad (60)$$

Also the relation holds

$$\delta\boldsymbol{\omega} = \mathbf{R}\dot{\hat{\mathbf{w}}}, \quad (61)$$
$$\boldsymbol{\omega}_{,i} = \mathbf{R}\dot{\hat{\mathbf{k}}}_i, \quad (62)$$
$$\mathbf{w}_{,i} = \mathbf{R}\delta\hat{\mathbf{k}}_i, \quad (63)$$

which relate the rates defined from the right to those defined from the left. Finally,

$$\delta\hat{\boldsymbol{\omega}} = \mathbf{R}^T\dot{\mathbf{w}}, \quad (64)$$
$$\hat{\boldsymbol{\omega}}_{,i} = \mathbf{R}^T\dot{\mathbf{k}}_i \quad (65)$$
$$\hat{\mathbf{w}}_{,i} = \mathbf{R}^T\delta\mathbf{k}_i. \quad (66)$$

Representation of the rotation tensor:

We recall the fundamental result that any rotation tensor can be expressed with the the help of the exponential map acted on a skew-symmetric tensor. Let $\boldsymbol{\Gamma} \in so(3)$, that is $\boldsymbol{\Gamma} = \boldsymbol{\Gamma}^T$. Then the map

$$\mathbf{R} = \exp(\boldsymbol{\Gamma}) = \mathbf{1} + \boldsymbol{\Gamma} + \frac{\boldsymbol{\Gamma}^2}{2!} + \frac{\boldsymbol{\Gamma}^3}{3!} + \cdots \quad (67)$$

is a rotation tensor. By *Rodrigues formula* the closed expression is valid

$$\mathbf{R} = \mathbf{1} + \frac{\sin|\gamma|}{|\gamma|}\boldsymbol{\Gamma} + \frac{1-\cos|\gamma|}{|\gamma|^2}\boldsymbol{\Gamma}^2, \quad (68)$$

where $\boldsymbol{\gamma}$ is the axial vector of $\boldsymbol{\Gamma}$ and $|\gamma|$ is its absolute value. $\boldsymbol{\gamma}$ is called the rotation vector of the rotation tensor.

For any $\mathbf{b} \in \mathbb{R}^3$ we have

$$\mathbf{R}\mathbf{b} = \mathbf{b} + \frac{\sin|\gamma|}{|\gamma|}(\gamma \times \mathbf{b}) + \frac{1-\cos|\gamma|}{|\gamma|^2}[\gamma \times (\gamma \times \mathbf{b})]. \quad (69)$$

If $\mathbf{b}$ coincides with $\boldsymbol{\gamma}$ we get

$$\mathbf{R}\gamma = \gamma. \quad (70)$$

That is, $\boldsymbol{\gamma}$ is an Eigenvector of $\mathbf{R}$. It is also convenient to use the notation $\mathbf{\Gamma} \equiv \boldsymbol{\gamma} \times \mathbf{1}$ with $\mathbf{1}$ being the identity tensor. The notation is motivated by the relation $\mathbf{\Gamma b} = \boldsymbol{\gamma} \times \mathbf{1b} = \boldsymbol{\gamma} \times \mathbf{b}$.

Now, between $\boldsymbol{\omega}$ and $\hat{\boldsymbol{\omega}}$, $\mathbf{w}$ and $\hat{\mathbf{w}}$, $\mathbf{k}_i$ and $\hat{\mathbf{k}}_i$ on the one hand and $\boldsymbol{\gamma}$ and its derivatives or variation on the other hand, certain relations must exist which are of importance. We give these relations first before motivating their derivatives. They read:

$$\boldsymbol{\omega} = \frac{\sin|\boldsymbol{\gamma}|}{|\boldsymbol{\gamma}|}\dot{\boldsymbol{\gamma}} - \frac{1-\cos|\boldsymbol{\gamma}|}{|\boldsymbol{\gamma}|^2}\dot{\boldsymbol{\gamma}} \times \boldsymbol{\gamma} + \left(\frac{1}{|\boldsymbol{\gamma}|} - \frac{\sin|\boldsymbol{\gamma}|}{|\boldsymbol{\gamma}|^2}\right)\frac{(\boldsymbol{\gamma}\cdot\dot{\boldsymbol{\gamma}})}{|\boldsymbol{\gamma}|}\boldsymbol{\gamma}, \tag{71}$$

$$\hat{\boldsymbol{\omega}} = \frac{\sin|\boldsymbol{\gamma}|}{|\boldsymbol{\gamma}|}\dot{\boldsymbol{\gamma}} + \frac{1-\cos|\boldsymbol{\gamma}|}{|\boldsymbol{\gamma}|^2}\dot{\boldsymbol{\gamma}} \times \boldsymbol{\gamma} + \left(\frac{1}{|\boldsymbol{\gamma}|} - \frac{\sin|\boldsymbol{\gamma}|}{|\boldsymbol{\gamma}|^2}\right)\frac{(\boldsymbol{\gamma}\cdot\dot{\boldsymbol{\gamma}})}{|\boldsymbol{\gamma}|}\boldsymbol{\gamma}, \tag{72}$$

$$\mathbf{w} = \frac{\sin|\boldsymbol{\gamma}|}{|\boldsymbol{\gamma}|}\delta\boldsymbol{\gamma} - \frac{1-\cos|\boldsymbol{\gamma}|}{|\boldsymbol{\gamma}|^2}\delta\boldsymbol{\gamma} \times \boldsymbol{\gamma} + \left(\frac{1}{|\boldsymbol{\gamma}|} - \frac{\sin|\boldsymbol{\gamma}|}{|\boldsymbol{\gamma}|^2}\right)\frac{(\boldsymbol{\gamma}\cdot\delta\boldsymbol{\gamma})}{|\boldsymbol{\gamma}|}\boldsymbol{\gamma} \tag{73}$$

,

$$\hat{\mathbf{w}} = \frac{\sin|\boldsymbol{\gamma}|}{|\boldsymbol{\gamma}|}\delta\boldsymbol{\gamma} + \frac{1-\cos|\boldsymbol{\gamma}|}{|\boldsymbol{\gamma}|^2}\delta\boldsymbol{\gamma} \times \boldsymbol{\gamma} + \left(\frac{1}{|\boldsymbol{\gamma}|} - \frac{\sin|\boldsymbol{\gamma}|}{|\boldsymbol{\gamma}|^2}\right)\frac{(\boldsymbol{\gamma}\cdot\delta\boldsymbol{\gamma})}{|\boldsymbol{\gamma}|}\boldsymbol{\gamma}, \tag{74}$$

$$\mathbf{k}_i = \frac{\sin|\boldsymbol{\gamma}|}{|\boldsymbol{\gamma}|}\boldsymbol{\gamma}_{,i} - \frac{1-\cos|\boldsymbol{\gamma}|}{|\boldsymbol{\gamma}|^2}\boldsymbol{\gamma}_{,i} \times \boldsymbol{\gamma} + \left(\frac{1}{|\boldsymbol{\gamma}|} - \frac{\sin|\boldsymbol{\gamma}|}{|\boldsymbol{\gamma}|^2}\right)\frac{(\boldsymbol{\gamma}\cdot\boldsymbol{\gamma}_{,i})}{|\boldsymbol{\gamma}|}\boldsymbol{\gamma}, \tag{75}$$

$$\hat{\mathbf{k}}_i = \frac{\sin|\boldsymbol{\gamma}|}{|\boldsymbol{\gamma}|}\boldsymbol{\gamma}_{,i} + \frac{1-\cos|\boldsymbol{\gamma}|}{|\boldsymbol{\gamma}|^2}\boldsymbol{\gamma}_{,i} \times \boldsymbol{\gamma} + \left(\frac{1}{|\boldsymbol{\gamma}|} - \frac{\sin|\boldsymbol{\gamma}|}{|\boldsymbol{\gamma}|^2}\right)\frac{(\boldsymbol{\gamma}\cdot\boldsymbol{\gamma}_{,i})}{|\boldsymbol{\gamma}|}\boldsymbol{\gamma}. \tag{76}$$

Obviously all the above relations have the same structure. Also, a dot between vectors denotes the scalar product. These relations have been repeatedly derived in the literature. Specifically the last two can be found in (Pietraszkiewicz and Badur, 1983). In what follows we would like to motivate their forms and exemplary derive the first relation in a way which may be different than the usual approach by using the concept of the Lie bracket introduced in (10). The starting point is the series representation of the rotation tensor in (67) from which immediately follows

$$\mathbf{R}^T = \mathbf{1} - \mathbf{\Gamma} + \frac{\mathbf{\Gamma}^2}{2!} - \frac{\mathbf{\Gamma}^3}{3!} + \cdots \tag{77}$$

and

$$\dot{\mathbf{R}} = \dot{\mathbf{\Gamma}} + \frac{1}{2!}\left(\mathbf{\Gamma}\dot{\mathbf{\Gamma}} + \dot{\mathbf{\Gamma}}\mathbf{\Gamma}\right) + \frac{1}{3!}\left(\mathbf{\Gamma}^2\dot{\mathbf{\Gamma}} + \mathbf{\Gamma}\dot{\mathbf{\Gamma}}\mathbf{\Gamma} + \dot{\mathbf{\Gamma}}\mathbf{\Gamma}^2\right) + \frac{1}{4!}\left(\mathbf{\Gamma}^3\dot{\mathbf{\Gamma}} + \mathbf{\Gamma}^2\dot{\mathbf{\Gamma}}\mathbf{\Gamma} + \mathbf{\Gamma}\dot{\mathbf{\Gamma}}\mathbf{\Gamma}^2 + \dot{\mathbf{\Gamma}}\mathbf{\Gamma}^3\right) + \cdots \tag{78}$$

The multiplication gives

$$\begin{aligned}\dot{\mathbf{R}}\mathbf{R}^T = {} & \dot{\mathbf{\Gamma}} + \frac{1}{2!}\left(\mathbf{\Gamma}\dot{\mathbf{\Gamma}} + \dot{\mathbf{\Gamma}}\mathbf{\Gamma}\right) - \dot{\mathbf{\Gamma}}\mathbf{\Gamma} + \frac{1}{3!}\left(\mathbf{\Gamma}^2\dot{\mathbf{\Gamma}} + \mathbf{\Gamma}\dot{\mathbf{\Gamma}}\mathbf{\Gamma} + \dot{\mathbf{\Gamma}}\mathbf{\Gamma}^2\right) + \frac{1}{2!}\dot{\mathbf{\Gamma}}\mathbf{\Gamma}^2 \\ & - \frac{1}{2!}\left(\mathbf{\Gamma}\dot{\mathbf{\Gamma}}\mathbf{\Gamma} + \dot{\mathbf{\Gamma}}\mathbf{\Gamma}^2\right) + \frac{1}{4!}\left(\mathbf{\Gamma}^3\dot{\mathbf{\Gamma}} + \mathbf{\Gamma}^2\dot{\mathbf{\Gamma}}\mathbf{\Gamma} + \mathbf{\Gamma}\dot{\mathbf{\Gamma}}\mathbf{\Gamma}^2 + \dot{\mathbf{\Gamma}}\mathbf{\Gamma}^3\right) - \frac{1}{3!}\dot{\mathbf{\Gamma}}\mathbf{\Gamma}^3 \\ & + \frac{1}{2!2!}\left(\mathbf{\Gamma}\dot{\mathbf{\Gamma}}\mathbf{\Gamma}^2 + \dot{\mathbf{\Gamma}}\mathbf{\Gamma}^3\right) - \frac{1}{3!}\left(\mathbf{\Gamma}^2\dot{\mathbf{\Gamma}}\mathbf{\Gamma} + \mathbf{\Gamma}\dot{\mathbf{\Gamma}}\mathbf{\Gamma}^2 + \dot{\mathbf{\Gamma}}\mathbf{\Gamma}^3\right) + \cdots\end{aligned} \tag{79}$$

Using the Lie bracket, this seemingly complex equation can be put in a nice form as follows

$$\dot{\mathbf{R}}\mathbf{R}^T = \mathbf{\Omega} = \dot{\mathbf{\Gamma}} + \frac{1}{2!}[\mathbf{\Gamma}, \dot{\mathbf{\Gamma}}] + \frac{1}{3!}\left[\mathbf{\Gamma}, [\mathbf{\Gamma}, \dot{\mathbf{\Gamma}}]\right] + \frac{1}{4!}\left[\mathbf{\Gamma}, \left[\mathbf{\Gamma}, [\mathbf{\Gamma}, \dot{\mathbf{\Gamma}}]\right]\right] + \cdots \tag{80}$$

where multiple applications of the Lie bracket are envisaged. Accordingly, one has $\left[\mathbf{\Gamma}, [\mathbf{\Gamma}, \dot{\mathbf{\Gamma}}]\right] = \mathbf{\Gamma}\left(\mathbf{\Gamma}\dot{\mathbf{\Gamma}} - \dot{\mathbf{\Gamma}}\mathbf{\Gamma}\right) - \left(\mathbf{\Gamma}\dot{\mathbf{\Gamma}} - \dot{\mathbf{\Gamma}}\mathbf{\Gamma}\right)\mathbf{\Gamma}$ etc. From the relation (80) it becomes clear that each involved term is skew-symmetric. The axial vectors are then

$$\boldsymbol{\omega} = \dot{\boldsymbol{\gamma}} + \frac{1}{2!}(\boldsymbol{\gamma} \times \dot{\boldsymbol{\gamma}}) + \frac{1}{3!}(\boldsymbol{\gamma} \times (\boldsymbol{\gamma} \times \dot{\boldsymbol{\gamma}})) + \frac{1}{4!}(\boldsymbol{\gamma} \times (\boldsymbol{\gamma} \times (\boldsymbol{\gamma} \times \dot{\boldsymbol{\gamma}}))) + \cdots \tag{81}$$

Standard evaluation of the cross product shows that three terms are involved. These are $\dot{\boldsymbol{\gamma}}$, $\boldsymbol{\gamma} \times \dot{\boldsymbol{\gamma}}$ and $(\boldsymbol{\gamma} \cdot \dot{\boldsymbol{\gamma}})\boldsymbol{\gamma}$. Moreover, $\boldsymbol{\omega}$ must have the form

$$\boldsymbol{\omega} = a\dot{\boldsymbol{\gamma}} + b\boldsymbol{\gamma} \times \dot{\boldsymbol{\gamma}} + c(\boldsymbol{\gamma} \cdot \dot{\boldsymbol{\gamma}})\boldsymbol{\gamma}\,, \tag{82}$$

where a, b, c are scalars given by the series

$$a = 1 - \frac{1}{3!}|\boldsymbol{\gamma}|^2 + \frac{1}{5!}|\boldsymbol{\gamma}|^4 - \frac{1}{7!}|\boldsymbol{\gamma}|^6 + \cdots = \frac{\sin|\boldsymbol{\gamma}|}{|\boldsymbol{\gamma}|} \tag{83}$$

$$b = \frac{1}{2!} - \frac{1}{4!}|\boldsymbol{\gamma}|^2 + \frac{1}{6!}|\boldsymbol{\gamma}|^4 - \frac{1}{8!}|\boldsymbol{\gamma}|^6 + \cdots = \frac{1 - \cos|\boldsymbol{\gamma}|}{|\boldsymbol{\gamma}|^2} \tag{84}$$

$$c = \frac{1}{3!} - \frac{1}{5!}|\boldsymbol{\gamma}|^2 + \frac{1}{7!}|\boldsymbol{\gamma}|^4 - \frac{1}{9!}|\boldsymbol{\gamma}|^6 + \cdots = \frac{1}{|\boldsymbol{\gamma}|^{\mathbf{2}}} - \frac{\sin|\boldsymbol{\gamma}|}{|\boldsymbol{\gamma}|^3} \tag{85}$$

in agreement with the relation in (71). All other relations from (72) to (76) can be derived similarly.

4 Classical formulations with an independent rotation tensor

4.1 A first formulation with a symmetric stretch tensor

Let $\mathcal{B} \subset \mathbb{R}^3$ define a body. The map

$$\varphi(t) : \mathcal{B} \to \mathbb{R}^3 \tag{86}$$

is an embedding depending on a well chosen parameter $t \in \mathbb{R}$. Hereby, $\varphi_0 = \varphi(t = t_0)$ defines a reference configuration, which enables the identification of the material points, and which, without loss of generality, we choose to be the identity map. Writing $\mathcal{B}$ for $\varphi_0\mathcal{B}$ and $\mathcal{B}_t$ for $\varphi(t)\mathcal{B}$ we get

$$\varphi(t) : \mathcal{B} \to \mathcal{B}_t. \tag{87}$$

For $\mathbf{X} \in \mathcal{B}$ and $\mathbf{x} \in \mathcal{B}_t$ we have

$$\mathbf{x}(t) = \varphi(\mathbf{X}, t) \qquad \text{and} \qquad \mathbf{X}(t) = \varphi^{-1}(\mathbf{x}, t). \tag{88}$$

Let further $\vartheta^i, i = 1, 2, 3$ be co-ordinate charts in $\mathcal{B}$, which we choose to be attached to the body (convected). With $\mathcal{T}B$, $\mathcal{T}B_t$ being the tangent spaces of $\mathcal{B}$ and $\mathcal{B}_t$ the covariant base vectors are

$$\mathbf{G}_i = \partial\mathbf{X}/\partial\vartheta^i \qquad \text{with} \qquad \mathbf{G}_i \in \mathcal{T}B \tag{89}$$

and

$$\mathbf{g}_i = \partial\mathbf{x}/\partial\vartheta^i \qquad \text{with} \qquad \mathbf{g}_i \in \mathcal{T}B_t. \tag{90}$$

The Riemannian metric in either configuration is given by

$$G_{ij} = \mathbf{G}_i \cdot \mathbf{G}_j \qquad \text{and} \qquad g_{ij} = \mathbf{g}_i \cdot \mathbf{g}_j \tag{91}$$

respectively. here again the scalar product of vectors is denoted by a dot. The determinants of the metrics are denoted by G and g respectively; their inverse as usual by G^{ij} and g^{ij} respectively. Further we denote the basic skew-symmetric three-dimensional *Levi-Civita* tensors (permutation tensors) by

$$e_{ijk} = \begin{cases} +1 & \text{for odd permutations of } i, j, k \\ -1 & \text{for even permutations of } i, j, k \end{cases} \tag{92}$$

and the Ricci tensor by

$$\varepsilon_{ijk} := \sqrt{G}\, e_{ijk}, \qquad \varepsilon^{ijk} := \frac{1}{\sqrt{G}}\, e^{ijk}, \tag{93}$$

where we have $e_{ijk} = e^{ijk}$ by its *Euclidean* structure. We make also use of the absolute notation

$$\boldsymbol{\epsilon} \equiv \varepsilon^{ijk} \mathbf{G}_i \otimes \mathbf{G}_j \otimes \mathbf{G}_k. \tag{94}$$

The deformation gradient is defined as the tangent map

$$\mathbf{F} = \mathbf{g}_i \otimes \mathbf{G}^i = \mathbf{x}_{,i} \otimes \mathbf{G}^i = \tag{95}$$

with the help of which suitable strain measures are defined. They are provided by splitting the deformation gradient into its rotational and stretching parts. Resting on the polar decomposition theorem one has

$$\mathbf{F} = \mathbf{Q}\mathbf{U}, \qquad \mathbf{Q} \in SO(3), \quad \mathbf{U} = \mathbf{U}^T. \tag{96}$$

The stretch tensor

$$\mathbf{U} = \mathbf{Q}^T\mathbf{F}, \qquad \mathbf{U} = \mathbf{U}^T \tag{97}$$

plays now the role of a strain measure. Although the rotations enter the formulation explicitly, the configuration space of the continuum is nevertheless determined by the displacement field alone since the rotation tensor $\mathbf{Q}$ is determined uniquely by the symmetry condition in $(96)_2$. In other words the rotation field is a dependent one.

The first possible extension of a classical theory of deformation as far as degrees of freedom are concerned is achieved by allowing for the rotation tensor to become an independent variable. A corresponding variational statement would then deliver as Euler-Lagrange equations both equilibrium equations: The force as well as the moment equilibrium equations, the latter expressed as a symmetry condition. First we define the restricted Cosserat continuum as a continuum with independent rotational degrees of freedom where the free (stored) energy function depends on the stretch tensor alone. The symmetry of $\mathbf{U}$ is first treated as a side condition and enforced by a Lagrangian multiplier (Bufler, 1985). Later a modified version will be considered as in (Sansour and Bednarczyk, 1995), where the restriction to symmetry is dropped to allow for a richer structure of the theory.

We recall first the force equilibrium equation in terms of the Cauchy stress tensor $\boldsymbol{\sigma}$ and the rotated one $\boldsymbol{\Sigma} = \mathbf{R}^T\boldsymbol{\sigma}\mathbf{R}$:

$$\text{div}\boldsymbol{\sigma} + \mathbf{b} = \mathbf{0}, \qquad \text{div}\mathbf{R}\boldsymbol{\Sigma}\mathbf{R}^T + \mathbf{b} = \mathbf{0}\,, \tag{98}$$

where $\mathbf{b}$ denote the field forces and div denote the divergence operator with respect to the actual configuration.

We consider now the free energy as a function of $\mathbf{U}$, $\mathbf{U} = \mathbf{U}^T$, and formulate the following variational statement:

$$\int_{\mathcal{B}_t} \rho\psi(\mathbf{U})dV - \int_{\mathcal{B}_t} \mathbf{b}\cdot\mathbf{x}dV - \int_{\partial\mathcal{B}_{t\sigma}} \mathbf{t}\cdot\mathbf{x}dA = stat, \tag{99}$$

where $\mathbf{t}$ is the boundary traction and dV, dA are the volume and area elements, respectively. For the independent variations of the displacements and of the rotations to take place, the symmetry condition on $\mathbf{U}$ has to be relaxed and is included in the functional statement as a side condition with $\mathbf{\Lambda}$ as a corresponding *Lagrange* multiplier. $\mathbf{\Lambda}$ is then skew-symmetric. Accordingly, the relaxed variational statement reads:

$$\int_{\mathcal{B}_t} (\rho\psi(\mathbf{U}) + \mathbf{\Lambda} : \mathbf{U})\, dV - \int_{\mathcal{B}_t} \mathbf{b}\cdot\mathbf{x}dV - \int_{\partial\mathcal{B}_{t\sigma}} \mathbf{t}\cdot\mathbf{x}dA = stat. \tag{100}$$

For simplicity while we write down $\mathbf{U}$, it is implicitly understood that $\mathbf{U}$ is to be replaced by $\mathbf{R}^T\mathbf{F}$.

The variation carried out reads

$$\int_{\mathcal{B}_t} \left(\rho\frac{\partial\psi(\mathbf{U})}{\partial\mathbf{U}} + \mathbf{\Lambda}\right) : \left(\delta\mathbf{R}^T\mathbf{F} + \mathbf{R}^T\delta\mathbf{F}\right) dV + \int_{\mathcal{B}_t} \delta\mathbf{\Lambda} : \mathbf{R}^T\mathbf{F}\, dV$$
$$- \int_{\mathcal{B}_t} \mathbf{b}\cdot\delta\mathbf{x}dV - \int_{\partial\mathcal{B}_{t\sigma}} \mathbf{t}\cdot\delta\mathbf{x}\, dA \quad = \tag{101}$$

Rewriting the last statement leads to

$$\int_{\mathcal{B}_t} \left(\rho\frac{\partial\psi(\mathbf{U})}{\partial\mathbf{U}} + \mathbf{\Lambda}\right)\mathbf{U}^T : \delta\mathbf{R}^T\mathbf{R}\, dv + \int_{\mathcal{B}_t} \mathbf{R}\left(\rho\frac{\partial\psi(\mathbf{U})}{\partial\mathbf{U}} + \mathbf{\Lambda}\right) : \delta\mathbf{F}\, dv$$
$$+ \int_{\mathcal{B}_t} \delta\mathbf{\Lambda} : \mathbf{R}^T\mathbf{F}\, dv - \int_{\mathcal{B}_t} \mathbf{b}\cdot\delta\mathbf{x}\, dv - \int_{\partial\mathcal{B}_{t\sigma}} \mathbf{t}\cdot\delta\mathbf{x}\, da \quad = \tag{102}$$

Noticing that $\delta\mathbf{R}\mathbf{R}^T = \mathbf{W} \in so(3)$ as well as $\mathbf{\Lambda}$ are skew-symmetric, making use of

$$\delta\mathbf{F} = \delta\mathbf{x}_{,i} \otimes \mathbf{G}^I = \delta\mathbf{x}_{,i} \otimes \mathbf{g}^i\mathbf{F} \tag{103}$$

and since all variations are independent, we end up with the following Euler-Lagrange equations:

$$\operatorname{div}\left[\mathbf{R}\left(\rho\frac{\partial\psi(\mathbf{U})}{\partial\mathbf{U}} + \mathbf{\Lambda}\right)\right] + \mathbf{b} \;=\; \mathbf{0} \tag{104}$$

$$\left(\rho\frac{\partial\psi(\mathbf{U})}{\partial\mathbf{U}} + \mathbf{\Lambda}\right)\mathbf{U}^T \;=\; symmetric \tag{105}$$

$$\mathbf{R}^T\mathbf{F} \;=\; symmetric \tag{106}$$

together with corresponding boundary conditions.

From (98) and (104) we conclude that

$$\boldsymbol{\Sigma} = \left(\rho\frac{\partial\psi(\mathbf{U})}{\partial\mathbf{U}} + \boldsymbol{\Lambda}\right)\mathbf{U}^T. \tag{107}$$

Thus, Eq. (105) is basically the statement that the rotated Cauchy and hence the Cauchy stress tensor itself is symmetric.

The symmetry condition employed on $\boldsymbol{\Sigma}$ leads to

$$\boldsymbol{\Lambda}\mathbf{U}^T + \mathbf{U}\boldsymbol{\Lambda} = \rho\left(\mathbf{U}\frac{\partial\psi(\mathbf{U})}{\partial\mathbf{U}^T} - \frac{\partial\psi(\mathbf{U})}{\partial\mathbf{U}}\mathbf{U}^T\right). \tag{108}$$

From this equation it is evident that the skew-symmetric tensor $\boldsymbol{\Lambda}$ is a measure of the non-commutativity of $\partial\psi(\mathbf{U})/\partial\mathbf{U}$ with $\mathbf{U}$. In the case of isotropic material behaviour, when both terms commute, $\boldsymbol{\Lambda}$ vanishes identically. It is also important to recall that the inclusion of an independent rotation field does not alter the material behaviour since the free energy function is still a function of symmetric $\mathbf{U}$. A classical formulation in terms of the right Cauchy-Green tensor $\mathbf{C}$ delivers same solutions for a problem (assuming exact ones).

4.2 A second formulation with a non-symmetric stretch tensor

A further extension of the theory is achieved by dropping the symmetry condition of $\mathbf{U}$ (Sansour and Bednarczyk, 1995). In this case no *Lagrange* multiplier is introduced and the independent variation of the displacements and rotations leads to the statement

$$\int_{\mathcal{B}_t} \left(\rho\frac{\partial\psi(\mathbf{U})}{\partial\mathbf{U}}\right)\mathbf{U}^T : \delta\mathbf{R}^T\mathbf{R}\, dV + \int_{\mathcal{B}_t} \mathbf{R}\left(\rho\frac{\partial\psi(\mathbf{U})}{\partial\mathbf{U}}\right) : \delta\mathbf{F}\, dV - \int_{\mathcal{B}_t} \mathbf{b}\cdot\delta\mathbf{x}\, dV - \int_{\partial\mathcal{B}_{t\sigma}} \mathbf{t}\cdot\delta\mathbf{x}\, dA = 0 \quad (109)$$

which results in the Euler-Lagrange equations

$$\mathrm{div}\left[\mathbf{R}\left(\rho\frac{\partial\psi(\mathbf{U})}{\partial\mathbf{U}}\right)\right] + \mathbf{b} = \mathbf{0} \quad (110)$$

$$\left(\rho\frac{\partial\psi(\mathbf{U})}{\partial\mathbf{U}}\right)\mathbf{U}^T = symmetric. \quad (111)$$

Within this formulation, $\mathbf{U}$ is not symmetric in general. The statement accounts for a general anisotropic formulation of the free energy function. However, in the special case of isotropic constitutive law, the angular momentum equation proved to be equivalent to a symmetry condition for $\mathbf{U}$. This is proved in the following proposition.

PROPOSITION: Let the free energy function be an isotropic function of $\mathbf{U}$ alone ($\psi = \psi(\mathbf{U})$). Let further the external power be a function of the placement field $\mathbf{x}$ alone ($W_{ext} = W_{ext}(\mathbf{x})$). Then $\mathbf{U}$ is symmetric.
Proof: The moment equilibrium (Eq. (111)) gives

$$\frac{\partial\psi(\mathbf{U})}{\partial\mathbf{U}}\mathbf{U}^T = symmetric. \quad (112)$$

To exploit this relation we split $\mathbf{U}$ in its symmetric part $\mathbf{U}^s$ and its skew-symmetric part $\mathbf{U}^a$:

$$\mathbf{U} = \mathbf{U}^s + \mathbf{U}^a. \quad (113)$$

Since $\psi(\mathbf{U})$ is an isotropic function of its argument $\mathbf{U}$, a complete representation of ψ is given in terms of the so called generators of the isotropic function which consist in our case of the products (see Smith (1971); Spencer (1971))

$$\mathbf{U}^s, \mathbf{U}^{s^2}, \mathbf{U}^{a^2}, (\mathbf{U}^s\mathbf{U}^a - \mathbf{U}^a\mathbf{U}^s), \mathbf{U}^a\mathbf{U}^s\mathbf{U}^a, (\mathbf{U}^{s^2}\mathbf{U}^a - \mathbf{U}^a\mathbf{U}^{s^2}), (\mathbf{U}^a\mathbf{U}^s\mathbf{U}^{a^2} - \mathbf{U}^{a^2}\mathbf{U}^s\mathbf{U}^a).$$

A tensor, say $\mathbf{\Sigma}$, as an isotropic function has then the representation

$$\begin{aligned}\mathbf{\Sigma} &= \alpha_1\mathbf{U}^s + \alpha_2\mathbf{U}^{s^2} + \alpha_3\mathbf{U}^{a^2} + \alpha_4(\mathbf{U}^s\mathbf{U}^a - \mathbf{U}^a\mathbf{U}^s) + \alpha_5\mathbf{U}^a\mathbf{U}^s\mathbf{U}^a \\ &\quad + \alpha_6(\mathbf{U}^{s^2}\mathbf{U}^a - \mathbf{U}^a\mathbf{U}^{s^2}) + \alpha_7(\mathbf{U}^a\mathbf{U}^s\mathbf{U}^{a^2} - \mathbf{U}^{a^2}\mathbf{U}^s\mathbf{U}^a) \qquad (114)\end{aligned}$$

where the α_i are functions of the irreducible integrity basis consisting of the following invariants:

$$\mathrm{tr}\mathbf{U}^s, \mathrm{tr}\mathbf{U}^{s^2}, \mathrm{tr}\mathbf{U}^{s^3}, \mathrm{tr}\mathbf{U}^{a^2}, \mathrm{tr}\mathbf{U}^s\mathbf{U}^{a^2}, \mathrm{tr}\mathbf{U}^{s^2}\mathbf{U}^{a^2}, \mathrm{tr}\mathbf{U}^{s^2}\mathbf{U}^{a^2}\mathbf{U}^s\mathbf{U}^a. \qquad (115)$$

The free energy is then a function of these invariants which we abbreviate by $I_1, I_2, ..., I_n$.

Eq. (112) renders

$$\left\{\sum_{i=1}^{n}\left[\frac{\partial\psi(I_1, I_2....I_n)}{\partial I_1}\frac{\partial I_1}{\partial \mathbf{U}} + \frac{\partial\psi(I_1, I_2....I_n)}{\partial I_2}\frac{\partial I_2}{\partial \mathbf{U}} + ...\right]\right\}\mathbf{U}^T = symmetric\,. \qquad (116)$$

Elaborating explicitly the last equation is very involved. In anyway, the outshot is an equation of the form

$$\sum_{i=1}^{n}\beta_i\mathbf{B}_i(\mathbf{U}^s, \mathbf{U}^a) + \sum_{j=1}^{m}\gamma_j\mathbf{C}_j(\mathbf{U}^a) = 0\,, \qquad (117)$$

where β_i, γ_j are some scalar functions of the integrity basis, n, m integers denoting the number of the different products involved in the equation, and $\mathbf{B}_i$ are tensors consisting of products of $\mathbf{U}^s$ and $\mathbf{U}^a$ while $\mathbf{C}_j$ are tensors consisting of products of $\mathbf{U}^a$ alone. It is important to note that in every tensor product $\mathbf{B}_i$ at least one $\mathbf{U}^a$ is involved. Since (116) (and consequently (117)) represents three scalar equations and since $\mathbf{U}^s \neq 0$ the last equation determines $\mathbf{U}^a$. By its homogeneity, $\mathbf{U}^a$ vanishes necessarily (otherwise the rotations would remain undetermined). Hence the statement of the proposition.

In the general anisotropic case the present formulation is a real extension from the classical theory. For $\mathbf{U}$ is non-symmetric in general and so the constitutive law in terms of $\mathbf{U}$ cannot be described in terms of the symmetric right Cauchy-Green tensor $\mathbf{C}$ as in the previous case.

4.3 Uniqueness of the decomposition $\mathbf{F} = \mathbf{RU}$

In the above sections the deformation gradient was decomposed according to $\mathbf{F} = \mathbf{RU}$, with $\mathbf{R} \in SO(3)$. $\mathbf{R}$ has been treated as an independent degree of freedom, and $\mathbf{U}$, symmetric or not, as a strain measure. The question arises as to whether such a decomposition is also possible using another group instead of $SO(3)$, e.g. $GL^+(3)$. In what follows we show that the above decomposition is in fact unique.

Proposition. For a function $\psi(\mathbf{U})$ with $\mathbf{U} = \mathbf{R}^{-1}\mathbf{F}, \mathbf{R} \in GL^+(3)$, the Euler-Lagrange equations of the functional

$$\int_{\mathcal{B}_t} \rho\dot{\psi}(\mathbf{U})\, dV - \int_{\mathcal{B}_t} \mathbf{b}\cdot\dot{\mathbf{x}}\, dV - \int_{\partial\mathcal{B}_{t_s}} \mathbf{t}\cdot\dot{\mathbf{x}}\, dA = 0$$

result in the statement that $\mathbf{R}$ can only be in $SO(3)$.

Proof. Again, the Euler-Lagrange equations read

$$\operatorname{div}\left(\rho\mathbf{R}\frac{\partial\psi(\mathbf{U})}{\partial\mathbf{U}}\mathbf{U}^T\mathbf{R}^{-1}\right) + \mathbf{b} = \mathbf{0}, \tag{118}$$

$$\rho\frac{\partial\psi(\mathbf{U})}{\partial\mathbf{U}}\mathbf{U}^T : \mathbf{Z} = 0, \tag{119}$$

$$\mathbf{Z} = \dot{\mathbf{R}}\mathbf{R}^{-1}. \tag{120}$$

$\mathbf{Z}$ is an element of the algebra of the Lie group, in our case $\mathbf{Z} \in gl^+(3)$.

The additive decomposition of $\mathbf{Z}$:

$$\mathbf{Z} = \mathbf{W} + \mathbf{D}, \qquad \mathbf{W}^T = -\mathbf{W}, \quad \mathbf{D} = \mathbf{D}^T \tag{121}$$

splits (119) in the two equations

$$\rho\frac{\partial\psi(\mathbf{U})}{\partial\mathbf{U}}\mathbf{U}^T : \mathbf{W} = 0, \tag{122}$$

$$\rho\frac{\partial\psi(\mathbf{U})}{\partial\mathbf{U}}\mathbf{U}^T : \mathbf{D} = 0. \tag{123}$$

We recall that $\mathbf{D}, \mathbf{W}$ are independent of each other. Hence, the first equation of the two will be a statement of symmetry

$$\rho\frac{\partial\psi(\mathbf{U})}{\partial\mathbf{U}}\mathbf{U}^T = symmetric. \tag{124}$$

Taking (118) into account, the second equation can only exist for

$$\mathbf{D} = \mathbf{0}, \tag{125}$$

that is $\mathbf{Z} \in so(3)$ and $\mathbf{R} \in SO(3)$. Hence the proof.

5 The Cosserat continuum

The following sections are organised as follows: In Sec. 5.1 the two *Cosserat strain tensors* are derived and their corresponding first variations are provided. The weak problem statement for the *Cosserat continuum* is outlined in Sec. 5.2 which also includes the corresponding equilibrium equations. The configuration space is discussed in detail in Sec. 5.3. Subsequently, Sec. 5.4 provides the details of the multiplicative rotation updating algorithm and investigates its applicability to the moving least square method. Finally, in Sec. 5.5 the proposed variational statement is used to model size-scale effects of bone and polymeric specimens.

5.1 Strain measures of the Cosserat continuum

The configuration space, as the set of all admissible degrees of freedom attached to every point of $\mathcal{B}$, is then defined as follows

$$\mathcal{C}(\mathcal{B}) = \{(\mathbf{x}, \mathbf{R}) \mid \mathbf{x} \in \mathbb{R}^3, \mathbf{R} \in SO(3)\}. \tag{126}$$

The construction of strain measures for the Cosserat continuum goes as follows. First we consider the pair of derivatives $(\mathbf{x}_{,i}, \mathbf{R}_{,i})$. The strain measures of the Cosserat continuum are then derived by modifying the above derivatives so as to arrive at expressions invariant with respect to the action of elements of $SO(3)$. Specifically one arrives at $\mathbf{R}^T\mathbf{x}_{,i}, \mathbf{R}^T\mathbf{R}_{,i}$. The second term is nothing but the skew-symmetric tensor $\hat{\mathbf{K}}_\mathbf{i}$ from the second chapter. With its axial vectors we end up with the pair of vectors $(\mathbf{R}^T\mathbf{x}_{,i}, \hat{\mathbf{k}}_i)$ as strain measures of the Cosserat continuum. In the following reference to ˆ will be omitted. From the context it should become clear which of the rates are considered.

Instead of the vector form, equivalently, one can consider the pair of tensors $(\mathbf{U}, \mathbf{K})$ defined by

$$\mathbf{U} = \mathbf{R}^T\mathbf{F}, \qquad \mathbf{K} = -\mathbf{k}_i \otimes \mathbf{G}^i. \tag{127}$$

$\mathbf{F}$ is the deformation gradient given by $\mathbf{F} = \mathbf{x}_{,i} \otimes \mathbf{G}^i$. The tensors $\mathbf{U}$ and $\mathbf{K}$ are called the first and the second Cosserat strain tensors.

The first variation (similarly time derivative) of these strain measures can immediately be written down as

$$\begin{aligned} \delta\mathbf{U} &= \delta\left(\mathbf{R}^T\mathbf{x}_{,i}\right) \otimes \mathbf{G}^i = \\ &= -\mathbf{R}^T\mathbf{W}\left(\mathbf{x}_{,i} \otimes \mathbf{G}^i\right) + \mathbf{R}^T\left(\delta\mathbf{x}_{,i} \otimes \mathbf{G}^i\right). \end{aligned} \tag{128}$$

$$\delta\mathbf{k}_i = \text{axial}\left[\delta\left(\mathbf{R}^T\mathbf{R}_{,i}\right)\right] = \mathbf{R}^T\mathbf{w}_{,i}. \tag{129}$$

From the above context it is clear that we have chosen $\mathbf{W}$ to be right invariant (spatial). Obviously, alternative choices are possible which, however, will not amend the physical meaning of the equations.

The difference now between an extended classical formulation, as discussed in the last chapter, and a Cosserat continuum, is that the tensor $\mathbf{K}$ (change of curvature tensor) is now assumed to have a conjugate stress measure. In the elastic case this statement is equivalent to the assumption that $\mathbf{K}$ will contribute to the internal potential.

In the following we want to confine ourselves to the quasi-static case and consider no time dependency.

5.2 The weak form and its corresponding equilibrium equations

Let us consider a non-linear boundary value problem on domain $\mathcal{B}$ with boundary $\partial\mathcal{B}$. Dirichlet boundary conditions are prescribed on $\partial\mathcal{B}_D \subset \partial\mathcal{B}$ and Neumann boundary conditions are prescribed on $\partial\mathcal{B}_N = \partial\mathcal{B} \setminus \partial\mathcal{B}_D$. Let us assume a hyper-elastic material behaviour and the existence of a free energy function $\rho_0\psi(\mathbf{U},\mathbf{K})$, function of the *first Cosserat strain tensor* $\mathbf{U}$ and the *second Cosserat strain tensor* $\mathbf{K}$ and defined per unit un-deformed volume. Now, the internal potential of the system in the Lagrangian form is written as

$$\Psi_{int} = \int_{\mathcal{B}} \left\{ \rho_0\, \psi\,(\mathbf{U},\mathbf{K}) \right\} dV\,. \tag{130}$$

Furthermore, we have conjugate to $\mathbf{U}$ and $\mathbf{K}$ the force stress tensor $\boldsymbol{n}$ and the couple stress tensor $\boldsymbol{m}$, respectively, which are defined as follows

$$\boldsymbol{n} = \rho_0\, \frac{\partial\psi\,(\mathbf{U},\mathbf{K})}{\partial\mathbf{U}} \qquad \text{and} \qquad \boldsymbol{m} = \rho_0 \frac{\partial\psi\,(\mathbf{U},\mathbf{K})}{\partial\mathbf{K}}\,. \tag{131}$$

Note that both stress tensors are generally non-symmetric. Corresponding to the internal potential let $\mathcal{W}_{ext}$ define the external virtual work in the Lagrangian form as follows

$$\mathcal{W}_{ext} = \int_{\mathcal{B}} \rho_0\, \mathbf{b}\cdot\delta\mathbf{x}\; dV + \int_{\mathcal{B}} \rho_0\, \mathbf{l}\cdot\mathbf{w}\; dV + \int_{\partial\mathcal{B}_N} \mathbf{t}^{(\mathbf{n})}\cdot\delta\mathbf{x}\; dA + \int_{\partial\mathcal{B}_N} \mathbf{q}^{(\mathbf{n})}\cdot\mathbf{w}\; dV\,, \tag{132}$$

where the vectors $\mathbf{b}$, $\mathbf{l}$ are the external body force and torque, respectively, and the vectors $\mathbf{t}^{(\mathbf{n})}$, $\mathbf{q}^{(\mathbf{n})}$ are the corresponding quantities on the boundary. The vector $\mathbf{w}$ is the rate (infinitesimal rotation) conjugate to the external torque.

For the static case and considering only mechanical processes the *first law of thermodynamics* provides the following variational statement

$$\delta\Psi - \mathcal{W}_{ext} = 0\,, \tag{133}$$

which can be written, with Eq. (302) and Eq. (296), as

$$\int_{\mathcal{B}} \left\{ \rho_0 \frac{\partial \psi_{int}(\mathbf{U})}{\partial \mathbf{U}} : \delta \mathbf{U} + \rho_0 \frac{\partial \psi_{int}(\mathbf{K})}{\partial \mathbf{K}} : \delta \mathbf{K} \right\} dV - \mathcal{W}_{ext} \quad = \quad 0 \qquad (134)$$

Making use of the expressions for the stress measures (Eq. 131) and the variational forms of both *Cosserat* strain tensors (Eq. 128) and (Eq. 129), respectively, the above functional takes its final form

$$\int_{\mathcal{B}} \left\{ \mathbf{R}n\mathbf{G}^i \cdot \delta \mathbf{x}_{,i} - \mathbf{R}n\mathbf{F}^T : \mathbf{W} + \mathbf{R}m\mathbf{G}^i \cdot \mathbf{w}_{,i} \right\} dV - \mathcal{W}_{ext} = 0 \,. \qquad (135)$$

Applying *Gauss's theorem* we can redraft the above functional as follows

$$\begin{aligned} -\int_{\mathcal{B}} \frac{1}{\sqrt{G}} \left(\sqrt{G}\, \mathbf{R}n\mathbf{G}^i \right)_{,i} \cdot \delta \mathbf{x} \, dV + \int_{\partial \mathcal{B}} \mathbf{R}n\, \mathbf{n} \cdot \delta \mathbf{x} \, dA \\ - \int_{\mathcal{B}} \mathbf{x}_{,i} \times \mathbf{R}n\mathbf{G}^i \cdot \mathbf{w} \, dV - \int_{\mathcal{B}} \frac{1}{\sqrt{G}} \left(\sqrt{G}\, \mathbf{R}m\mathbf{G}^i \right)_{,i} \cdot \mathbf{w} \, dV \\ + \int_{\partial \mathcal{B}} \mathbf{R}m\, \mathbf{n} \cdot \mathbf{w} \, dA - \mathcal{W}_{ext} = 0 \,. \end{aligned} \qquad (136)$$

where G is the determinate of the *Riemannian* metric coefficients (Eq. 91) and where use has been made of the following algebraic manipulations

$$\begin{aligned} -\mathbf{R}n\mathbf{F}^T : \mathbf{W} &= -\mathbf{R}n\mathbf{G}^i \otimes \mathbf{x}_{,i} : \mathbf{W} = -\mathbf{R}n\mathbf{G}^i \cdot \mathbf{W}\mathbf{x}_{,i} = \\ &= -\mathbf{R}n\mathbf{G}^i \cdot \mathbf{w} \times \mathbf{x}_{,i} = -\mathbf{x}_{,i} \times \mathbf{R}n\mathbf{G}^i \cdot \mathbf{w} \,. \end{aligned} \qquad (137)$$

Note that here and throughout this section the basis vectors $\mathbf{G}_i$ are considered to be independent of ϑ^k. Since $\delta \mathbf{x}$ and $\mathbf{w}$ are arbitrary functions, the local statements of the governing equations of this functional, which are the equilibrium equations, are expressed by

$$\frac{1}{\sqrt{G}} \left(\sqrt{G}\, \mathbf{R}n\mathbf{G}^i \right)_{,i} + \rho_0 \, \mathbf{b} = \mathbf{0} \,, \quad \text{in } \mathcal{B} \qquad (138)$$

$$\mathbf{x}_{,i} \times \mathbf{R}n\mathbf{G}^i + \frac{1}{\sqrt{G}} \left(\sqrt{G}\, \mathbf{R}m\mathbf{G}^i \right)_{,i} + \rho_0 \, \mathbf{l} = \mathbf{0} \,, \quad \text{in } \mathcal{B} \,. \qquad (139)$$

The corresponding natural boundary conditions are written as

$$\mathbf{R}n\, \mathbf{n} = \mathbf{t}^{(\mathbf{n})} \qquad \text{and} \qquad \mathbf{R}m\, \mathbf{n} = \mathbf{q}^{(\mathbf{n})} \,, \quad \text{on } \partial \mathcal{B}_N \,, \qquad (140)$$

whereas the essential boundary conditions are given by

$$\mathbf{x} = \mathbf{h}_{\mathbf{x}} \qquad \text{and} \qquad \gamma = \mathbf{h}_{\gamma} \,, \quad \text{on } \partial \mathcal{B}_D \,. \qquad (141)$$

Note that the first term in (139) is nothing but the axial vector of the skew symmetric part of the Kirchhoff stress tensor. Hence, the equation is another way to write the moment equilibrium and so is equivalent to that written down in tensor form where then the skew-symmetric part of the Kirchhoff tensor would be the first term. In a transition to a classical formulation, $\boldsymbol{m}$ becomes zero and the mentioned axial vector becomes zero as well, which is equivalent to the symmetry of the Kirchhoff stress tensor itself.

5.3 The configuration space

Configuration space with a *Killing* metric The deformation of the *Cosserat* continuum is completely described in terms of the pair $(\mathbf{x}, \mathbf{R})$ attached to every point of the continuum. This motivates the definition of the configuration space as the set consisting of all admissible configurations of the body $\mathcal{B}$. A precise definition of it, which we denote by $\mathcal{C}$, is given by

$$\mathcal{C}(B) = \{\mathcal{U} = (\mathbf{x}, \mathbf{R}) \,|\mathcal{U} : \mathcal{B} \longrightarrow \mathbb{R}^3 \times SO(3)\} \tag{142}$$

with $\mathbf{R} = \exp(\mathbf{\Gamma})$ and $\mathbf{\Gamma} \in so(3)$. For two elements $\mathcal{U}, V$ of $\mathcal{C}$ with $\mathcal{U} = (\mathbf{x}, \mathbf{R})$ and $\mathcal{V} = (\mathbf{v}, \mathbf{Q})$ the group operation is then defined by

$$\mathcal{U} + V = (\mathbf{x}, \mathbf{R}) + (\mathbf{v}, \mathbf{Q}) = (\mathbf{x} + \mathbf{v}, \mathbf{QR}). \tag{143}$$

Clearly the group structure is not *Abelian* and its neutral element is $(\mathbf{0}, \mathbf{1})$.

The following elaborations deal with the formulation and definition of rates and second derivatives. As to the rotations one has to choose between material (right, left invariant) rates or spatial (left, or right invariant) ones. We develop the ideas for the first case but it is clear that same considerations are valid for the second class of rates as well.

A curve $\mathcal{V}(t)$ on $\mathcal{C}$ passing through $\mathcal{U} = (\mathbf{x}, \mathbf{R})$ with $\mathcal{V}(t = 0) = \mathcal{U}$ is given as the parameter subgroup

$$\mathcal{V}(t) = [\mathbf{x} + t\mathbf{v}, \exp(t\mathbf{W})\mathbf{R}]. \tag{144}$$

Its tangent at $\mathcal{U}$ is defined in the usual manner by

$$D\mathcal{U} = \frac{d}{dt}\mathcal{V}_{|t=0} = \frac{d}{dt}[\mathbf{x}+t\mathbf{v}, \exp(t\mathbf{W})\mathbf{R}]_{|t=0} = (\mathbf{v}, \mathbf{WR}) \quad \text{with} \quad \mathbf{W} \in so(3). \tag{145}$$

The pair $(\mathbf{v}, \mathbf{W}) = \delta\mathcal{U}$ defines the infinitesimal deformation to be superimposed on a given admissible state. That is, a neighbourhood of $\mathcal{U}$ is given as

$$\mathcal{U} + \delta\mathcal{U} = (\mathbf{x}, \mathbf{R}) + (\mathbf{v}, \mathbf{W}) = (\mathbf{x} + \mathbf{v}, \mathbf{WR}). \tag{146}$$

From physical point of view $D\mathcal{U}$ defines the way to take variations of a physical quantity (e.g. free energy) which depends on the configuration space.

In nonlinear analysis one is interested, mostly always, in the second variations of certain functionals describing a physical process (e.g. for a *Newton-Raphson* iteration). Since the variations of the rotation part of $\mathcal{C}(B)$ is defined by tensor multiplication which does not commute, the tangent operators of such functionals are non-symmetric in general, even then, when the physical problem does posses a potential. To see this, consider any function depending on $\mathcal{C}(B)$. For simplicity we choose a function $\psi(\mathbf{K})$ depending only on $\mathbf{K}$. If necessary one can associate with this function that part of the free energy function related to the bending. First we have

$$D\psi(\mathbf{K};\delta\mathbf{K}) = \frac{\partial\psi(\mathbf{K})}{\partial\mathbf{K}} : \delta\mathbf{K} \tag{147}$$

$$D^2\phi(\mathbf{K};\delta\mathbf{K},\Delta\mathbf{K}) = \frac{\partial^2\psi(\mathbf{K})}{\partial\mathbf{K}^2}\Delta\mathbf{K} : \delta\mathbf{K} + \frac{\partial\phi(\mathbf{K})}{\partial\mathbf{K}} : \delta^2\mathbf{K}. \tag{148}$$

Now in view of (57) we have

$$\delta\mathbf{K} = D\mathbf{K}(\mathbf{R};\mathbf{w}) = -\left(\mathbf{R}^T\mathbf{w}_{,i}\right) \otimes \mathbf{G}^i. \tag{149}$$

Now, the second variation reads

$$\delta^2\mathbf{K} = D^2\mathbf{K}(\mathbf{R};\mathbf{y},\mathbf{w}) = \left(\mathbf{R}^T(\mathbf{y}\times\mathbf{w}_{,i}\right) \otimes \mathbf{G}^i \tag{150}$$

with $\mathbf{y}$ being the axial vector of $\mathbf{Y} \in so(3)$. It is not difficult to see that the last equation renders the second term in Eq. (150) non-symmetric.

To arrive at a symmetric tangent operator a more structured configuration space equipped with a metric is needed. This was proposed first by Simo (1992), who treated the first variation of the potential as a co-tangent vector. Instead we make here direct use of the idea of vector fields defined on $SO(3)$. One can then show directly that the tangent operator resulting from the variation of the weak form of equilibrium retains symmetry. We exemplify the ideas by considering $R_{\mathbf{W}}(\mathbf{R}) = \mathbf{WR}$ which we called a right-invariant tangent vector field on $SO(3)$.

At an arbitrary point on $SO(3)$, say at $\mathbf{R}$, a scalar product of the tangent vectors can be defined by

$$\langle R_{\mathbf{W}}(\mathbf{R}), R_{\mathbf{Y}}(\mathbf{R})\rangle = tr(\mathbf{WR}(\mathbf{YR})^T) = tr(\mathbf{WY}^T). \tag{151}$$

With this scalar product, a metric, the *Killing* metric, is defined on the space of tangent vectors on $SO(3)$ (see again Dubrovin et al. (1984)). Further, with the operation given by the *Lie* bracket $[\mathbf{W},\mathbf{Y}]$:

$$[\mathbf{W},\mathbf{Y}] = \mathbf{WY} - \mathbf{YW}, \qquad \mathbf{W},\mathbf{Y} \in so(3), \tag{152}$$

the tangent space is furnished with the structure of a *Lie* algebra. On the space of tangent vectors there exists a connection given by the relation (Dubrovin et al., 1984; Brickell and Clark, 1970; Choquet-Bruhat et al., 1982).

$$\begin{aligned} 2\langle\nabla_{\mathbf{U}}\mathbf{V},\mathbf{W}\rangle &= \mathbf{U}(\langle\mathbf{V},\mathbf{W}\rangle)+\mathbf{V}(\langle\mathbf{W},\mathbf{U}\rangle)-\mathbf{W}(\langle\mathbf{U},\mathbf{V}\rangle)- \\ &\quad \langle\mathbf{U},[\mathbf{V},\mathbf{W}]\rangle+\langle\mathbf{V},[\mathbf{W},\mathbf{U}]\rangle+\langle\mathbf{W},[\mathbf{U},\mathbf{V}]\rangle- \\ &\quad \langle\mathbf{U},\mathbf{T}(\mathbf{V},\mathbf{W})\rangle+\langle\mathbf{V},\mathbf{T}(\mathbf{W},\mathbf{U})\rangle+\langle\mathbf{W},\mathbf{T}(\mathbf{U},\mathbf{V})\rangle \end{aligned} \quad (153)$$

where the torsion tensor $\mathbf{T}$ is defined as

$$\mathbf{T}(\mathbf{U},\mathbf{V})=\nabla_{\mathbf{U}}\mathbf{V}-\nabla_{\mathbf{V}}\mathbf{U}-[\mathbf{U},\mathbf{V}].$$

For a symmetric connection the torsion $\mathbf{T}$ vanishes and we can elaborate from (153) the expression

$$\nabla_{R_{\mathbf{Y}}}R_{\mathbf{W}}=\frac{1}{2}R_{[\mathbf{Y},\mathbf{W}]}. \quad (154)$$

Hereby one has to make use of the metric as defined in Eq. (151) as well as of the idea that tangent vectors can be understood as derivatives operating on function spaces. By the latter fact, the first line in (153) vanishes identically since the metric is independent of the particular point on the curve in $\mathcal{C}$.

Dealing with vector fields, it makes sense to define with the help of this connection a covariant derivative (or covariant variation). In fact such a derivative is given on the right invariant vector fields as

$$\begin{aligned} \nabla_{\mathbf{Y}}R_{\mathbf{W}}(\mathbf{R}) &= R_{\mathbf{WY}}(\mathbf{R})+\nabla_{R_{\mathbf{Y}}}R_{\mathbf{W}}(\mathbf{R}) \\ &= \mathbf{WYR}+\frac{1}{2}[\mathbf{Y},\mathbf{W}]\mathbf{R} \\ &= \frac{1}{2}(\mathbf{YW}+\mathbf{WY})\mathbf{R}. \end{aligned} \quad (155)$$

Remarkably, this symmetric result for the second (covariant) derivative can be obtained by considering the second derivatives of the two-parameter subgroup $\exp(t\mathbf{W}+s\mathbf{Y})\mathbf{R}$, where s is a second parameter, according to

$$\frac{1}{2}(\mathbf{YW}+\mathbf{WY})\mathbf{R}=\left(\frac{\partial^2}{\partial s\partial t}[\exp(t\mathbf{W}+s\mathbf{Y})\mathbf{R}]\right)_{t=0,s=0}. \quad (156)$$

Note that one has to make use of the definition of the exponential map as given in Eq. (5).

Now we recall Eq. (149) in its modified form given in (34):

$$\delta_W\mathbf{K}_i=(\mathbf{WR})^T\mathbf{R}_{,i}+\mathbf{R}^T(\mathbf{WR})_{,i} \quad (157)$$

(where we have written δ_W to denote that the first variation is taken in the direction of $\mathbf{W}$). In this equation the term $\mathbf{WR}$ define a vector field on $SO(3)$. The covariant variation of Eq. (157) then renders

$$\begin{aligned}\nabla_{\mathbf{Y}}(\delta_W \mathbf{K}_i) &= \frac{1}{2}\{[(\mathbf{YW}+\mathbf{WY})\mathbf{R}]^T\mathbf{R}_{,i}+\mathbf{R}^T[(\mathbf{YW}+\mathbf{WY})\mathbf{R}]_{,i}\}+\\ &\quad (\mathbf{WR})^T\mathbf{YR}_{,i}+(\mathbf{YR})^T\mathbf{WR}_{,i}. \end{aligned} \tag{158}$$

Obviously the formula is symmetric with respect to $\mathbf{Y}$ and $\mathbf{W}$.

To complete our discussion, we give the formula for the first variation of the stretch tensor as well as for its second covariant variation. We recall $\mathbf{U}=\mathbf{R}^T\mathbf{F}$. Hence

$$\delta\mathbf{U}=\delta\mathbf{R}^T\mathbf{F}+\mathbf{R}^T\delta\mathbf{F}=(\mathbf{WR})^T\mathbf{F}+\mathbf{R}^T\delta\mathbf{F}. \tag{159}$$

With the usual second variation (*Gâuteaux* derivative) we have

$$\delta^2\mathbf{U}=(\mathbf{WYR})^T\mathbf{F}+(\mathbf{WR})^T\Delta\mathbf{F}+(\mathbf{YR})^T\delta\mathbf{F}, \tag{160}$$

where again $\mathbf{W},\mathbf{Y}\in so(3)$. The first term on the right side renders the second variation non-symmetric. The second variation carried out co-variantly leads to

$$\nabla\delta\mathbf{U}=\frac{1}{2}[(\mathbf{YW}+\mathbf{WY})\mathbf{R}]^T\mathbf{F}+(\mathbf{WR})^T\Delta\mathbf{F}+(\mathbf{YR})^T\delta\mathbf{F} \tag{161}$$

which is again the symmetrisation of Eq. (160).

Configuration space with an *Euclidean* metric In addition to the already discussed configuration space equipped with a *Killing* metric, it is possible to define the configuration space in a way preserving the structure of a vector group. The configuration space may be redefined by considering the rotation vector instead of the rotation tensor itself:

$$\mathcal{C}(B)=\{\mathcal{U}=(\mathbf{u},\boldsymbol{\gamma})\ |\mathcal{U}:\mathcal{B}\longrightarrow\mathbb{R}^3\times\mathbb{R}^3\longrightarrow\mathbb{R}^3\times SO(3)\}. \tag{162}$$

This definition allows now to equip the space $\mathcal{C}(B)$ with a vector group operation. By letting again $\mathcal{U},V\in C$ with $\mathcal{U}=(\mathbf{u},\boldsymbol{\gamma})$, $\mathcal{V}=(\mathbf{v},\boldsymbol{\psi})$, the group operation is defined as

$$\mathcal{U}+V=(\mathbf{u},\boldsymbol{\gamma})+(\mathbf{v},\boldsymbol{\psi})=(\mathbf{u}+\mathbf{v},\boldsymbol{\gamma}+\boldsymbol{\psi}) \tag{163}$$

with $(\mathbf{0},\mathbf{0})$ defining the neutral element of the group. A curve $\mathcal{V}(t)$ on $\mathcal{C}$ passing through $\mathcal{U}=(\mathbf{u},\boldsymbol{\gamma})$ with $\mathcal{V}(t=0)=\mathcal{U}$ is given by

$$\mathcal{V}(t)=(\mathbf{u}+t\delta\mathbf{u},\boldsymbol{\gamma}+t\delta\boldsymbol{\gamma}) \tag{164}$$

and its tangent at $\mathcal{U}$ by

$$\frac{d}{dt}\mathcal{U}_{|t=0} = \frac{d}{dt}[\mathbf{u} + t\delta\mathbf{u}, \boldsymbol{\gamma} + t\delta\boldsymbol{\gamma}]_{|t=0} = (\delta\mathbf{u}, \delta\boldsymbol{\gamma}). \tag{165}$$

Alternatively one may define the curve as

$$\mathcal{V}(t) = \{\mathbf{u} + t\delta\mathbf{u}, \exp[(\boldsymbol{\gamma} + t\delta\boldsymbol{\gamma}) \times \mathbf{1}]\} \quad \text{with} \quad t \in \mathbb{R}. \tag{166}$$

The tangent is then

$$\frac{d}{dt}\mathcal{V}_{|t=0} = \frac{d}{dt}\{\mathbf{u} + t\delta\mathbf{u}, \exp[(\boldsymbol{\gamma} + t\delta\boldsymbol{\gamma}) \times \mathbf{1}]\}_{|t=0}. \tag{167}$$

By writing $\delta\mathbf{R}$ for $d/dt \exp(\boldsymbol{\gamma} + t\delta\boldsymbol{\gamma})_{|t=0}$ one may elaborate the lengthy expression

$$\begin{aligned}
\delta\mathbf{R} \quad = \quad & \left(\frac{\cos|\boldsymbol{\gamma}|}{|\boldsymbol{\gamma}|} - \frac{\sin|\boldsymbol{\gamma}|}{|\boldsymbol{\gamma}|^2}\right)\frac{(\boldsymbol{\gamma}\cdot\delta\boldsymbol{\gamma})}{|\boldsymbol{\gamma}|}(\boldsymbol{\gamma}\times\mathbf{1}) + \frac{\sin|\boldsymbol{\gamma}|}{|\boldsymbol{\gamma}|}(\delta\boldsymbol{\gamma}\times\mathbf{1}) \\
& + \left[\frac{\sin|\boldsymbol{\gamma}|}{|\boldsymbol{\gamma}|^2} - \frac{2(1-\cos|\boldsymbol{\gamma}|)}{|\boldsymbol{\gamma}|^3}\right]\frac{(\boldsymbol{\gamma}\cdot\delta\boldsymbol{\gamma})}{|\boldsymbol{\gamma}|}[\boldsymbol{\gamma}\times(\boldsymbol{\gamma}\times\mathbf{1})] \\
& + \frac{1-\cos|\boldsymbol{\gamma}|}{|\boldsymbol{\gamma}|^2}[\delta\boldsymbol{\gamma}\times(\boldsymbol{\gamma}\times\mathbf{1}) + \boldsymbol{\gamma}\times(\delta\boldsymbol{\gamma}\times\mathbf{1})].
\end{aligned} \tag{168}$$

It is clear that tangent operators of functionals depending on an internal potential with a configuration space $\mathcal{C}(B)$, by virtue of its *Euclidean* structure, are always symmetric. What about the contribution to the tangent operator from the external work? While the contribution of the part related to the forces and tractions $\mathbf{b}$ and $\mathbf{t}$ is independent of the configuration space, that related to the external moments $\mathbf{l}$ and $\mathbf{q}$ is not.

Assuming constant external moments, the external work is linear in $\mathbf{w}$ and so the second derivative would be zero. Accordingly, in the case of a configuration space with a Killing metric the symmetry of the tangent operator can be granted when constant external moments apply (as far as the external forces themselves are conservative).

In the case of a configuration space with an *Euclidean* metric, $\delta\boldsymbol{\gamma}$ does not constitute the conjugate quantity for the external moments. We recall that, as formulated in (296), this quantity is $\mathbf{w}$. Already in (73) $\mathbf{w}$ was related to $\delta\boldsymbol{\gamma}$. Hence, for the external power one has the contribution

$$\mathbf{l}\cdot\mathbf{w} = \mathbf{g}(\boldsymbol{\gamma})\cdot\delta\boldsymbol{\gamma} \tag{169}$$

with

$$\mathbf{g}(\boldsymbol{\gamma}) = \frac{\sin|\gamma|}{|\gamma|}\mathbf{l} - \frac{1-\cos|\gamma|}{|\gamma|^2}\boldsymbol{\gamma}\times\mathbf{l} + \left(\frac{1}{|\gamma|} - \frac{\sin|\gamma|}{|\gamma|^2}\right)\frac{\boldsymbol{\gamma}\cdot\mathbf{l}}{|\gamma|}\boldsymbol{\gamma}. \tag{170}$$

The contribution to the tangent operator is now given by the derivative of the above expression with respect to $\boldsymbol{\gamma}$. For a constant moment vector $\mathbf{l}$ one has

$$\begin{aligned}\frac{\partial\mathbf{g}(\boldsymbol{\gamma})}{\partial\boldsymbol{\gamma}} &= \left(\frac{\cos|\gamma|}{|\gamma|^2} - \frac{\sin|\gamma|}{|\gamma|^3}\right)\mathbf{l}\otimes\boldsymbol{\gamma} \qquad (171)\\ &+ \left(\frac{\sin|\gamma|}{|\gamma|^3} - \frac{2(1-\cos|\gamma|)}{|\gamma|^4}\right)(\boldsymbol{\gamma}\times\mathbf{l})\otimes\boldsymbol{\gamma} + \\ &\left(\frac{-(2+\cos|\gamma|)}{|\gamma|^3} + \frac{3\sin|\gamma|}{|\gamma|^4}\right)\frac{\boldsymbol{\gamma}\cdot\mathbf{l}}{|\gamma|}\boldsymbol{\gamma}\otimes\boldsymbol{\gamma} + \\ &\left(\frac{1}{|\gamma|^2} - \frac{\sin|\gamma|}{|\gamma|^3}\right)(\boldsymbol{\gamma}\otimes\mathbf{l} + (\boldsymbol{\gamma}\cdot\mathbf{l})\mathbf{1}))\,. \qquad (172)\end{aligned}$$

Obviously the expression is not symmetric in general, which reveals that external moments do not constitute conservative loading. The above is clearly a disadvantage of an *Euclidean* configuration space.

The case of a follower moment. For completion we include here the special case of a moment which is not constant but of the follower type. In this case the moment is given by

$$\mathbf{l} = \mathbf{R}\mathbf{l}_0\,, \qquad \mathbf{l}\cdot\mathbf{w} = \mathbf{R}\mathbf{l}_0\cdot\mathbf{w} = \mathbf{l}_0\cdot\mathbf{R}^T\mathbf{w}\,, \tag{173}$$

where $\mathbf{l}_0$ is a constant vector which could define a moment at the initial configuration. Taking derivatives of this quantity gives

$$\mathbf{l}_0\cdot\mathbf{R}^T\mathbf{Y}^T\mathbf{w} = \mathbf{R}\mathbf{l}_0\cdot(\mathbf{w}\times\mathbf{y})\,. \tag{174}$$

Accordingly, the contribution of the above term to the tangent operator is neither zero nor symmetric but skew-symmetric by the very feature of the vector product. What if we now switch to an Euclidean configuration space? For this purpose we rewrite Eq. (73) in the form

$$\mathbf{w} = \mathbf{A}\delta\boldsymbol{\gamma}\,, \qquad \mathbf{A} = \frac{\partial\mathbf{w}}{\partial\delta\boldsymbol{\gamma}}\,, \tag{175}$$

which is possible because of the linear dependency of $\mathbf{w}$ on $\delta\gamma$. $\mathbf{A}$ is then the corresponding operator which relates both quantities to each other. Hence we can write

$$\mathbf{Rl_0} \cdot (\mathbf{w} \times \mathbf{y}) = \mathbf{Rl_0} \cdot (\mathbf{A}\delta\gamma) \times (\mathbf{A}\Delta\gamma) . \tag{176}$$

Finally, using indices the contribution to the tangent operator is

$$e_{krs}(\mathbf{Rl_0})_k(\mathbf{A})_{ri}(\mathbf{A})_{sj}\delta(\gamma)_i(\Delta\gamma)_j . \tag{177}$$

5.4 Path-independent updating of the rotation tensor

Multiplicative updating At a certain loading step i we have a stage of equilibrium defined by the displacement field $\mathbf{u}_i$ and the rotation field $\mathbf{R}_i$. By changing the external loading, the body will attain from this state of equilibrium a new one characterized by the new kinematical fields $\mathbf{u}_{i+1}$ and $\mathbf{R}_{i+1}$.

Within an iteration process in a finite element analysis the resulting field equations are to be linearized leading to linear equations for incremental quantities of the kinematical fields. Computationally, to get from a load step i to another one $i+1$, several linear computations (iterations) are performed which lead in each iteration to incremental displacements $\Delta\mathbf{u}^j$ and incremental rotations $\Delta\mathbf{W}^j$, where j denotes the iteration step. While it is evident how to construct the displacement field at the end of an iteration step by simply relying on the additive structure of the displacement vectors to obtain

$$\mathbf{u}_{i+1}^j = \mathbf{u}_{i+1}^{(j-1)} + \Delta\mathbf{u} , \tag{178}$$

it is by no means clear how to build up a rotation tensor, given the incremental rotation. Note that according to the above notation we must have $\mathbf{u}_{i+1}^{(0)} = \mathbf{u}_i$.

The first idea is the following. Let the linearisation of the rotation tensor be carried out multiplicatively either by left or right multiplications. That is, the linearisation is of the form $\Delta\mathbf{R} = \Delta\mathbf{W}\mathbf{R} = \mathbf{R}\Delta\mathbf{Y}$ (by simply replacing the variations by the increments), where the rotational increments are also skew symmetric. Then, it is natural to build up the rotation tensor according to

$$\mathbf{R}_{i+1}^j = \exp(\Delta\mathbf{W})\,\mathbf{R}_{i+1}^{(j-1)} = \mathbf{R}_{i+1}^{(j-1)}\exp(\Delta\mathbf{Y}) . \tag{179}$$

In other words, given a rotation tensor $\mathbf{R}_{i+1}^{(j-1)}$ corresponding to the $(j-1)$-iteration at the $(i+1)$-load step, one builds an incremental rotation tensor $\Delta\mathbf{R} = \exp(\Delta\mathbf{W})$ and multiplies it with the rotation tensor $\mathbf{R}_{i+1}^{(j-1)}$.

The above multiplicative scheme pertains to the general structure of the rotational group, but it does not tell us how to interpolate rotational parameters. Recall that the rotation tensor itself cannot be interpolated, only rotational parameters can. The rotation vector of $\Delta\mathbf{R}$ is $\Delta\mathbf{w}$, that of any rotation tensor $\mathbf{R}$ will be the corresponding $\boldsymbol{\gamma}$, and only these quantities can be interpolated.

A first method for multiplicative updating was assigned in Simo and Vu-Quoc (1986) (see also Simo et al. (1995) where the elaborations on the method are more explicit), which is based on the idea that only $\Delta\mathbf{w}$ is interpolated such that the incremental rotation vector is defined at every *Gauss* point. Given this rotation vector, one constructs the corresponding rotation tensor at the *Gauss* points which is then multiplied with that existing from the previous step and the result is stored as a history variable. In other words, the operation documented in (179) is carried out at *Gauss* points and $\mathbf{R}_{i+1}^{(j)}$ is stored as a history variable. At the nodes, only $\Delta\mathbf{W}$ is defined and, hence, only the increment of the rotation vector is interpolated. Note that at the reference configuration we have $\mathbf{R} = \mathbf{1}$.

The above method was widely accepted and used but discovered to be path-dependent (Jelenic and Crisfield (1999)). In contrast, we will device in the following a path-independent multiplicative scheme for rotational parameters which is actually based on the first author's work in Sansour and Bednarczyk (1995). The key to making a scheme path-independent is the observation that the method must interpolate total rotational parameters, not only the incremental ones. In order to preserve the multiplicative structure and not end up in an additive one, a special scheme is needed in order to construct the total rotational parameters which have to be defined at the nodal points. The concept now is therefore as follows. Let the rotation vector $\boldsymbol{\gamma}_{j+1}^{i-1}$ be given at nodal points (at the reference configuration we have simply $\boldsymbol{\gamma} = \mathbf{0}$). The linear iteration step provides us with an incremental rotation vector $\Delta\mathbf{w}$. Accordingly, we have $\Delta\mathbf{R}$ and $\mathbf{R}_{i+1}^{j-1}$ defined at nodal points. Now we build up multiplicatively $\mathbf{R}_{i+1}^{j} = \Delta\mathbf{R}\mathbf{R}_{i+1}^{(j-1)}$, also at nodal points, and extract from $\mathbf{R}_{i+1}^{j}$ the corresponding rotation vector $\boldsymbol{\gamma}_{i+1}^{j}$. The latter can now be interpolated to arrive at the rotation vector, and hence rotation tensor, at *Gauss* points. Accordingly, both quantities $\Delta\mathbf{w}$ as well as $\boldsymbol{\gamma}_{i+1}^{j}$ are defined at nodal points and can be interpolated in the classical way. The method is simple and efficient as well, since one does not need to carry out the tensor multiplications explicitly. Based on the notion of quaternions, one can directly generate the vector $\boldsymbol{\gamma}_{i+1}^{j}$ from given $\Delta\mathbf{w}$ and $\boldsymbol{\gamma}_{i+1}^{j-1}$. In the following we briefly describe the basic computational steps.

Updating method using quaternions For a readable amount on quaternions we refer to Hestens Hestens (1990), see also McRobie and Lasenby McRobie and Lasenby (1999), who entirely reformulated the rod equations in terms of quaternions. Let $\mathbf{a}$ be any vector in $\mathbb{R}^3$ and let i be defined as a complex number with $i^2 = -1$. The '$\wedge$'-product (wedge product) of two *Euclidean* vectors $\mathbf{a}$ and $\mathbf{b}$ is defined as

$$\mathbf{a} \wedge \mathbf{b} = i(\mathbf{a} \times \mathbf{b}), \tag{180}$$

where $\mathbf{a} \times \mathbf{b}$ is the usual cross product of vectors.

The geometric product of two vectors is now defined as

$$\mathbf{ab} = \mathbf{a} \cdot \mathbf{b} + \mathbf{a} \wedge \mathbf{b} = \mathbf{a} \cdot \mathbf{b} + i(\mathbf{a} \times \mathbf{b}). \tag{181}$$

Note that the result consists of a scalar and a vector. The operation (+) is not to be confused with the usual addition of scalars or vectors.

Using the above terminology any rotation tensor, say $\mathbf{R}$ with corresponding rotation vector $\boldsymbol{\gamma}$, can be described as

$$\mathbf{R} = \alpha + i\boldsymbol{\beta}, \tag{182}$$

where α and $\boldsymbol{\beta}$ are defined according to

$$\alpha = \cos(\frac{|\boldsymbol{\gamma}|}{2}), \tag{183}$$

$$\boldsymbol{\beta} = \sin(\frac{|\boldsymbol{\gamma}|}{2})\frac{\boldsymbol{\gamma}}{|\boldsymbol{\gamma}|}. \tag{184}$$

The relations hold

$$\mathbf{R}^T = \alpha - i\boldsymbol{\beta} \tag{185}$$

and

$$\mathbf{R}^T\mathbf{R} = 1, \tag{186}$$

since we have, using (182) and (185),

$$\mathbf{R}^T\mathbf{R} = \alpha^2 + \boldsymbol{\beta}^2 = 1, \tag{187}$$

which shows that the four parameters α and $\boldsymbol{\beta}$ are not independent but have to fulfill the side condition (187). From a computational point of view the real advantage of the above lies in the simple multiplication of rotations.

Now, given two rotation vectors $\boldsymbol{\gamma}_i$ and $\Delta\mathbf{w}$, what is the rotation vector which corresponds to the multiplicative product of the corresponding rotation tensors? The computation goes as follows. From (183) and (184) one has

$$\alpha_1 = \cos(\frac{|\Delta\mathbf{w}|}{2}), \tag{188}$$

and

$$\boldsymbol{\beta}_1 = \sin(\frac{|\Delta\mathbf{w}|}{2})\frac{\Delta\mathbf{w}}{|\Delta\mathbf{w}|}\,. \tag{189}$$

Similarly, the relations are valid

$$\alpha_2 = \cos(\frac{|\boldsymbol{\gamma}_i|}{2})\,, \tag{190}$$

and

$$\boldsymbol{\beta}_2 = \sin(\frac{|\boldsymbol{\gamma}_i|}{2})\frac{\boldsymbol{\gamma}_i}{|\boldsymbol{\gamma}_i|}\,. \tag{191}$$

The geometric product of both quaternions reads:

$$\begin{aligned}
\mathbf{R}_{i+1} &= \Delta\mathbf{R}\mathbf{R}_i \\
&= (\alpha_1 + i\boldsymbol{\beta}_1)(\alpha_2 + i\boldsymbol{\beta}_2) \\
&= \alpha_1\alpha_2 + \alpha_1 i\boldsymbol{\beta}_2 + \alpha_2 i\boldsymbol{\beta}_1 + i^2(\boldsymbol{\beta}_1\boldsymbol{\beta}_2) \\
&= \alpha_1\alpha_2 + \alpha_1 i\boldsymbol{\beta}_2 + \alpha_2 i\boldsymbol{\beta}_1 + i^2(\boldsymbol{\beta}_1\cdot\boldsymbol{\beta}_2 + i(\boldsymbol{\beta}_1\times\boldsymbol{\beta}_2)) \\
&= (\alpha_1\alpha_2 - \boldsymbol{\beta}_1\cdot\boldsymbol{\beta}_2) + i(\alpha_1\boldsymbol{\beta}_2 + \alpha_2\boldsymbol{\beta}_1 - \boldsymbol{\beta}_1\times\boldsymbol{\beta}_2) \\
&= \alpha_{i+1} + i\boldsymbol{\beta}_{i+1}\,.
\end{aligned} \tag{192}$$

From the last equation we read

$$\boldsymbol{\beta}_{i+1} = (\alpha_1\boldsymbol{\beta}_2 + \alpha_2\boldsymbol{\beta}_1 - \boldsymbol{\beta}_1\times\boldsymbol{\beta}_2)$$

Using (184), the rotation vector corresponding to $\mathbf{R}_{i+1}$ is recovered as

$$\boldsymbol{\gamma}_{i+1} = \boldsymbol{\beta}_{i+1}\frac{2\arccos\alpha_{i+1}}{|\boldsymbol{\beta}_{i+1}|}\,, \tag{193}$$

which can now be classically interpolated.

5.5 Numerical examples

Before addressing the examples an initial remark is made with regard to the used numerical pre-conditioner and solver. Generally, the system of algebraic equations which is the result of the evaluation of the problem functional (Eq. 306) is usually not adequately conditioned and also asymmetric. Therefore, the right choice of a suitable pre-conditioning and solving method is indispensable. The *Generalized Minimal Residual* method together with the *Gram-Schmidt Orthogonalization* and left pre-conditioning is found to perform excellent. All numerical examples within this section make use of the hyperelastic *Saint-Venant-Kirchhoff* constitutive model which involves as material parameters *Young's modulus* E and *Poisson's ratio* ν.

Epoxy polymeric beams

Bending experiments on epoxy (bisphenol-A epichlorohydrin resin) polymeric micro-beams revealed a size-scale dependency of the resulting deformation (Lam *et al.* 2003). The authors demonstrated that in contrast to the classical elasticity theory the incorporation of strain gradients enabled to approximate size-dependent behaviour of these small-scale structures. Apparently, the nano-structure influences the macroscopic response of the polymeric specimens so that the heterogeneity of the material has to be addressed in order not to underestimate stress and rigidity. Epoxy polymers consist of molecule chains which are partly crosslinked to a network-type structures, if the side branches of such chains are joined up with other polymer chains. Their morphology is semi-crystalline, that is they form mixtures of small crystals and amorphous material. The amorphous or glass-like part though, exhibits no long range order, and the chains are tangled. The experiment consisted of four different micro-cantilever beams of

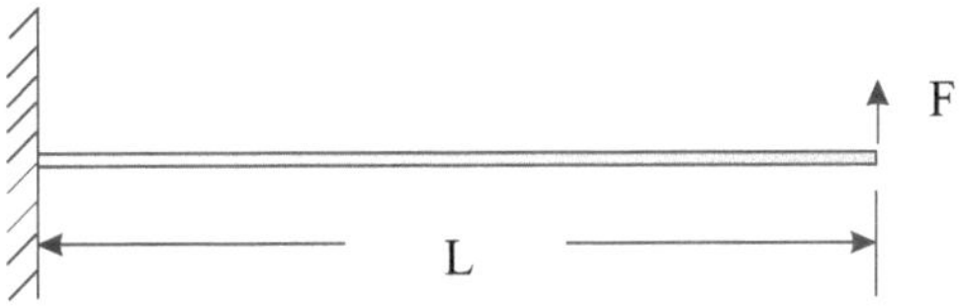

Figure 1. *problem configuration*

length $L = 1150\,\mu m$, $750\,\mu m$, $380\,\mu m$ and $200\,\mu m$ depicted in Fig. 1. The width $b = 235\,\mu m$ and the ratio length to height $L/h = 10$ was kept constant for all four beams. This configuration allowed to compare directly the results of all beams despite their different scaling levels. *Young's modulus* was determined as $E = 1.55\,GPa$, whereas the transversal contraction was found negligible. The experiments did not show strictly linear deformation behaviour, but exhibited some kind of softening in the last third of the deformation process. The relation between different scaling levels to changing rigidity of the beams was clearly non-linear. The strain gradient formulation developed by Lam *et al.* was based on the bending theory of plain strain beams and involved one additional material parameter, a so-called *length scale parameter* l. Their approach provided a fairly accurate match with the experimental data, if this length scale parameter was chosen to

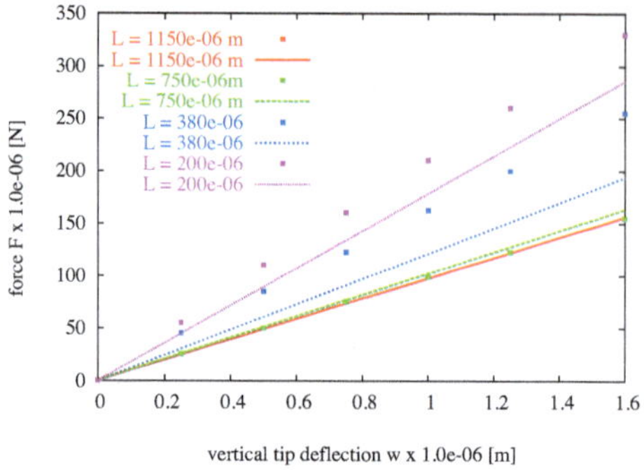

Figure 2. *load displacement with diagram* $l = 21\,\mu\,m$ *and* $\nu = 0.3$

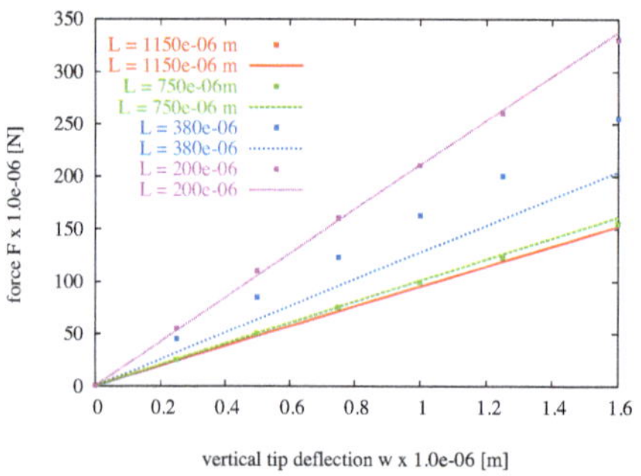

Figure 3. *load displacement diagram with* $l = 25\,\mu\,m$ *and* $\nu = 0$

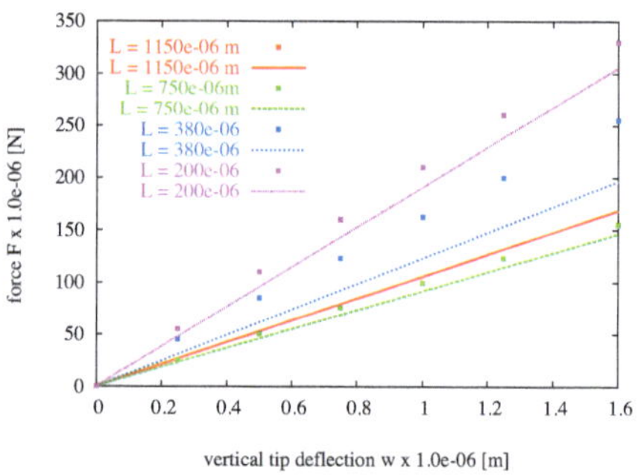

Figure 4. *load displacement diagram with* $l = 25\,\mu\,m$ *and* $\nu = 0.3$

$l = 24\,\mu m$.

This bending experiment is now modelled with $31 \times 3 \times 3$ particles in longitudinal and thickness direction, respectively, making use of the variational formulation (Eq. 306) which also involves a length scale parameter l. In Fig. 2 - 4 the load-deflection graphs of all four scaling levels of the micro-cantilever beam are combined for micro-length scale parameters $l = 21\,\mu m$ and $l = 25\,\mu m$, respectively. The classical elastic plain strain relation between load and deflection is expressed by

$$w = \frac{1}{3}\frac{FL^3}{bh^3}\frac{12\nu}{E}\,, \tag{194}$$

which agrees with the red curves illustrated in Fig. 2 to 4. Each of the remaining colors represents a different scaling level. The dotted curves are the experimental results and the continuous curves illustrate the results which are simulated. Note that the experimental data indicate no difference for the beams with $L = 1150\,\mu m$ and $750\,\mu m$. In Fig. 2 and 4 modelling with transversal contraction is illustrated setting *Poisson's ratio* $\nu = 0.3$, while in Fig. 3 it is neglected by setting $\nu = 0$. All simulated load-deflection diagrams do not show any non-linearity, which is plausible as deformation and strain are fairly small. The relation between the scaling and the changing rigidity however, is non-linear as it is in the experiment too. The simulated results vary for different length scale parameters as expected, but also the influence of the transversal contraction can not be neglected. Furthermore, there is a difference between linear and non-linear modelling. That is, a preliminary linear simulation actually indicates that the parameter setting $l = 25\,\mu m$ and $\nu = 0.3$ should provide a fairly accurate match with experimental data for the beam with $L = 1150\,\mu m$ and also for $L = 200\,\mu m$. In contrast, the non-linear simulation depicted in Fig. 4 shows that this is not the case. Neglecting the transversal contraction illustrated in Fig. 3 significantly improves the agreement of simulation with the experiment for the beams with $L = 1150\,\mu m$, $750\,\mu m$ and $200\,\mu m$. But then again, this result is not predicted by the preliminary linear run with $\nu = 0$, which indicates less accuracy than with $\nu = 0.3$. It can be generally stated that the transversal contraction affects the least the beam with $L = 1150\,\mu m$ and most significantly the beam with $L = 200\,\mu m$.

The simulation deviates the most from the experiment for the beam with length $L = 380\,\mu m$. Overall, no parameter setting for ν and l can be found which provides an accurate match between modelled and experimentally obtained tip-deflection for all scaling levels. This might indicate that the micro-structural behaviour is not fully characterized by rotations only.

Micro-films

The second example is a thin square sheet which is clamped at all its four edges depicted in Fig. 5. Due to the symmetry only one quarter of the sheet is modelled with $11 \times 11 \times 3$ particles in longitudinal and thickness direction, respectively, applying the appropriate symmetry conditions for displacement and rotation. The material is chosen to be the same as for the micro-cantilever beam, except that the *Poisson ratio* is kept constant at $\nu = 0.3$. Four different scaling levels are simulated with $L = 6.0\,mm$, $3.0\,mm$, $1.2\,mm$ and $0.6\,mm$ using the variational formulation (Eq. 306). The ratio $L/h = 300$ is fixed for all four plates so that all results are directly comparable. The classical solution denoted by the red line in all graphs Fig. 6 - 9 is obtained using the *Green* strain tensor based model (Skatulla and Sansour 2005). The results depicted in Fig. 6 and Fig. Fig.

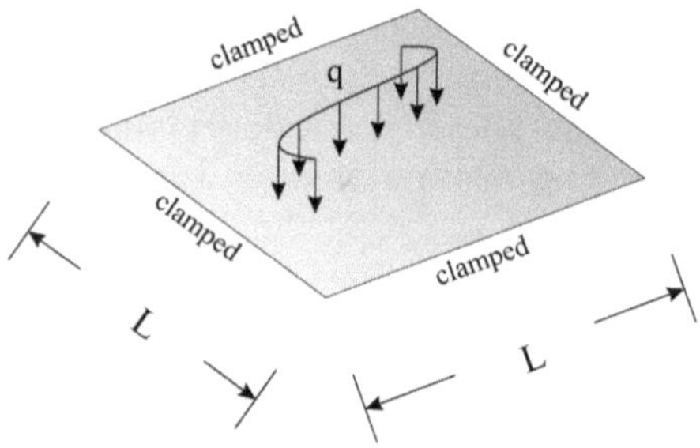

Figure 5. *problem configuration*

7 are achieved by fixing only all displacement but not the rotational degrees of freedom at the outer edges of the plate. The load-deflection graph in Fig. 6 illustrates that in the first stage of the simulation the rigidity of the clamped plate is increasing for smaller scaling levels. In the subsequent modelling process displayed in Fig. 7 the increased rigidity relative to the classical solution is again decreasing and eventually turning into a rigidity regime which is of lower magnitude than the classical reference solution. The crossing point however, where the curves of the non-classical and classical solution meet, is increasingly delayed for the smaller scaling levels.

Holding also all three rotational degrees of freedom at the outer edges

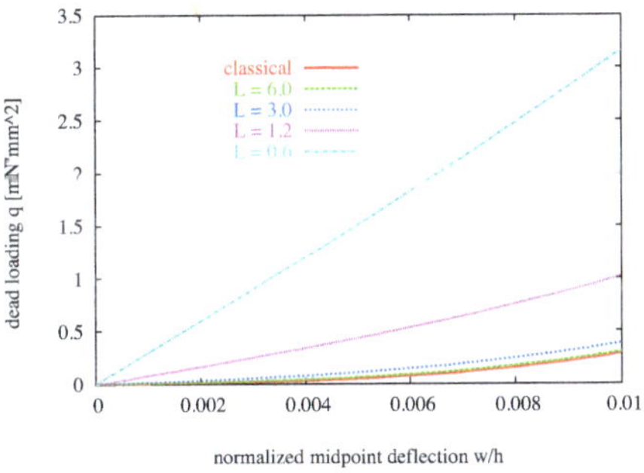

Figure 6. *displacement diagram with displacement boundary conditions*

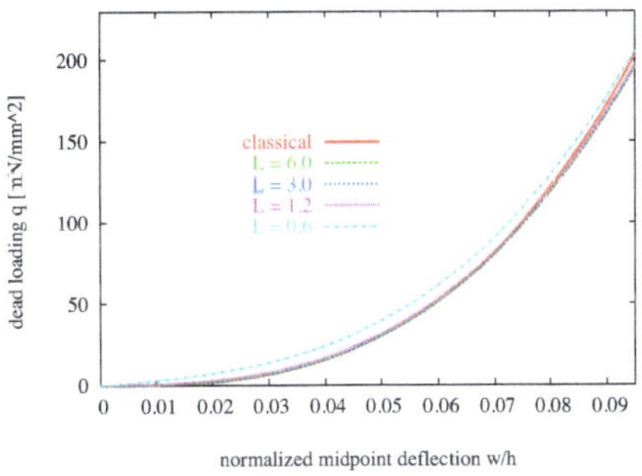

Figure 7. *displacement diagram with displacement boundary conditions*

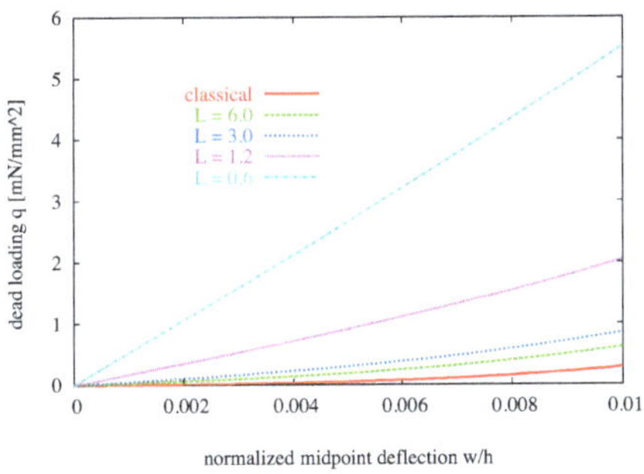

Figure 8. *displacement diagram with displacement and rotation boundary conditions*

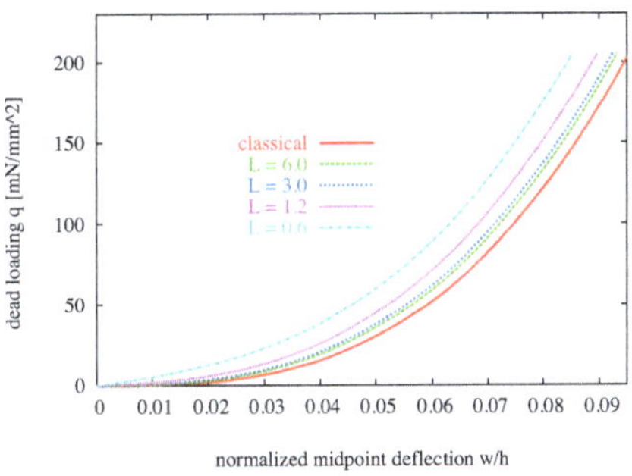

Figure 9. *displacement diagram with displacement and rotation boundary conditions*

of the plate leads somewhat different results depicted in Fig. 8 and Fig. 9. The additional rotational boundary conditions have the effect that the stiffness increase for smaller scaling levels is much higher than without these rotational boundary conditions. Furthermore, the load-deflection curves illustrated in Fig. 9 do not indicate a declining rigidity relative to the classical solution.

Torsion experiments on human bone specimens

Materials such as foam and bones show a porous micro-structure which consists of osteons of $0.15\,mm - 0.25\,mm$ in diameter. Torsion experiments on human bone specimens exhibited a size-dependent elastic response (Yang and Lakes 1981). The authors employed the linear elastic isotropic couple stress theory to model experimental results which involved two additional material parameters. These material parameters were related to a so-called internal length scale parameter l which had about the same magnitude as the bone pores. Later on, Aifantis (1999) demonstrated that also his strain gradient approach was able to predict these experimental results. His approach incorporated, besides the normal shear strain the second order gradient of the same, the latter of which was associated with another material parameter.

In the following, the experiment is idealized as a set of thin beams with a square cross-section with side length $a = 5.0\,mm$, $2.0\,mm$, $1.0\,mm$ and $0.5\,mm$ depicted in Fig. 10. It is not the aim of this study to predict the experimental data, but to show qualitatively that size-scale effects in torsion of thin beams can be modelled applying the proposed variational principle (Eq. 306) to a MLS-based meshfree method. The ratio of beam length to

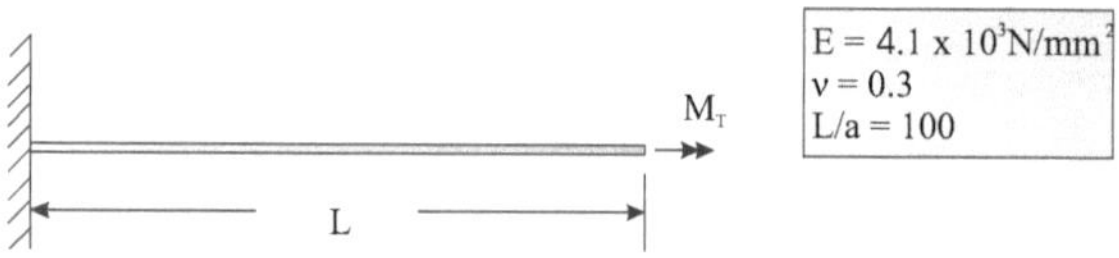

Figure 10. *problem configuration*

side length $L/a = 100$ is kept constant all four beams so that the results for the twist θ and the corresponding torsion moment M_T are comparable. The torsion moment is applied to the tip of the beam. The opposite end of the beam is clamped, that is all degree displacement degrees of freedom are held over the entire cross-section. The problem is simulated with $21 \times 4 \times 4$ particles in longitudinal and thickness direction, respectively. The internal

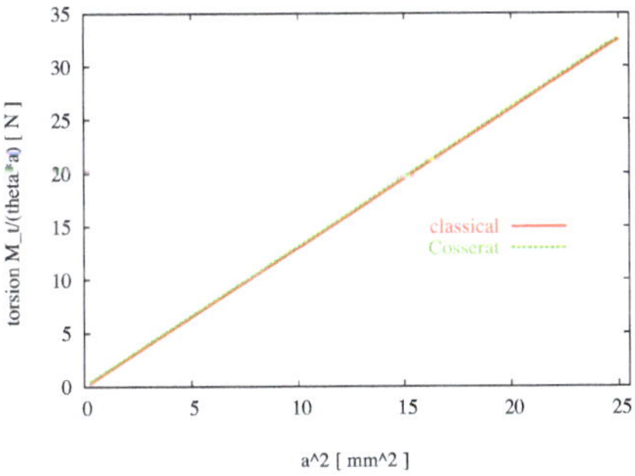

Figure 11. *diagram: normalized torsion vs. cross-section size*

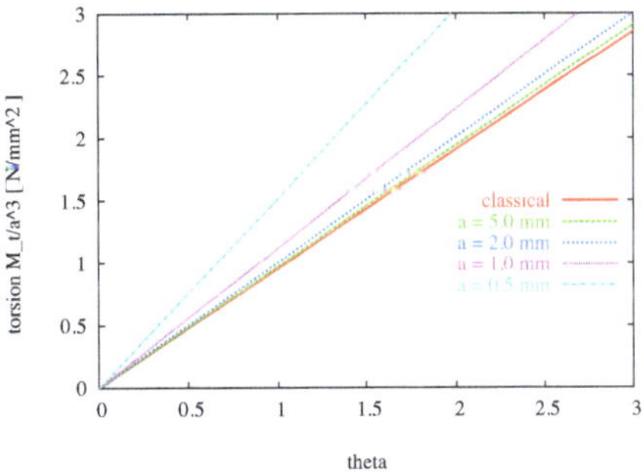

Figure 12. *diagram: normalized torsion vs. twist*

length scale parameter is assumed as $l = 0.25\,mm$.

If the twist is considered to be small, then the analytical relation of the applied torsion moment to the resulting twist θ is given by

$$M_T = C\,G\,a^4\,\frac{\theta}{L}\,, \tag{195}$$

where C is a constant and $L = 100\,a$. Aifantis as well Yang and Lakis very closely fitted the experimental data with their linear non-classical theories. Plotting the results in graphs with $\frac{M_T}{\theta\,a}$ versus cross-section size a^2 for various bone specimens a slight but distinct difference between classical and non-classical theory was recognizable. Qualitatively equivalent results achieved with the variational principle based on the *Cosserat theory* (Eq. 306) are depicted in Fig. 11, where for all sizes of the beam cross-section, also for the largest, the predictions of classical and *Cosserat* theory indicate a clear difference. Nonlinear modelling however, makes the size-scale effects much more visible as displayed in Fig. 12. Here, the classical theory predicts that curves should fall for all side lengths a into one line which is the red curve. In fact, the beams behave stiffer with decreasing cross-section size.

6 The micromorphic continuum

6.1 A first formulation

In the following we critically discuss the passage to the micromorphic continuum essentially as proposed by Eringen and co-workers (Eringen and Kafadar, 1976). Starting again from an adequate definition of the configuration space as

$$\mathcal{C}(\mathcal{B}) = \{(\mathbf{x}, \mathbf{A}) \mid \mathbf{x} \in \mathbb{E}^3, \mathbf{A} \in GL^+(3)\} \tag{196}$$

where $GL^+(3)$ is the general linear group the elements of which have positive determinants. In Eringen and Kafadar (1976) the following strain measures are formulated

$$\mathbf{U} = \mathbf{A}^{-1}\mathbf{F}, \tag{197}$$

$$\mathbf{K}_1 = \mathbf{A}^{-1}\mathbf{A}_{,i} \otimes \mathbf{G}^i, \tag{198}$$

$$\mathbf{K}_2 = \mathbf{A}^T\mathbf{A}. \tag{199}$$

In case $\mathbf{A} \in SO(3)$, the above equations reduce to the strain measures of the Cosserat continuum. In some sense the strain measures ((197))-((199)) are a generalization of those of the Cosserat ones for the case of $\mathbf{A}$ being an element of a general group not necessary the orthogonal one. Now, the first two strain measures can be interpreted in analogy to those of the Cosserat continuum. One considers the derivatives $\mathbf{x}_{,i}$ and $\mathbf{A}_{,i}$ and modify them resulting in expressions invariant with respect to the action of an element of the general group $GL^+(3)$.

The question which arises now is whether such an invariance is of a physical background. To answer this question it is completely sufficient to consider the simple case of a hyperelastic material response. We assume, accordingly, the existence of a stored energy function depending on the three strain measures of the micromorphic continuum and derive the corresponding field equations. Thus, we consider the function $\psi(\mathbf{U}, \mathbf{K}_1, \mathbf{K}_2)$ together with the functional

$$\int_{\mathcal{B}_t} (\rho\delta\psi(\mathbf{U}, \mathbf{K}_1, \mathbf{K}_2) - \mathbf{b} \cdot \delta\mathbf{x})\, dv = 0. \tag{200}$$

Here, $\mathbf{b}$ denotes a field of external loading, dv is the volume element, $\mathcal{B}_t$ is the actual configuration of $\mathcal{B}$, and ρ is the actual mass density. For simplicity, inertia terms, boundary traction, as well as higher order external loading are suppressed. One has first

$$\int_{\mathcal{B}_t} \rho\delta\psi(\mathbf{U}, \mathbf{K}_1, \mathbf{K}_2)\, dv = \int_{\mathcal{B}_t} \rho\left(\frac{\partial\psi}{\partial\mathbf{U}} : \delta\mathbf{U} + \frac{\partial\psi}{\partial\mathbf{K}_1} : \delta\mathbf{K}_1 + \frac{\partial\psi}{\partial\mathbf{K}_2} : \delta\mathbf{K}_2\right) dv. \tag{201}$$

With (:) we denote the scalar product of two tensors. The consideration of (197)-(199) gives

$$\begin{aligned}\delta\mathbf{U} &= \delta\mathbf{A}^{-1}\mathbf{F}+\mathbf{A}^{-1}\delta\mathbf{F}=\delta\mathbf{A}^{-1}\mathbf{A}\mathbf{U}+\mathbf{A}^{-1}\delta\mathbf{F}, &(202)\\ \delta\mathbf{K}_1 &= \delta\mathbf{A}^{-1}\mathbf{A}_{,i}\otimes\mathbf{G}^i+\mathbf{A}^{-1}\delta\mathbf{A}_{,i}\otimes\mathbf{G}^i, &(203)\\ \delta\mathbf{K}_2 &= \delta\mathbf{A}^T\mathbf{A}+\mathbf{A}^T\delta\mathbf{A}. &(204)\end{aligned}$$

For (201) one gets, accordingly,

$$\int_{\mathcal{B}_t}\left(\rho\mathbf{A}^{-T}\frac{\partial\psi}{\partial\mathbf{U}}\mathbf{U}^T\mathbf{A}^T:\delta\mathbf{F}\mathbf{F}^{-1}+\rho\mathbf{A}^{-T}\frac{\partial\psi}{\partial\mathbf{U}}\mathbf{U}^T\mathbf{A}^T:\mathbf{A}\delta\mathbf{A}^{-1}+\right.$$
$$\rho\mathbf{A}^{-T}\frac{\partial\psi}{\partial\mathbf{K}_1}\mathbf{G}^i\mathbf{A}^T_{,i}:\mathbf{A}\delta\mathbf{A}^{-1}+\frac{1}{\sqrt{g}}\left[\sqrt{g}\rho\mathbf{A}^{-T}\frac{\partial\psi}{\partial\mathbf{K}_1}\mathbf{G}^i\right]_{,i}\mathbf{A}^T:\mathbf{A}\delta\mathbf{A}^{-1}-$$
$$\left.2\rho\mathbf{A}\frac{\partial\psi}{\partial\mathbf{K}_2}\mathbf{A}^T:\mathbf{A}\delta\mathbf{A}^{-1}\right)dv-\int_{\mathcal{B}_t}\mathbf{b}\cdot\delta\mathbf{x}\,dv = 0.(205)$$

With g we denoted the metric at the actual configuration. Note that $\delta\mathbf{F}\mathbf{F}^{-1}$ as well as $\mathbf{A}\delta\mathbf{A}^{-1}$ are elements of the Lie Algebra $gl^+(3)$. The Euler-Lagrange equations now read

$$div\left(\rho\mathbf{A}^{-T}\frac{\partial\psi}{\partial\mathbf{U}}\mathbf{U}^T\mathbf{A}^T\right)+\mathbf{b} = \mathbf{0}, \qquad (206)$$

$$\rho\mathbf{A}^{-T}\frac{\partial\psi}{\partial\mathbf{U}}\mathbf{U}^T\mathbf{A}^T+\rho\mathbf{A}^{-T}\frac{\partial\psi}{\partial\mathbf{K}_1}\mathbf{G}^i\mathbf{A}^T_{,i}+$$
$$\frac{1}{\sqrt{g}}\left[\sqrt{g}\rho\mathbf{A}^{-T}\frac{\partial\psi}{\partial\mathbf{K}_1}\mathbf{G}^i\right]_{,i}\mathbf{A}^T-2\rho\mathbf{A}\frac{\partial\psi}{\partial\mathbf{K}_2}\mathbf{A}^T = \mathbf{0}, \qquad (207)$$

together with corresponding boundary conditions.

The first term at the left hand side of (206) is evidently nothing but the Cauchy stress tensor. Now a great disadvantage of these equations lies in the fact that the Cauchy stress tensor is present in the second equation (Eq. (207). That is, using (207) one may eliminate the Cauchy stress tensor completely from (206). The external forces must then be in equilibrium with the higher order stress quantities. This stands in contradiction to the fact that higher order stresses or moments could be as small as wished and may be even negligible in passing to a standard continuum.

6.2 An alternative formulation

To avoid this contradiction we conclude that invariance of the strain measures is to be enforced with respect to elements of $SO(3)$ alone and not with respect to a more general group. In view of this we suggest the following strain measures for a general micromorphic continuum

$$\begin{aligned} \mathbf{U} &= \mathbf{R}^T \mathbf{x}_{,i} \otimes \mathbf{G}^i, & (208)\\ \mathbf{K}_1 &= \mathbf{R}^T \mathbf{A}_{,i} \otimes \mathbf{G}^i, & (209)\\ \mathbf{K}_2 &= \mathbf{R}^T \mathbf{A}, & (210) \end{aligned}$$

where we have $\mathbf{A} \in GL^+(3)$ but $\mathbf{R} \in SO(3)$. To show that the resulting field equations are useful we consider again the functional

$$\int_{\mathcal{B}_t} (\rho \delta\psi(\mathbf{U}, \mathbf{K}_1, \mathbf{K}_2) - \mathbf{b} \cdot \delta\mathbf{x})\, dv = 0, \tag{211}$$

together with the variations

$$\begin{aligned} \delta\mathbf{U} &= \delta\mathbf{R}^T \mathbf{F} + \mathbf{R}^T \delta\mathbf{F} = \delta\mathbf{R}^T \mathbf{R}\mathbf{U} + \mathbf{R}^T \delta\mathbf{F}, & (212)\\ \delta\mathbf{K}_1 &= \delta\mathbf{R}^T \mathbf{A}_{,i} \otimes \mathbf{G}^i + \mathbf{R}^T \delta\mathbf{A}_{,i} \otimes \mathbf{G}^i, & (213)\\ \delta\mathbf{K}_2 &= \delta\mathbf{R}^T \mathbf{A} + \mathbf{R}^T \delta\mathbf{A}. & (214) \end{aligned}$$

Noting that the product $\delta\mathbf{R}^T\mathbf{R}$ is an element of the Lie algebra $so(3)$, that is, it is skew symmetric, the resulting field equations read

$$\begin{aligned} div\left(\rho\mathbf{R}\frac{\partial\psi}{\partial\mathbf{U}}\mathbf{U}^T\mathbf{R}^T\right) + \mathbf{b} &= \mathbf{0}, & (215)\\ \rho\frac{\partial\psi}{\partial\mathbf{U}}\mathbf{U}^T + \rho\frac{\partial\psi}{\partial\mathbf{K}_1}\mathbf{G}^i(\mathbf{R}^T\mathbf{A}_{,i})^T + \rho\frac{\partial\psi}{\partial\mathbf{K}_2}\mathbf{K}_2^T &= symmetric, & (216)\\ \frac{1}{\sqrt{g}}\left[\sqrt{g}\rho\mathbf{R}\frac{\partial\psi}{\partial\mathbf{K}_1}\mathbf{G}^i\right]_{,i}\mathbf{A}^T + \rho\mathbf{R}\frac{\partial\psi}{\partial\mathbf{K}_2}\mathbf{A}^T &= \mathbf{0}. & (217) \end{aligned}$$

The equations are to be supplemented with corresponding boundary conditions. Remarkably, and in contrast to (207), only the skew symmetric part of $\boldsymbol{\sigma}$ enters the second equation. This fact allows for a natural passing to a classical continuum.

7 A unified concept of generalised continua

This chapter is dedicated to the development of generalized continuum models based on the theoretical framework proposed by Sansour (1998a). Its basic principles are outlined Sec. 7.1. It is demonstrated that by specific definition of the topology of the generalized vector and tensor space different types of generalized deformation, strain and stress measures can be formulated suitable to describe the physical problem at hand. Corresponding variational principles are derived analog to the classical case and the governing equations are deduced. Here, we want to illustrate the potential of the proposed generalized continuum framework starting firstly with a micromorphic continuum outlined in Sec. 7.2. The micromorphic deformation description is based on a specific quadratic ansatz for the micro which opens the door for inelastic formulations of such continua at finite strains as introduced in Sec. 7.3 but also for scale effects in specimen experiencing homogenous deformation. The theory will be used to model scale effects and shear band formation using a finite element method. Secondly, in Sec. 7.4 we want to consider generalized micropolar continuum defined such that the microcontinuum only undergoes rigid body rotations only. This is achieved by the use of a macroscopic rotation field which is assumed to be independent of the macroscopic displacement field so that the formulation incorporates besides displacement three additional rotational degrees of freedom. The rotation field is included in this generalized deformation description by means of the macroscopic rotation tensor. The procedure leads to a generalized micropolar continuum which is closely related to the *Cosserat Continuum*. Unlike the classical *Cosserat Continuum* however, there is only one strain measure needed which implicitly accounts for the conventional and higher order strain contribution. Thirdly, two different strain gradient approaches are outlined in Secs. 7.5 and 7.5. In analogy to (Ericksen and Truesdell, 1957; Mindlin, 1964) the incorporation of additional degrees of freedom, besides displacement ones, are avoided by assuming the micro-deformation as to depend on the derivatives of the macroscopic deformation which, in essence, constitutes a higher gradient continuum. In particular, it is shown that the *moving least square*-approximation (MLS) proves to be suitable to provide the continuity and consistency needed by this kind of generalized formulation.

7.1 Generalized deformation

The basic idea is that a generalized continuum $\mathcal{G}$ can be assumed to inherit the mathematical structure of a fibre bundle (see e.g. Choquet-Bruhat et al. (1982)). In the simplest case, this is the Cartesian product of

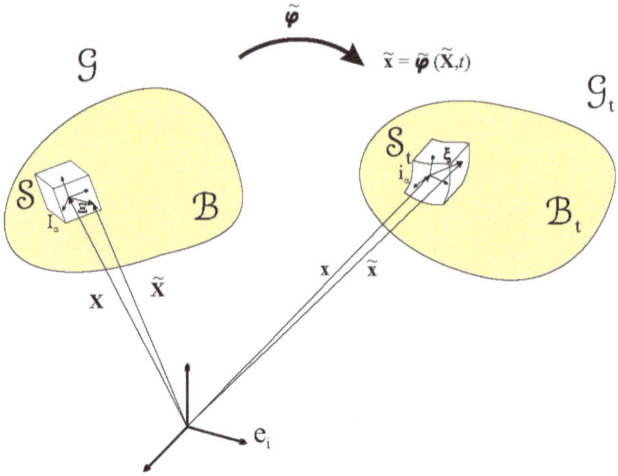

Figure 13. Configuration spaces

a macro space $\mathcal{B} \subset \mathbb{E}(3)$ and a micro space $\mathcal{S}$ which we write as

$$\mathcal{G} := \mathcal{B} \times \mathcal{S}\,. \tag{218}$$

This definition assumes an additive structure of $\mathcal{G}$ which implies that the integration over the macro- and the micro-continuum can be performed separately. The macro-space $\mathcal{B}$ is parameterized by the curvilinear coordinates ϑ^i, $i = 1, 2, 3$ and the micro-space or micro-continuum $\mathcal{S}$ by the curvilinear coordinates ζ^α. Here, and in what follows, Greek indices take the values 1, ... or n. The dimension of $\mathcal{S}$ denoted by n is arbitrary, but finite. Furthermore, we want to exclude that the dimension and topology of the micro-space is dependent on ϑ^i. Each material point $\tilde{\mathbf{X}} \in \mathcal{G}$ is related to its spatial placement $\tilde{\mathbf{x}} \in \mathcal{G}_t$ at time $t \in \mathbb{R}$ by the mapping

$$\tilde{\varphi}\,(t) : \mathcal{G} \longrightarrow \mathcal{G}_t\,. \tag{219}$$

For convenience but without loss of generality we identify $\mathcal{G}$ with the undeformed reference configuration at a fixed time t_0 in what follows. The generalized space can be projected to the macro-space in its reference and its current configuration by

$$\pi_0(\tilde{\mathbf{X}}) = \mathbf{X} \qquad \text{and} \qquad \pi_t\,(\tilde{\mathbf{x}}) = \mathbf{x} \tag{220}$$

respectively, where π_0 as well as π_t represent projection maps, and $\mathbf{X} \in \mathcal{B}$ and $\mathbf{x} \in \mathcal{B}_t$. The tangent space $\mathcal{TG}$ in the reference configuration is defined by the pair $(\tilde{\mathbf{G}}_i \times \mathbf{I}_\alpha)$ given by

$$\tilde{\mathbf{G}}_i = \frac{\partial \tilde{\mathbf{X}}}{\partial \vartheta^i} \qquad \text{and} \qquad \mathbf{I}_\alpha = \frac{\partial \tilde{\mathbf{X}}}{\partial \zeta^\alpha}\,, \tag{221}$$

where the corresponding dual contra-variant vectors are denoted by $\mathbf{G}^i$ and $\mathbf{I}^\alpha$, respectively. A corresponding tangent space in the current configuration $\mathcal{T}\mathcal{G}_t$ is spanned by the pair $(\tilde{\mathbf{g}}_i \times \mathbf{i}_\alpha)$ given by

$$\tilde{\mathbf{g}}_i = \frac{\partial \tilde{\mathbf{x}}}{\partial \vartheta^i} \qquad \text{and} \qquad \mathbf{i}_\alpha = \frac{\partial \tilde{\mathbf{x}}}{\partial \zeta^\alpha}\,. \tag{222}$$

The generalized tangent space can also be projected to its corresponding macro-space by

$$\pi_0^*(\tilde{\mathbf{G}}_i) = \mathbf{G}_i \qquad \text{and} \qquad \pi_t^*\,(\tilde{\mathbf{g}}_i) = \mathbf{g}_i \tag{223}$$

respectively, where the tangent vectors $\mathbf{I}_\alpha$ are assumed to be constant throughout $\mathcal{S}$ for simplicity. Note that the definition of a projection map is not trivial. The tangent of the projection map defines the geometry of the extra space and so the metric which is to be used to evaluate the integral over the generalized space. The concept is rich in its structure.

Now, we assume that the placement vector $\tilde{\mathbf{x}}$ of a material point P $(\tilde{\mathbf{X}} \in \mathcal{G})$ is of an additive nature and is the sum of its position in the macro-continuum $\mathbf{x} \in \mathcal{B}_t$ and in the micro-continuum $\boldsymbol{\xi} \in \mathcal{S}_t$ as follows

$$\tilde{\mathbf{x}}\left(\vartheta^k, \zeta^\beta, t\right) = \mathbf{x}\left(\vartheta^k, t\right) + \boldsymbol{\xi}\left(\vartheta^k, \zeta^\beta, t\right)\,. \tag{224}$$

Thereby, the macro-placement vector $\mathbf{x}$ defines the origin of the micro co-ordinate system such that the micro-placement $\boldsymbol{\xi}$ is assumed to be relative to the macro-placement. Accordingly, for $t = t_0$ the placement of a material point, Eq. (224), takes

$$\tilde{\mathbf{X}} = \mathbf{X}\left(\vartheta^k\right) + \boldsymbol{\Xi}\left(\vartheta^k, \zeta^\beta\right)\,, \qquad \text{with } \boldsymbol{\Xi} \in \mathcal{S}\,. \tag{225}$$

The definition of the generalized continuum and so of the extra degrees of freedom depends directly on the choices to be made for the micro deformation $\boldsymbol{\xi}\left(\vartheta^k, \zeta^\beta, t\right)$. The theory is based on the fact that the dependency on the micro co-ordinates ζ^β must be determined apriori. Specific choices define specific continua.

The simplest case is provided by the linear ansatz, also suggested by Eringen (1999) albeit in a different context,

$$\tilde{\mathbf{x}} = \mathbf{x}\left(\vartheta^k, t\right) + \zeta^\alpha\,\mathbf{a}_\alpha\left(\vartheta^k, t\right)\,. \tag{226}$$

The vector functions $\mathbf{a}_\alpha\left(\vartheta^k, t\right)$, with their corresponding micro co-ordinates ζ^α, are independent degrees of freedom. The number α must be chosen according to the specific topology of the micro-space as well as depending on the physical properties of the material due to its intrinsic structure. Generally, the vector functions $\mathbf{a}_\alpha\left(\vartheta^k, t\right)$ can be described with the help of a second order tensor $\mathbf{A}$ as follows

$$\mathbf{a}_\alpha\left(\vartheta^k, t\right) = \mathbf{A}\mathbf{I}_\alpha\,. \tag{227}$$

Note that if the dimension of $\mathcal{S}$ is three, then we have $\mathbf{A} \in GL^+(3, \mathbb{R})$ which can be restricted to subgroups of $GL^+(3, \mathbb{R})$ as well. Here $GL^+(3, \mathbb{R})$ defines the general linear group of 3×3 matrices (defined over the body of real numbers) with positive determinants (simply all invertible 3×3 matrices with positive determinants).

As it will be shown later, the above linear ansatz underlies certain deficiencies, because it can result in a second strain measure which is not of full rank. To achieve full rank the ansatz is extended as to additionally include quadratic terms. Of interest is the following form

$$\tilde{\mathbf{x}} = \mathbf{x}\left(\vartheta^k, t\right) + \zeta^\alpha \mathbf{a}_\alpha\left(\vartheta^k, t\right) + \zeta^\alpha \zeta^\beta \mathbf{b}_{\alpha\beta}\left(\vartheta^k, t\right)\,. \tag{228}$$

In addition to the classical three degrees of freedom, the above function introduces the vectors $\mathbf{a}_\alpha$ as well as the vectors $\mathbf{b}_{\alpha\beta}$ as extra degrees of freedom. However, the mentioned full rank can be achieved using a simpler form of the function as follows

$$\tilde{\mathbf{x}} = \mathbf{x}\left(\vartheta^k, t\right) + \zeta^\alpha \left(1 + \zeta^\beta \chi_\beta\left(\vartheta^k, t\right)\right) \mathbf{a}_\alpha\left(\vartheta^k, t\right)\,. \tag{229}$$

Accordingly, with each micro co-ordinate we have now four extra degrees of freedom instead of the larger number above. These are the three components of the vector $\mathbf{a}_\alpha$ as well as the independent displacement-like functions χ_α. Note, however, $\mathbf{a}_\alpha$ as well as χ_α are constant over $\mathcal{S}$. While the functions χ_α contribute to the definition of the strains, their special importance lies in the fact that they allow for the complete definition of linear distribution of strain in the extra dimensions. Also, it is important to realize that the dimension of the micro-space does not have to coincide with the dimension of the macro-space.

Discussion. At this stage it is worthwhile to have a discussion regarding the above approach. At its heart, the deformation is viewed as embedded in an extended space of $3+n$ dimensions. This definition takes into account the consideration that loading and deformation takes place in the $3+n$ dimensional space and contributes to all entities of $\mathbf{x}$ as well as $\boldsymbol{\xi}$, and so to $\mathbf{a}_\alpha$, in general. One is immediately inclined to look at the pair $(\mathbf{x}, \boldsymbol{\xi})$ as defining an extended placement field. A realisation of such a field would be to look at a new vector, say $\tilde{\mathbf{x}}$, simply as having $3+n$ entities, where n is the dimension of the additional space. Close inspection, however, reveals that such an approach cannot produce the results we are looking for, as it does not provide us with higher order stress and strain measures. In fact, we would end up with a strain measure, say the right *Cauchy-Green* deformation tensor which is simply defined in a $(3+n)\times(3+n)$ space; the field equation will not change in appearance from the classical case albeit the stress tensor being defined in a higher space and so the divergence operator. This approach does not lead to modified equilibrium equations (e.g. angular momentum equation) with higher order stress tensors (moments) modifying its form. Crucial is to get the spaces where $\mathbf{x}$ and $\boldsymbol{\xi}$ live to interact with each other. Accordingly, the importance does not lie in simply extending the space, it is vital to define the means of interaction between the classical degrees of freedom and the additional ones that specify the generalized continuum at hand. For the interaction to be physically meaningful it must ensure that the stress quantities which result from the new degrees of freedom are going to modify the Euler-Lagrange equations (e.g. the angular momentum equation) already at the linear level. One may think of different approaches to achieve such an interaction, the above presented one views $\boldsymbol{\xi}(\vartheta^k,\zeta^\beta,t)$ as the complement to $\mathbf{x}(\vartheta^k,t)$. Decisive now is to assume that $\boldsymbol{\xi}(\vartheta^k,\zeta^\beta,t)$ depends on both co-ordinate sets ϑ^k and ζ^β. This provides us with the interaction we are seeking. When it comes to computations, obviously all vectors must be treated as living in some space, to be decided as appropriate to the problem at hand, and which is large enough to accommodate all of them. As mentioned above, in general the placement vectors at the current configuration, i.e. $\mathbf{x}$ and $\mathbf{a}_\alpha$, are of $3+n$ dimension. From a computational point of view it makes sense to reduce this dimension and deal only with the physically most relevant degrees of freedom. Since we are interested in the equilibrium equations in the three-dimensional *Euclidean* space, obviously the restrictions to only the first three components of the mentioned vectors is not only physically sound but makes also perfect sense in reducing the computational cost. In other words, the embedding of $\boldsymbol{\xi}$ in the three-dimensional space is especially attractive, because it significantly saves computational time, while delivers as far as the physics is concerned. In our case the vectors $\mathbf{x}$ and $\mathbf{a}_\alpha$ as given in Eq. (226) can be treated as belonging to the same three-dimensional vector space albeit their physical interpretation and background. The approach is most effective when it comes to inelastic material behaviour, where, classically, either the strain measures or their rates are split in elastic and inelastic parts.

In the following sections strain gradient, micropolar and micromorphic formulations are shown to be derived from the generalized continuum framework. For each a generalized *Cauchy-Green* deformation tensor, a corresponding variational principle and its govering equations are formulated. The differences and their suitability for specific applications are illustrated.

7.2 Generalized micromorphic continuum

So far formulations of generalized continua are faced with the problem relating to the observation of scale effects also in a homogeneously deforming specimen. Cosserat and higher gradient theories cannot predict such scale effects because the extra strain measures are identically zero for homogeneous deformation.

Here, a micromorphic continuum is derived from the generalized deformation description outlined in Sec. 7.1 which can address scale effects in specimen experiencing homogenous deformation as well as inelastic material behaviour at finite strains in a very straightforward manner.

The approach not only considers a geometrically exact description of finite deformations within the macro-continuum, but can accommodate also higher-order deformations, within the micro-continuum. Associated with the non-local deformation are internal length scale parameters relevant to describe size-scale effects. In this paper the micromorphic deformation will be based on a specific quadratic ansatz for the micro deformation. Most notably, the difference in the number of material parameters in comparison to a classical formulation is kept to a minimum and is confined to purely geometric ones which describe the micro continuum.

Generalized deformation and strain measures For a micromorphic deformation description, the components of micro-directors $\mathbf{a}_\alpha$ (Eq. 227) are independent degrees freedom and additional to the displacements unlike in the strain gradient approach (Eq. 314). It is also meaningful in order to avoid rank deficiency in the higher order deformation tensor to consider quadratic terms in ζ^α as follows:

$$\tilde{\mathbf{x}} = \mathbf{x}\left(\vartheta^k, t\right) + \zeta^\alpha \left(1 + \zeta^\beta \chi_\beta \left(\vartheta^k, t\right)\right) \mathbf{a}_\alpha \left(\vartheta^k, t\right) . \tag{230}$$

Accordingly, with each micro co-ordinate we have the three components of the vector $\mathbf{a}$ and additionally the displacement-like functions χ as independent degrees of freedom.

Now we proceed to define the strain measures. Taking the derivatives of

$\tilde{\mathbf{x}}$ with respect to the macro-coordinates ϑ^i

$$\tilde{\mathbf{x}}_{,i} = \frac{\partial \tilde{\mathbf{x}}}{\partial \vartheta^i} = \mathbf{x}_{,i}\left(\vartheta^k, t\right) + \zeta^\alpha \zeta^\beta \chi_{\beta,i}\left(\vartheta^k, t\right) \mathbf{a}_\alpha\left(\vartheta^k, t\right) + \zeta^\alpha \left(1 + \zeta^\beta \chi_\beta\left(\vartheta^k, t\right)\right) \mathbf{a}_{\alpha,i}\left(\vartheta^k, t\right), \quad (231)$$

as well as with respect to the micro co-ordinates ζ^α,

$$\tilde{\mathbf{x}}_{,\alpha} = \frac{\partial \tilde{\mathbf{x}}}{\partial \zeta^\alpha} = \mathbf{a}_\alpha\left(\vartheta^k, t\right) + \zeta^\beta \left(\chi_\beta\left(\vartheta^k, t\right) \mathbf{a}_\alpha\left(\vartheta^k, t\right) + \chi_\alpha\left(\vartheta^k, t\right) \mathbf{a}_\beta\left(\vartheta^k, t\right)\right), \quad (232)$$

the generalized deformation gradient tensor can be expressed as follows

$$\begin{aligned}\tilde{\mathbf{F}} = & \Big[\mathbf{x}_{,i}\left(\vartheta^k, t\right) + \zeta^\alpha \zeta^\beta \chi_{\beta,i}\left(\vartheta^k, t\right) \mathbf{a}_\alpha\left(\vartheta^k, t\right) \\ & + \zeta^\alpha \left(1 + \zeta^\beta \chi_\beta\left(\vartheta^k, t\right)\right) \mathbf{a}_{\alpha,i}\left(\vartheta^k, t\right)\Big] \otimes \tilde{\mathbf{G}}^i + \\ & \Big[\mathbf{a}_\alpha\left(\vartheta^k, t\right) + \zeta^\beta \left(\chi_\beta\left(\vartheta^k, t\right) \mathbf{a}_\alpha\left(\vartheta^k, t\right) + \chi_\alpha\left(\vartheta^k, t\right) \mathbf{a}_\beta\left(\vartheta^k, t\right)\right)\Big] \otimes \mathbf{I}^\alpha .\end{aligned} \quad (233)$$

As for approaches before, a generalized right *Cauchy-Green* deformation tensor based on Eq. (233) is formulated as follows

$$\tilde{\mathbf{C}} = \tilde{\mathbf{F}}^T \tilde{\mathbf{F}} . \quad (234)$$

and neglecting higher order terms in ζ^α and extracting only the dominant parts of $\tilde{\mathbf{C}}$ (constant and linear in ζ^α) we arrive at

$$\begin{aligned}\tilde{\mathbf{C}} = & \Big(\mathbf{x}_{,k} \cdot \mathbf{x}_{,l} + \zeta^\alpha \left(\mathbf{a}_{\alpha,k} \cdot \mathbf{x}_{,l} + \mathbf{x}_{,k} \cdot \mathbf{a}_{\alpha,l}\right)\Big) \tilde{\mathbf{G}}^k \otimes \tilde{\mathbf{G}}^l \\ & + \Big(\mathbf{x}_{,k} \cdot \mathbf{a}_\beta + \zeta^\alpha \mathbf{a}_{\alpha,k} \cdot \mathbf{a}_\beta + \zeta^\alpha \mathbf{x}_{,k} \cdot \left(\chi_\alpha \mathbf{a}_\beta + \chi_\beta \mathbf{a}_\alpha\right)\Big) \\ & \qquad \Big(\tilde{\mathbf{G}}^k \otimes \mathbf{I}^\beta + \mathbf{I}^\beta \otimes \tilde{\mathbf{G}}^k\Big) \\ & + \Big(\zeta^\alpha \left(\chi_\gamma \mathbf{a}_\alpha \cdot \mathbf{a}_\beta + \chi_\beta \mathbf{a}_\alpha \cdot \mathbf{a}_\gamma\right) + 2 \zeta^\alpha \chi_\alpha \mathbf{a}_\beta \cdot \mathbf{a}_\gamma + \mathbf{a}_\beta \cdot \mathbf{a}_\gamma\Big) \mathbf{I}^\beta \otimes \mathbf{I}^\gamma = \\ = & \ \mathbf{C} + \zeta^\alpha \mathbf{K}_\alpha .\end{aligned} \quad (235)$$

Explicitly one has

$$\begin{aligned}\mathbf{C} = & \ \mathbf{x}_{,k} \cdot \mathbf{x}_{,l} \mathbf{G}^k \otimes \mathbf{G}^l + \mathbf{x}_{,k} \cdot \mathbf{a}_\beta \left(\mathbf{G}^k \otimes \mathbf{I}^\beta + \mathbf{I}^\beta \otimes \mathbf{G}^k\right) \\ & + \mathbf{a}_\beta \cdot \mathbf{a}_\gamma \mathbf{I}^\beta \otimes \mathbf{I}^\gamma\end{aligned} \quad (236)$$

and

$$
\begin{aligned}
\mathbf{K}_\alpha &= (\mathbf{a}_{\alpha,k}\cdot\mathbf{x}_{,l}+\mathbf{x}_{,k}\cdot\mathbf{a}_{\alpha,l})\ \mathbf{G}^k\otimes\mathbf{G}^l + \\
&\Big(\mathbf{a}_{\alpha,k}\cdot\mathbf{a}_\beta+\mathbf{x}_{,k}\cdot(\chi_\alpha\,\mathbf{a}_\beta+\chi_\beta\,\mathbf{a}_\alpha)\Big)\ (\mathbf{G}^k\otimes\mathbf{I}^\beta+\mathbf{I}^\beta\otimes\mathbf{G}^k) + \\
&\Big(\chi_\gamma\,\mathbf{a}_\alpha\cdot\mathbf{a}_\beta+\chi_\beta\,\mathbf{a}_\alpha\cdot\mathbf{a}_\gamma+2\,\chi_\alpha\,\mathbf{a}_\beta\cdot\mathbf{a}_\gamma\Big)\ \mathbf{I}^\beta\otimes\mathbf{I}^\gamma\,. \qquad (237)
\end{aligned}
$$

Note that for the transition from Eq. (235) to Eqs. (236) and (237) to take place, the geometry of the micro space must be specified. Specifically, one has to decide about the projection map $\pi_0^*(\tilde{\mathbf{G}}_i)$ which defines the transition from the tangent vectors $\tilde{\mathbf{G}}^i$, defined in the generalized space, to the tangent vectors $\mathbf{G}^i$, defined in $\mathcal{T}B$. Generally, $\tilde{\mathbf{G}}^i$ are functions of the micro coordinates ζ^α. In the above transition the projection map of the tangent space is assumed to be trivial, which means that $\tilde{\mathbf{G}}^i$ coincide with $\mathbf{G}^i$. It is noted, however, that the presented theoretical framework is rich and allows for the micro space to be equipped with further geometric structures such as curvature, should this be deemed relevant from a physical point of view. In such cases the projection map no longer can be treated as trivial.

From the last equation it becomes clear that for $\chi_\alpha = 0$, the tensor $\mathbf{K}_\alpha$ will be deficient. This fact is relevant when it comes to the definition of constitutive laws.

Generalized principle of virtual work Similar as in Sec. 7.4, a micromorphic variational principle is established based on the generalized strain tensor $\tilde{\mathbf{C}}$ (Eq. 235). From a non-linear boundary value problem in the domain $\mathcal{B}\times\mathcal{S}$ considering the static case and considering only mechanical processes, the *first law of thermodynamics* provides the following variational statement

$$
\delta\Psi - \mathcal{W}_{ext} = 0\,. \qquad (238)
$$

The external virtual work $\mathcal{W}_{ext}$ is defined in the Lagrangian form as follows

$$
\begin{aligned}
\mathcal{W}_{ext}(\mathbf{u}) = &\int_{\mathcal{B}} \mathbf{b}\cdot\delta\mathbf{u}\;dV + \int_{\mathcal{B}} \mathbf{l}^\alpha\cdot\delta\mathbf{a}_\alpha\;dV \\
&+ \int_{\partial\mathcal{B}_N} \mathbf{t}^{(\mathbf{n})}\cdot\delta\mathbf{u}\;dA + \int_{\partial\mathcal{B}_N} \mathbf{q}^{(\mathbf{n})\,\alpha}\cdot\delta\mathbf{a}_\alpha\;dA \qquad (239)
\end{aligned}
$$

where the external body force and moment $\mathbf{b}$ and $\mathbf{l}$, respectively, acting in $\mathcal{B}$ and the external traction and surface moment $\mathbf{t}^{(\mathbf{n})}$ and $\mathbf{q}^{(\mathbf{n})}$, respectively, acting on the Neumann boundary $\partial\mathcal{B}_N$ are obtained by integrating over the micro-space $\mathcal{S}$:

$$\mathbf{b}\left(\vartheta^k\right) = \frac{1}{V_{\mathcal{S}}} \int_{\mathcal{S}} \tilde{\rho}_0 \, \tilde{\mathbf{b}}\left(\vartheta^k, \zeta^\beta\right) \, dS \tag{240}$$

$$\mathbf{l}\left(\vartheta^k\right) = \frac{1}{V_{\mathcal{S}}} \int_{\mathcal{S}} \tilde{\rho}_0 \, \zeta^\alpha \, \mathbf{R}\mathbf{I}_\alpha \times \tilde{\mathbf{b}}\left(\vartheta^k, \beta^\alpha\right) \, dS \tag{241}$$

$$\mathbf{t}^{(\mathbf{n})}\left(\eta^k\right) = \frac{1}{V_{\mathcal{S}}} \int_{\mathcal{S}} \tilde{\rho}_0 \, \tilde{\mathbf{t}}^{(\mathbf{n})}\left(\eta^k, \zeta^\beta\right) \, dS \tag{242}$$

$$\mathbf{q}^{(\mathbf{n})}\left(\eta^k\right) = \frac{1}{V_{\mathcal{S}}} \int_{\mathcal{S}} \tilde{\rho}_0 \, \zeta^\alpha \, \mathbf{R}\mathbf{I}_\alpha \times \tilde{\mathbf{t}}^{(\mathbf{n})}\left(\eta^k, \zeta^\beta\right) \, dS \, . \tag{243}$$

The internal virtual power in the Lagrangian form is given by

$$\delta\Psi = \int_{\mathcal{B}} \int_{\mathcal{S}} \tilde{\rho}_0 \, \frac{\partial\psi(\tilde{\mathbf{C}})}{\partial\tilde{\mathbf{C}}} \, dS \, dV \, , \tag{244}$$

and with Eq. (236) and Eq. (237) we arrive at

$$\delta\Psi = \int_{\mathcal{B}} \frac{1}{2} \Big\{ \mathbf{S} : \delta\mathbf{C} + \mathbf{M}^\alpha : \delta\mathbf{K}_\alpha \Big\} \, dV \, , \tag{245}$$

with the force stress

$$\mathbf{S}\left(\vartheta^k\right) = \frac{1}{V_{\mathcal{S}}} \int_{\mathcal{S}} 2 \, \tilde{\rho}_0 \, \frac{\partial\psi(\tilde{\mathbf{C}})}{\partial\tilde{\mathbf{C}}} \, dS \tag{246}$$

and higher-order and size-scale relevant stress

$$\mathbf{M}\left(\vartheta^k\right) = \frac{1}{V_{\mathcal{S}}} \int_{\mathcal{S}} 2 \, \zeta^\alpha \, \tilde{\rho}_0 \, \frac{\partial\psi(\tilde{\mathbf{C}})}{\partial\tilde{\mathbf{C}}} \, dS \, . \tag{247}$$

Then, substituting Eqs. (245) and (239) into Eq. (238) we end up with a micromorphic variational principle:

$$\int_{\mathcal{B}} \Big\{ \mathbf{S} : \delta\mathbf{C} + \mathbf{M}^\alpha : \delta\mathbf{K}_\alpha \Big\} \, dV - \mathcal{W}_{ext} = 0 \, . \tag{248}$$

The generalized principle of virtual work is supplemented by essential boundary conditions, the so-called Dirichlet boundary conditions

$$\mathbf{u} = \mathbf{h}_u \quad \text{on } \partial\mathcal{B}_D \tag{249}$$

$$\mathbf{a}_\alpha = \mathbf{h}_{\gamma,\alpha} \quad \text{on } \partial\mathcal{B}_D \, , \tag{250}$$

where $\mathbf{h}_u$ and $h_{\gamma,\alpha}$ are prescribed values at the boundary $\partial\mathcal{B}_D$.

Conclusion A general framework for a micromorphic continuum has been developed which is especially attractive for non-linear material behaviour. This approach motivates research into experimental verification of the mentioned extra degrees of freedom which is still elusive at large. While it is clear that generalized degrees of freedom and the internal lengths associated with them as well as their scale effects are related to the internal structure of the material, the direct mechanisms of deformations at the micro-scale which give rise to such degrees of freedom are widely subject to intensive research in many areas of mechanics and physics with many open questions. It is very likely that more than one mechanism could lead to a certain type of degrees of freedom. While these questions are beyond the scope of the present work we do acknowledge their importance. Multi-scale modelling and experimentation will be at the heart of any answer.

7.3 Inelastic micromorphic approach

Classically, inelastic formulations are based on decompositions of strain measures. Since generalized continua exhibit more than one strain or deformation measure the question arises as to how these can be decomposed into elastic and inelastic parts. Few suggestions were made in (Steinmann, 1994; Forest et al., 1997; Sansour, 1998b; Forest and Sievert, 2006; Chambon et al., 2004). These formulations remained, however, less satisfactory since the decomposition of the two deformation measures were, strictly speaking, independent of each other, which raises many questions regarding the adequate formulation of evolution laws for the inelastic parts.

The strength and generality of the micromorphic approach introduced in the previous section (Sec. 7.2) becomes apparent when fully non-linear formulations including inelastic deformations at finite strains are to be dealt with. One just needs to generalize the classical decomposition of the deformation gradient and carry it over to the generalized continuum. The theory will be used to model scale effects and shear band formation using a finite element method. The used inelastic material law can be any classical one which is now to be defined at the level of the micro-continuum. In what follows we adopt and tailor to our purposes the formulation of finite strain inelasticity based on unified constitutive models as developed in (Sansour and Kollmann (1997)). While the choice is convenient we stress that any alternative inelastic law could serve the purpose as well. Indeed, it exemplifies that any conventional materials can be incorporated in a very straightforward fashion.

Generalized kinematics of the elastic-inelastic body A point of departure for an inelastic formulation constitutes the multiplicative decomposition (Besseling, 1968; Kröner, 1960; Lee, 1969) of the generalized deformation gradient Eq. (233) into an elastic and an inelastic part

$$\tilde{\mathbf{F}} = \tilde{\mathbf{F}}_e \tilde{\mathbf{F}}_p. \tag{251}$$

For metals, the above decomposition is accompanied with the assumption $\tilde{\mathbf{F}}_p \in SL^+(3, \mathbb{R})$ which reflects the incompressibility of the inelastic deformations, where $SL^+(3, \mathbb{R})$ denotes the special linear group with determinant equal one.

The following generalized right *Cauchy-Green*-type deformation tensors are defined

$$\tilde{\mathbf{C}} := \tilde{\mathbf{F}}^T \tilde{\mathbf{F}}\,, \tag{252}$$

$$\tilde{\mathbf{C}}_e := \tilde{\mathbf{F}}_e^T \tilde{\mathbf{F}}_e\,, \tag{253}$$

$$\tilde{\mathbf{C}}_p := \tilde{\mathbf{F}}_p^T \tilde{\mathbf{F}}_p\,. \tag{254}$$

Since the deformation gradient $\tilde{\mathbf{F}}$ is also an element of $GL^+(3, \mathbb{R})$ with positive determinant, we can attribute to its time derivative a left and right rate

$$\dot{\tilde{\mathbf{F}}} = \tilde{\mathbf{l}} \tilde{\mathbf{F}}\,, \tag{255}$$

$$\dot{\tilde{\mathbf{F}}} = \tilde{\mathbf{F}} \tilde{\mathbf{L}}\,. \tag{256}$$

Both rates are mixed tensors (contravariant-covariant). They are related by means of the equation

$$\tilde{\mathbf{L}} = \tilde{\mathbf{F}}^{-1} \tilde{\mathbf{l}} \tilde{\mathbf{F}}\,. \tag{257}$$

Since $\tilde{\mathbf{F}}_p \in SL^+(3, \mathbb{R})$ we can define a right rate according to

$$\dot{\tilde{\mathbf{F}}}_p = \tilde{\mathbf{F}}_p \tilde{\mathbf{L}}_p \tag{258}$$

which proves more appropriate for a numerical treatment in a purely material context.

The constitutive model

General considerations Let $\tilde{\boldsymbol{\tau}}$ be the generalized *Kirchhoff* stress tensor. Consider the expression of the internal power

$$\mathcal{W} = \tilde{\boldsymbol{\tau}} : \tilde{\mathbf{l}}, \tag{259}$$

where $\tilde{\mathbf{l}}$ is defined in Eq. (255). The expression is rewritten using material tensors as

$$\mathcal{W} = \tilde{\mathbf{\Gamma}} : \tilde{\mathbf{L}}. \tag{260}$$

The comparison of Eq. (259) with Eq. (260) leads with the aid of Eq. (257) to the definition equation of the material stress tensor $\tilde{\mathbf{\Gamma}}$:

$$\tilde{\mathbf{\Gamma}} = \tilde{\mathbf{F}}^{\mathrm{T}} \tilde{\boldsymbol{\tau}} \tilde{\mathbf{F}}^{-T}. \tag{261}$$

The tensor $\tilde{\mathbf{\Gamma}}$ is, accordingly, the mixed variant pull-back of the generalized *Kirchhoff* tensor. It coincides with Noll's intrinsic stress tensor and determines up to a spherical part the Eshelby stress tensor.

A common feature of inelastic constitutive models is the introduction of phenomenological internal variables. We denote a typical internal variable as $\boldsymbol{Z}$. Assuming the existence of a free energy function according to $\psi = \psi(\tilde{\mathbf{C}}_e, \boldsymbol{Z})$, the localized form of the dissipation inequality for an isothermal process takes

$$\begin{aligned} \mathcal{D} &= \tilde{\boldsymbol{\tau}} : \tilde{\mathbf{l}} - \tilde{\rho}_{\mathrm{ref}} \dot{\psi} \\ &= \tilde{\mathbf{\Gamma}} : \tilde{\mathbf{L}} - \tilde{\rho}_{\mathrm{ref}} \dot{\psi} \geq 0, \end{aligned} \tag{262}$$

where ρ_{ref} is the density at the reference configuration.

Making use of the relation

$$\dot{\tilde{\mathbf{C}}}_e = \tilde{\mathbf{F}}_p^{-T} \tilde{\mathbf{L}}^T \tilde{\mathbf{C}} \tilde{\mathbf{F}}_p^{-1} + \tilde{\mathbf{F}}_p^{-T} \tilde{\mathbf{C}} \mathbf{L} \tilde{\mathbf{F}}_p^{-1} - \tilde{\mathbf{F}}_p^{-T} \tilde{\mathbf{L}}_p^T \tilde{\mathbf{C}} \tilde{\mathbf{F}}_p^{-1} - \tilde{\mathbf{F}}_p^{-T} \tilde{\mathbf{C}} \tilde{\mathbf{L}}_p \tilde{\mathbf{F}}_p^{-1} \tag{263}$$

one may derive

$$\dot{\psi} = 2 \tilde{\mathbf{C}} \tilde{\mathbf{F}}_p^{-1} \frac{\partial \psi}{\partial \tilde{\mathbf{C}}_e} \tilde{\mathbf{F}}_p^{-T} : (\tilde{\mathbf{L}} - \tilde{\mathbf{L}}_p) + \frac{\partial \psi}{\partial \boldsymbol{Z}} \cdot \dot{\boldsymbol{Z}}. \tag{264}$$

The insertion of Eq. (264) into Eq. (262) leads to

$$\begin{aligned} \mathcal{D} = &\left(\tilde{\mathbf{\Gamma}} - 2 \tilde{\rho}_{\mathrm{ref}} \tilde{\mathbf{C}} \tilde{\mathbf{F}}_p^{-1} \frac{\partial \psi(\tilde{\mathbf{C}}_e, \boldsymbol{Z})}{\partial \tilde{\mathbf{C}}_e} \tilde{\mathbf{F}}_p^{-T} \right) : \tilde{\mathbf{L}} \\ &+ 2 \tilde{\rho}_{\mathrm{ref}} \tilde{\mathbf{C}} \tilde{\mathbf{F}}_p^{-1} \frac{\partial \psi(\tilde{\mathbf{C}}_e, \boldsymbol{Z})}{\partial \tilde{\mathbf{C}}_e} \tilde{\mathbf{F}}_p^{-T} : \tilde{\mathbf{L}}_p - \tilde{\rho}_{\mathrm{ref}} \frac{\partial \psi(\tilde{\mathbf{C}}_e, \boldsymbol{Z})}{\partial \boldsymbol{Z}} \cdot \dot{\boldsymbol{Z}} \;\geq\; 0. \end{aligned}$$

By defining $\boldsymbol{Y}$ as the thermodynamical force conjugate to the internal variable $\boldsymbol{Z}$

$$\boldsymbol{Y} := -\tilde{\rho}_{\text{ref}} \frac{\partial \psi(\tilde{\mathbf{C}}_e, \boldsymbol{Z})}{\partial \boldsymbol{Z}}, \tag{265}$$

and making use of standard thermodynamical arguments, from Eq. (265) follows the elastic constitutive equation

$$\tilde{\boldsymbol{\Gamma}} = 2\tilde{\rho}_{\text{ref}} \tilde{\mathbf{C}} \tilde{\mathbf{F}}_p^{-1} \frac{\partial \psi(\tilde{\mathbf{C}}_e, \boldsymbol{Z})}{\partial \tilde{\mathbf{C}}_e} \tilde{\mathbf{F}}_p^{-T} = 2\tilde{\rho}_{\text{ref}} \tilde{\mathbf{F}}_p^T \tilde{\mathbf{C}}_e \frac{\partial \psi(\tilde{\mathbf{C}}_e, \boldsymbol{Z})}{\partial \tilde{\mathbf{C}}_e} \tilde{\mathbf{F}}_p^{-T} \tag{266}$$

as well as the reduced local dissipation inequality

$$\mathcal{D}_p := \tilde{\boldsymbol{\Gamma}} : \tilde{\mathbf{L}}_p + \boldsymbol{Y} \cdot \dot{\boldsymbol{Z}} \geq 0\,, \tag{267}$$

where Eq. (265) has been considered. $\mathcal{D}_p$ is the plastic dissipation function. From Eq. (267) follows as an essential result that the stress tensor $\tilde{\boldsymbol{\Gamma}}$ and the plastic rate $\tilde{\mathbf{L}}_p$ are conjugate variables. Observe that the tensor $\tilde{\mathbf{L}}_p$ is defined in Eq. (258).

The elastic constitutive model Further we assume that the elastic potential can be decomposed additively into one part depending only on the elastic generalized right *Cauchy-Green* deformation tensor $\tilde{\mathbf{C}}_e$ and the other one depending only on the internal variable $\boldsymbol{Z}$

$$\psi = \psi_{\text{e}}(\tilde{\mathbf{C}}_e) + \psi_{\text{Z}}(\boldsymbol{Z})\,. \tag{268}$$

Defining the logarithmic strain measure

$$\boldsymbol{\alpha} := \ln \tilde{\mathbf{C}}_e\,, \qquad \tilde{\mathbf{C}}_e = \exp \boldsymbol{\alpha} \tag{269}$$

and assuming that the material is elastically isotropic, one can prove that the relation holds

$$\tilde{\mathbf{C}}_e \frac{\partial \psi_{\text{e}}(\tilde{\mathbf{C}}_e)}{\partial \tilde{\mathbf{C}}_e} = \frac{\partial \psi_{\text{e}}(\boldsymbol{\alpha})}{\partial \boldsymbol{\alpha}}\,, \tag{270}$$

where $\psi_{\text{e}}(\boldsymbol{\alpha})$ is the potential expressed in the logarithmic strain measure $\boldsymbol{\alpha}$. Eq. (266) results then in

$$\tilde{\boldsymbol{\Gamma}} = 2\rho_{ref}\, \tilde{\mathbf{F}}_p^T \frac{\partial \psi_{\text{e}}(\boldsymbol{\alpha})}{\partial \boldsymbol{\alpha}} \tilde{\mathbf{F}}_p^{-T}\,. \tag{271}$$

Note that ψ_e is an isotropic function of $\boldsymbol{\alpha}$. The last equation motivates the introduction of a modified logarithmic strain measure

$$\bar{\boldsymbol{\alpha}} := \tilde{\mathbf{F}}_p^{-1}\boldsymbol{\alpha}\tilde{\mathbf{F}}_p\,. \tag{272}$$

Since the following relation for the exponential map holds

$$\tilde{\mathbf{F}}_p^{-1}(\exp\boldsymbol{\alpha})\tilde{\mathbf{F}}_p = \exp\bar{\boldsymbol{\alpha}}, \tag{273}$$

Eq. (271) takes

$$\tilde{\boldsymbol{\Gamma}} = 2\rho_{ref}\,\frac{\partial\psi(\bar{\boldsymbol{\alpha}})}{\partial\bar{\boldsymbol{\alpha}}}\,. \tag{274}$$

It is interesting to note that Eq. (273) together with Eqs. (269), (252), and (254) lead to a direct definition of $\bar{\boldsymbol{\alpha}}$. The relation holds

$$\bar{\boldsymbol{\alpha}} = \ln(\tilde{\mathbf{C}}_p^{-1}\tilde{\mathbf{C}}). \tag{275}$$

For computational simplicity a linear relation is assumed and therefore the elastic constitutive model Eq. (274) takes its final form

$$\tilde{\boldsymbol{\Gamma}} = K\,\mathrm{tr}\,\bar{\boldsymbol{\alpha}}^T\,\mathbf{1} + \mu\,\mathrm{dev}\bar{\boldsymbol{\alpha}}^T \tag{276}$$

where

$$\bar{\boldsymbol{\alpha}}^T = \ln(\tilde{\mathbf{C}}\tilde{\mathbf{C}}_p^{-1})\,, \tag{277}$$

and K is the bulk modulus and μ the shear modulus.

It should be stressed that the reduction of the elastic constitutive law to that given by Eq. (274) results in a considerable simplification of the computations necessary for the formulation of the weak form of equilibrium and its corresponding linearisation. The only assumption we used was the very natural one of having an internal potential depending on $\tilde{\mathbf{C}}_e$. The following reduction is carried out systematically.

Inelastic constitutive model The presented framework of generalized continua allows for the application of any set of classical constitutive laws. In what follows we confine ourselves to a unified constitutive law of the Bodner and Partom type as generalized in the first author's previous work (see e.g. Sansour and Kollmann (1997); Sansour and Wagner (2001)). We concluded from Eq. (267) that the tensors $\tilde{\boldsymbol{\Gamma}}$ and $\tilde{\mathbf{L}}_p$ are conjugate. Essentially we have to consider the stress tensor $\tilde{\boldsymbol{\Gamma}}$ as the driving stress quantity, while the plastic rate for which an evolution equation is to be formulated is taken to be $\tilde{\mathbf{L}}_p$. This leads to the following set of evolution equations

$$\tilde{\mathbf{L}}_p = \dot{\phi}\boldsymbol{\nu}^{\mathrm{T}}, \tag{278}$$

$$\dot{Z} = \frac{M}{Z_0}(Z_1 - Z)\dot{W}_p, \tag{279}$$

$$\dot{W}_p = \Pi_{\mathrm{dev}\tilde{\boldsymbol{\Gamma}}}\dot{\phi}(\Pi_{\mathrm{dev}\tilde{\boldsymbol{\Gamma}}}, Z), \qquad \Pi_{\mathrm{dev}\tilde{\boldsymbol{\Gamma}}} = \sqrt{\frac{3}{2}\mathrm{dev}\tilde{\boldsymbol{\Gamma}} : \mathrm{dev}\tilde{\boldsymbol{\Gamma}}}, \tag{280}$$

$$\dot{\phi} = \frac{2}{\sqrt{3}}D_0 \exp\left[-\frac{1}{2}\frac{N+1}{N}\left(\frac{Z}{\Pi_{\mathrm{dev}\tilde{\boldsymbol{\Gamma}}}}\right)^{2N}\right], \tag{281}$$

$$\boldsymbol{\nu} = \frac{3}{2}\frac{\mathrm{dev}\tilde{\boldsymbol{\Gamma}}}{\Pi_{\mathrm{dev}\tilde{\boldsymbol{\Gamma}}}}. \tag{282}$$

Here, Z_0, Z_1, D_0, N, M are material parameters. The choice of the transposed quantity in Eq. (278) reflects the form given by associative viscoplasticity, when the classical flow functions are generalized and formulated in terms of nonsymmetric quantities.

Numerical examples In this section two numerical examples are presented to demonstrate the applicability of the micromorphic theory. In this specific case the micro-deformation is assumed to be one-dimensional and of the following form:

$$\tilde{\mathbf{x}} = \mathbf{x}\left(\vartheta^k, t\right) + \zeta\,\left(1 + \zeta\,\chi\left(\vartheta^k, t\right)\right)\mathbf{a}\left(\vartheta^k, t\right). \tag{283}$$

Obviously the micro-continuum is assumed one-dimensional. The vector $\mathbf{a}$ is defined at the actual configuration and it is the map of a one-dimensional vector defined in the generalized reference configuration. We take this direction to be a unit vector which lies in parallel to the $x-$axis. The material parameters, typical for metals, are chosen as follows:

$$\begin{aligned}
K &= 1.64206E02\ \mathrm{N/mm^2},\\
\mu &= 1.6194E02\ \mathrm{N/mm^2},\\
D_0 &= 10000\ 1/\mathrm{sec},\\
Z_0 &= 1150\ \mathrm{N/mm^2},\\
Z_1 &= 1400\ \mathrm{N/mm^2},\\
N &= 1,\\
M &= 100.
\end{aligned}$$

The inelastic parameters are reported in the literature for titanium, e.g. Bodner and Partom (1975).

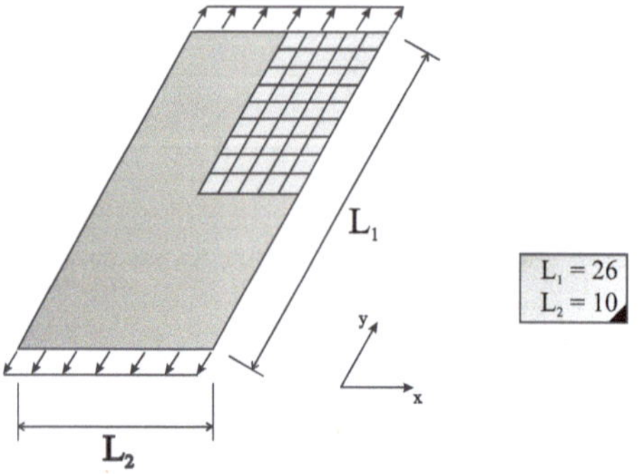

Figure 14. problem definition of a sheet under homogenous tension

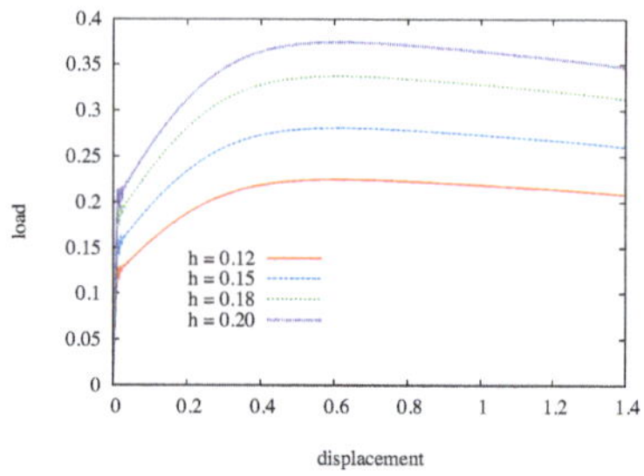

Figure 15. load displacement graph illustrating size-scale effects for different values of the internal length parameter h

Simple tension The first example is a thin sheet subjected to simple tension as shown in Fig. 14. One quarter of the sheet is discretised using 5×5 enhanced 4-node finite elements of the type developed in Sansour and Kollmann (1998), which, in this specific case, are equivalent to three-dimensional enhanced 8-node elements with thickness 1. The aim here is to illustrate size-scale effects in the viscoplastic regime at homogenous deformations. We consider four different internal length-scale parameters, denoted by h, which are nothing but the size/length of the microspace $\mathcal{S}$, and take them to be of the values 0.12, 0.15, 0.18 and $0.20\,mm$. The time step used is $0.1sec$ for the displacement at the top increasing by a velocity of $0.02N/sec$. While the specimen is under force loading with no prescribed displacements at the loading side, the computations are carried out displacement-controlled with the value of the external loading being scaled and determined to provide the prescribed displacement velocity.

The corresponding load-displacement graphs are depicted in Fig. 15. With increasing internal length-scale parameter it can be clearly seen that the onset of the plastic deformation takes place at larger loading values. During the plastic deformation the relative loading offset between the curves is maintained.

Now, this case of simple tension particularly illustrates the attractiveness of the proposed generalized theory as it predicts scale effects also in a homogenously deforming specimen. In fact alternative theories, such as micropolar (Cosserat) or strain gradient approaches, lack the means to predict this kind of scale effect. This is clear, because the former necessitates the rotation gradient and the latter the deformation gradient of higher order not to vanish. In this example, however, both of them do not arise and consequently, no scale effects would be observed.

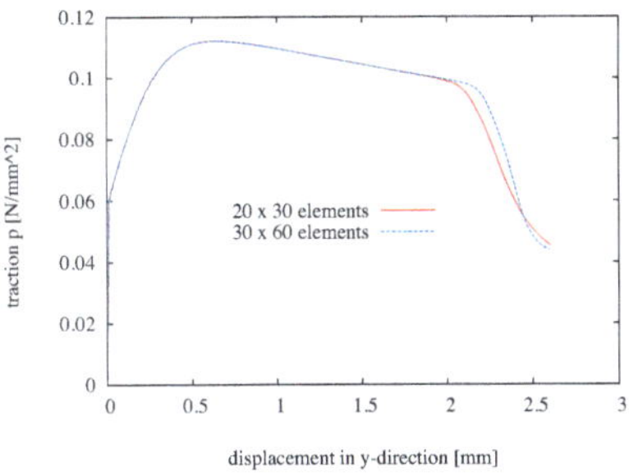

Figure 16. load displacement graph illustrating computations using 20×30 elements and 30×60 elements

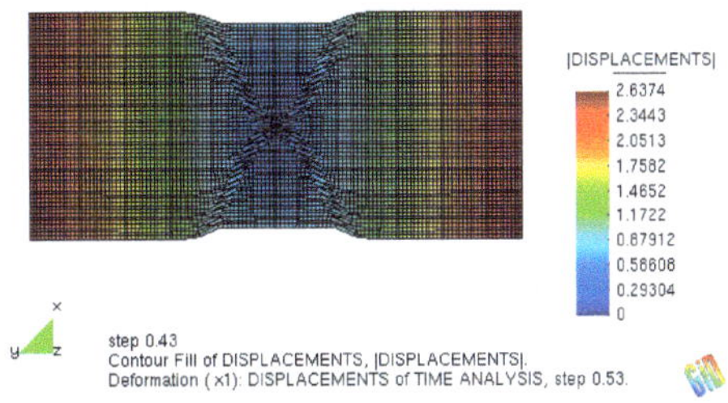

Figure 17. final deformed configuration displaying the shearband formation using 30×60 elements

Shearband formation The second numerical example is the same as before in terms of geometry, loading, and time step - a thin sheet under tension as displayed in Fig. 14. Shearbanding is initialized by decreasing the material parameter Z_0 by 10% within the first element (at the centre of the specimen). The internal length is considered to be $h = 0.1\,mm$. One quarter of the sheet is modeled using 20×30 and 30×60 4-node elements of the type described above. From the load-displacement curve in Fig. (16) it is clear that heavy softening related to the shearband formation takes place. This softening is independent of the mesh since both meshes give essentially the same results. The deformed configuration is pictured in Fig. (17).

Note that the constitutive law is of the viscoplastic type. As demonstrated by (Sluys, 1992) viscous effects have regularisation effect too. However, the Bodner-Partom model covers in the limit the time-independent case as well. The present choice of material parameters together with the applied loading velocity renders the time-dependent effect rather very small. Also, from the previous example we can conclude that the scale effect due to the micromorphic formulation is dominant here.

Conclusion In Sec. 7.3 the delicate issue of incorporating dissipative inelastic material behaviour into the framework of generalized continua is addressed. The presented method is attractive as it allows for any well established constitutive law to be considered without major modifications.

Numerical examples showed clearly that the formulation is capable of describing scale effects also in the case of homogeneous deformation - so far an open issue in the area of micro-mechanics.

7.4 Generalized micropolar continuum

The generalized continuum presented in Secs. 7.5 and 7.5 feature both a generalized deformation which is able to describe the rotation of the micro-continuum attached to each material point. These approaches hold the benefit that the displacement field is the only unknown. This is achieved by considering the deformation of the microspace to be dependent on the first-order derivatives of the macro-deformation field. Rigid body rotations of the micro-continuum, however, can also be described by making use of a rotation tensor (Sansour, 1998a). This will be outlined in the following.

Generalized deformation and strain measures The directors of the micro-space $\mathbf{a}_\alpha\left(\vartheta^k, t\right)$ (Eq. 226) are now associated with a macroscopic rotation tensor $\mathbf{Q}$ which is considered to be independent of the map $\varphi(\mathbf{X}, t)$ (Eq. 88) and element of the proper orthogonal transformation group $SO(3)$. Therefore, the micro-directors only undergo rotations and the micro-continuum behaves as a rigid body. The use of the rotation tensor as another unknown field induces three extra rotational degrees of freedom, i.e. in contrast to the strain gradient-based approach in Sec. 7.5, the rotation of the micro-directors is not hidden within the formulation.

This is achieved substituting the rotation tensor $\mathbf{Q}$ for tensor $\mathbf{A}$ in Eq. (227) and the generalized deformation (Eq. 226) takes the following form

$$\tilde{\mathbf{x}}\left(\vartheta^k, \zeta^\alpha, t\right) = \mathbf{x}\left(\vartheta^k, t\right) + \zeta^\alpha\, \mathbf{Q}\left(\vartheta^k, t\right) \mathbf{I}_\alpha \,. \tag{284}$$

This establishes the analogy to the classical *Cosserat* continuum and refers to a *micropolar* deformation description leaning on the definition of Eringen (1999). Consequently, each material point P is equipped with extra three rotational degrees of freedom additional to displacements, where the micro-deformation is restricted to rigid body motion.

A variable orientation of the microspace with respect to the macrospace can be modelled by re-defining Eq. (284) as follows

$$\tilde{\mathbf{x}} = \mathbf{x}\left(\vartheta^k, t\right) + \zeta^\alpha\, H_{\beta\alpha}\, \mathbf{Q}\left(\vartheta^k, t\right) \mathbf{I}_\beta \,, \tag{285}$$

where in general $H_{\beta\alpha}$ denotes a constant rotation matrix of dimension (3×3).

Assuming $\mathbf{I}_\alpha$ to be independent of ϑ^i and ζ^α the spatial derivatives of the position vector in the current configuration with respect to the macro-coordinates ϑ^i are given by

$$\tilde{\mathbf{x}}_{,i}\left(\vartheta^k,\zeta^\alpha,t\right) = \frac{\partial\tilde{\mathbf{x}}}{\partial\vartheta^i} = \mathbf{x}_{,i}\left(\vartheta^k,t\right) + \zeta^\alpha\,\mathbf{Q}_{,i}\left(\vartheta^k,t\right)\mathbf{I}_\alpha\,, \tag{286}$$

and with respect to the micro-coordinates ζ^α by

$$\tilde{\mathbf{x}}_{,\alpha}\left(\vartheta^k,t\right) = \frac{\partial\tilde{\mathbf{x}}}{\partial\zeta^\alpha} = \mathbf{Q}\left(\vartheta^k,t\right)\mathbf{I}_\alpha\,. \tag{287}$$

The generalized deformation gradient which describes the mapping of the tangent space $\mathcal{TG}$ in the reference configuration to a corresponding tangent space in the current configuration $\mathcal{TG}_t$ is with Eq. (286) and Eq. (287) expressed by

$$\tilde{\mathbf{F}}\left(\vartheta^k,\zeta^\alpha,t\right) = \Big(\mathbf{x}_{,i} + \zeta^\alpha\mathbf{Q}_{,i}\mathbf{I}_\alpha\Big)\otimes\tilde{\mathbf{G}}^i + \mathbf{Q}\mathbf{I}_\alpha\otimes\mathbf{I}^\alpha\,. \tag{288}$$

A natural passing to a classical continuum can be achieved, if, in the following, the strain measures are required to be invariant with respect to the group $SO(3)$, very much as in the *Cosserat continuum* (see Secs. 4.3 and 5.1). Following the above requirement and leaning on the classical *Cosserat continuum* a stretch-type strain measure is chosen involving $\tilde{\mathbf{F}}$ (Eq. 288) as well as an independent macroscopical rotation tensor $\mathbf{R}$:

$$\begin{aligned}\tilde{\mathbf{U}}\left(\vartheta^k,\zeta^\alpha,t\right) \;&:=\; \mathbf{R}^T\left(\vartheta^k,t\right)\tilde{\mathbf{F}}\left(\vartheta^k,\zeta^\alpha,t\right) = \\ &=\; \Big(\mathbf{R}^T\mathbf{x}_{,i} + \zeta^\alpha\mathbf{R}^T\mathbf{Q}_{,i}\mathbf{I}_\alpha\Big)\otimes\tilde{\mathbf{G}}^i + \mathbf{I}_\alpha\otimes\mathbf{I}^\alpha\,,\end{aligned} \tag{289}$$

where $\mathbf{R}$ is given by Eq. (68). The first term in Eq. (289) is an analogy to the classical stretch tensor and is denoted by

$$\mathbf{U}\left(\vartheta^k,t\right) = \mathbf{R}^T\mathbf{x}_{,i}\otimes\tilde{\mathbf{G}}^i\,, \tag{290}$$

whereas the second term is a set of curvature tensors

$$\mathbf{K}_\alpha\left(\vartheta^k,t\right) = \mathbf{R}^T\mathbf{Q}_{,i}\mathbf{I}_\alpha\otimes\tilde{\mathbf{G}}^i\,. \tag{291}$$

Since the the third term is constant, Eq. (289) is re-defined as

$$\tilde{\mathbf{U}}\left(\vartheta^k,\zeta^\alpha,t\right) = \mathbf{U} + \zeta^\alpha\,\mathbf{K}_\alpha\,. \tag{292}$$

In the following it is assumed $\mathbf{Q} = \mathbf{R}$. With $\mathbf{R} \in SO(3)$ the tensor products $\mathbf{R}^T \mathbf{R}_{,i}$ are skew-symmetric, and with Eq. (129) as axial vector, Eq. (291) can be alternatively expressed as

$$\mathbf{K}_\alpha \left(\vartheta^k, t\right) = \mathbf{k}_i \left(\vartheta^k, t\right) \times \mathbf{I}_\alpha \otimes \mathbf{G}^i \,, \tag{293}$$

or

$$\mathbf{K}_\alpha \left(\vartheta^k, t\right) = -\frac{1}{2} \boldsymbol{\epsilon} : \mathbf{R}^T \left(\vartheta^k, t\right) \mathbf{R}_{,i} \left(\vartheta^k, t\right) \mathbf{I}_\alpha \otimes \tilde{\mathbf{G}}^i \,. \tag{294}$$

Due the analogy to a micropolar continuum and the approach presented in this section may be therefore called a generalized micropolar continuum.

Generalized variational formulation and the equilibrium equations Analog to the strain gradient approach before, a micropolar variational principle is established based on the generalized strain tensor $\tilde{\mathbf{C}}$ for the static case and considering only mechanical processes. The external virtual work $\mathcal{W}_{ext}$ is defined in the Lagrangian form as follows

$$\mathcal{W}_{ext} = \int_{\mathcal{B}} \mathbf{b} \left(\vartheta^k\right) \cdot \delta \mathbf{u} \left(\vartheta^k\right) \, dV + \int_{\mathcal{B}} \mathbf{l} \left(\vartheta^k\right) \cdot \mathbf{w} \left(\vartheta^k\right) \, dV \tag{295}$$

$$+ \int_{\partial \mathcal{B}_N} \mathbf{t}^{(\mathbf{n})} \left(\eta^k\right) \cdot \delta \mathbf{u} \, dA + \int_{\partial \mathcal{B}_N} \mathbf{q}^{(\mathbf{n})} \left(\eta^k\right) \cdot \mathbf{w} \, dA \tag{296}$$

where the external body force and moment $\mathbf{b}$ and $\mathbf{l}$, respectively, acting in $\mathcal{B}$ and the external traction and surface moment $\mathbf{t}^{(\mathbf{n})}$ and $\mathbf{q}^{(\mathbf{n})}$, respectively, acting on the Neumann boundary $\partial \mathcal{B}_N$ are obtained by integrating over the micro-space $\mathcal{S}$:

$$\mathbf{b} \left(\vartheta^k\right) = \frac{1}{V_{\mathcal{S}}} \int_{\mathcal{S}} \tilde{\rho}_0 \, \tilde{\mathbf{b}} \left(\vartheta^k, \zeta^\beta\right) \, dS \tag{297}$$

$$\mathbf{l} \left(\vartheta^k\right) = \frac{1}{V_{\mathcal{S}}} \int_{\mathcal{S}} \tilde{\rho}_0 \, \zeta^\alpha \, \mathbf{R} \mathbf{I}_\alpha \times \tilde{\mathbf{b}} \left(\vartheta^k, \beta^\alpha\right) \, dS \tag{298}$$

$$\mathbf{t}^{(\mathbf{n})} \left(\eta^k\right) = \frac{1}{V_{\mathcal{S}}} \int_{\mathcal{S}} \tilde{\rho}_0 \, \tilde{\mathbf{t}}^{(\mathbf{n})} \left(\eta^k, \zeta^\beta\right) \, dS \tag{299}$$

$$\mathbf{q}^{(\mathbf{n})} \left(\eta^k\right) = \frac{1}{V_{\mathcal{S}}} \int_{\mathcal{S}} \tilde{\rho}_0 \, \zeta^\alpha \, \mathbf{R} \mathbf{I}_\alpha \times \tilde{\mathbf{t}}^{(\mathbf{n})} \left(\eta^k, \zeta^\beta\right) \, dS \,. \tag{300}$$

The internal virtual power in the Lagrangian form is given by

$$\delta\Psi = \int_{\mathcal{B}}\int_{\mathcal{S}} \tilde{\rho}_0 \frac{\partial\psi(\tilde{\mathbf{U}})}{\partial\tilde{\mathbf{U}}}\, dS\, dV\,, \tag{301}$$

and with Eq. 290 and Eq. 294 it is further redrafted to

$$\delta\Psi = \int_{\mathcal{B}} \frac{1}{2}\left\{\boldsymbol{n} : \delta\mathbf{U} + \boldsymbol{m}\,\mathbf{G}^i \cdot \delta\mathbf{k}_i\right\} dV \tag{302}$$

where

$$\boldsymbol{n}\left(\vartheta^k\right) = \frac{1}{V_{\mathcal{S}}}\int_{\mathcal{S}} 2\,\tilde{\rho}_0 \frac{\partial\psi\left(\tilde{\mathbf{U}}\right)}{\partial\tilde{\mathbf{U}}}\, dS \tag{303}$$

$$\boldsymbol{m}\left(\vartheta^k\right) = \frac{1}{V_{\mathcal{S}}}\int_{\mathcal{S}} 2\tilde{\rho}_0\,\zeta^\alpha\,\mathbf{I}_\alpha \times \frac{\partial\psi\left(\tilde{\mathbf{U}}\right)}{\partial\tilde{\mathbf{U}}}\mathbf{G}^i\, dS\,. \tag{304}$$

Note that $\boldsymbol{n}$ is an analogy to the so-called *Biot-Lurje* stress tensor which is energy conjugate to the classical stretch tensor $\mathbf{U}$ and with the *first Piola-Kirchhoff stress tensor* $\mathbf{P}$ defined as

$$\boldsymbol{n} = \mathbf{R}^T\mathbf{P}\,. \tag{305}$$

$\boldsymbol{m}$ may be interpreted as a couple stress tensor. For details on the variations $\delta\mathbf{U}$ and $\delta\mathbf{k}_i$ the reader is referred to Sec. 5.1.

Then, with Eqs. (296) and (302) a generalized variational formulation with free rotations is expressed as

$$\int_{\mathcal{B}} \frac{1}{2}\left\{\boldsymbol{n} : \delta\mathbf{U} + \boldsymbol{m}\,\mathbf{G}^i \cdot \delta\mathbf{k}_i\right\} dV - \mathcal{W}_{ext} = 0\,, \tag{306}$$

which is supplemented by following Dirichlet boundary conditions for displacement and rotation field, respectively:

$$\mathbf{u} = \mathbf{h}_u\,, \quad \text{on } \partial\mathcal{B}_D \tag{307}$$

$$\gamma = \mathbf{h}_\gamma\,, \quad \text{on } \partial\mathcal{B}_D\,. \tag{308}$$

The governing equations of this functional can be derived, if we apply *Gauss's* divergence theorem to Eq. (306) and we obtain

$$-\frac{1}{\sqrt{G}}\left(\sqrt{G}\,\mathbf{R}\boldsymbol{n}\mathbf{G}^i\right)_{,i} - \mathbf{b} = \mathbf{0}\,, \qquad \text{in} \quad \mathcal{B} \tag{309}$$

$$-\mathbf{x}_{,i} \times \mathbf{R}\boldsymbol{n} - \frac{1}{\sqrt{G}}\left(\sqrt{G}\,\mathbf{R}\boldsymbol{m}\mathbf{G}^i\right)_{,i} - \mathbf{l} = \mathbf{0}\,, \qquad \text{in} \quad \mathcal{B}\,, \tag{310}$$

where, as in the previous sections G denotes the determinate of the *Riemannian* metric coefficients $G_{ij} = \mathbf{G}_i \cdot \mathbf{G}_j$ and the projection map $\pi_0^*(\tilde{\mathbf{G}}_i)$ is defined to be the identity, such that $\tilde{\mathbf{G}}_i := \mathbf{G}_i$.

The corresponding natural boundary conditions are

$$\mathbf{R}\boldsymbol{n}\mathbf{n} = \mathbf{t}^{(\mathbf{n})}, \qquad \text{on} \quad \mathcal{B}_N \tag{311}$$

$$\mathbf{R}\boldsymbol{m}\mathbf{n} = \mathbf{q}^{(\mathbf{n})}, \qquad \text{on} \quad \mathcal{B}_N\,. \tag{312}$$

Numerical examples As in Sec. 7.5, the numerical examples in this section are modelled making use of a *moving least square*-based meshfree method (Skatulla and Sansour, 2008) and the numerical integration over the micro-continuum $\mathcal{S}$ is carried out with the help of the *Gauss quadrature*, the order of which is chosen to be second according to the used second-order basis polynomial of the meshfree approximation. The enforcement of the Dirichlet boundary conditions is achieved using the penalty method. The

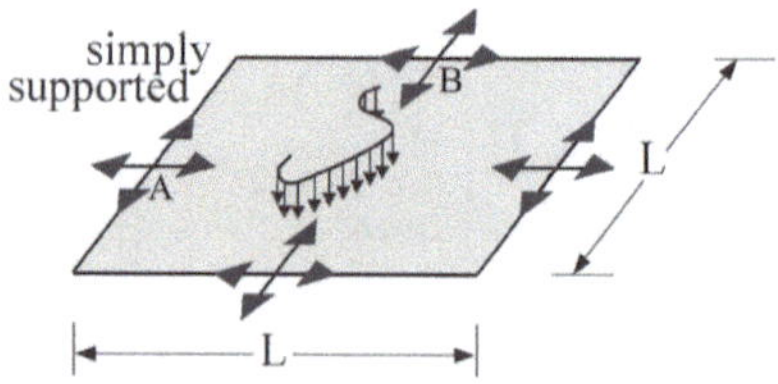

Figure 18. *problem definition*

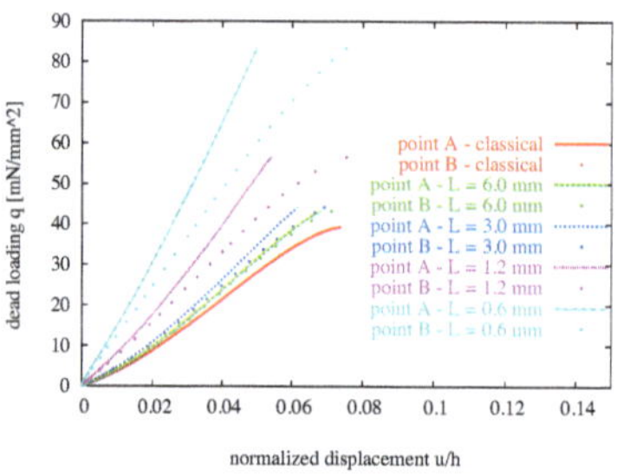

Figure 19. *load displacement diagram with $l_1 = 21\,\mu m$*

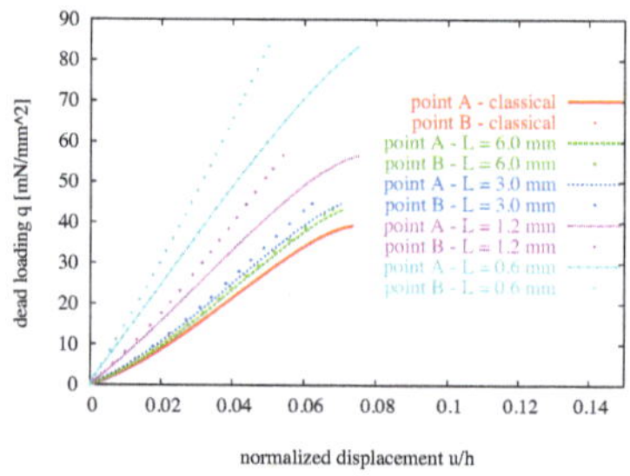

Figure 20. *load displacement diagram with $l_2 = 21\,\mu m$*

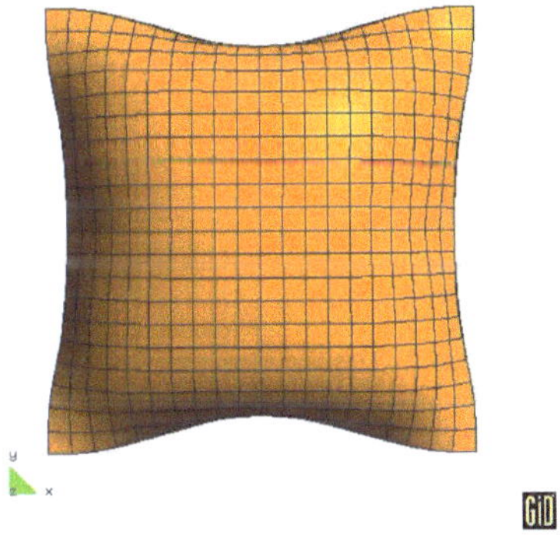

Figure 21. *deformed configuration at loading parameter* $83.41\, mN/mm^2$ *with* $l_1 = 21\, \mu m$

Figure 22. *deformed configuration at loading parameter* $83.41\, mN/mm^2$ *with* $l_2 = 21\, \mu m$

problem studied in the following is a square sheet which is considered in four different scaling levels with lengths $L = 6.0\, mm$, $L = 3.0\, mm$, $L = 1.2\, mm$ and $L = 0.6\, mm$. The ratio of the plate's length to its height $L/h = 300$ is kept constant for all four sheets so that the results are comparable. The sheets are simply supported at their outer edges in vertical direction and subjected to a dead loading depicted in Fig. 18. Only one quarter of the sheet is modelled with $11 \times 11 \times 3$ particles in longitudinal and thickness direction, respectively, making use of the appropriate symmetry conditions.

The material behaviour is assumed to be hyperelastic and the *Saint-Venant-Kirchhoff* constitutive model is utilized which involves as material parameters *Young's modulus* $E = 1.55\, GPa$ and *Poission's ratio* $\nu = 0.3$.

Making use of the generalized micropolar variational formulation (Eq. 306) it is now the aim to illustrate oriented deformation behaviour due to a particular specification of the micro-continuum $\mathcal{S}$. That is, the micro-continuum is first assumed to be one-dimensional and its basis vector in the reference configuration is taken to be $\mathbf{I}_1 = \mathbf{G}_1$. The corresponding internal length scale parameter is determined as $l_1 = 21\, \mu m$. The final deformed configuration of the smallest sheet with $L = 0.6\, mm$ displayed in Fig. 21 clearly shows that the magnitude of the displacement of point A in x-direction is larger than the one of point B in y-direction which the classical *Green strain tensor*-based model (Skatulla and Sansour, 2008) predicts for all scaling levels to be equal, illustrated in Fig. 19. It can also be seen that the difference in displacement of point A in x-direction denoted by the continuous graphs and of point B in y-direction denoted by the dotted line is increasing for the smaller sheets. The curves of each scaling level are indicated by choice of different colors, where the red graphs represent the classical solutions.

Subsequently, the basis vector of $\mathcal{S}$ in the reference configuration is now alternated to $\mathbf{I}_1 = \mathbf{G}_2$ and the corresponding internal length scale-parameter is chosen to be $l_2 = 21\,\mu m$. Again, the final deformed configuration of the smallest sheet with $L = 0.6\,mm$ illustrated in Fig. 22 indicates a distinct difference in the displacement of point A in x-direction and of point B in y-direction. Now however, the displacement magnitude of point B is larger than the one of point A shown also in Fig. 20.

7.5 A strain gradient formulation

In the previous section (Sec. 7.5) it is demonstrated how to formulate a generalized deformation description which incorporates an oriented micro-continuum, without adding extra degrees of freedom beyond displacements. However, due to the use of the vector cross product the resulting computational algorithms are rather involved. In the following an alternative formulation is outlined which also utilizes higher-order displacement field derivatives to address non-local material behaviour.

Generalized deformation and strain measures To avoid the incorporation of additional degrees of freedom, other than the displacement degrees of freedom, but to still maintain the simplicity of the formulation we redefine the directors of the micro-continuum $\mathbf{a}_\alpha$ (227) as follows

$$\mathbf{a}_\alpha = \frac{\partial \mathbf{x}}{\partial \vartheta^\alpha} = \mathbf{g}_\alpha\left(\vartheta^k, t\right) , \tag{313}$$

where $\mathbf{g}_\alpha$ are the basis vectors of the macro-space in the current configuration $\mathcal{B}_t$. Then the generalized deformation field Eq. (226) takes the following form

$$\tilde{\mathbf{x}} = \mathbf{x}\left(\vartheta^k, t\right) + \zeta^\alpha\,\mathbf{g}_\alpha\left(\vartheta^k, t\right) . \tag{314}$$

In the reference configuration we have:

$$\tilde{\mathbf{X}} = \mathbf{X}\left(\vartheta^k\right) + \zeta^\alpha\,\mathbf{G}_\alpha\left(\vartheta^k\right) \qquad \text{with } \mathbf{G}_\alpha\left(\vartheta^k\right) = \frac{\partial \mathbf{X}}{\partial \vartheta^\alpha} , \tag{315}$$

where $\mathbf{G}_\alpha$ are the basis vectors of the macro-space in the reference configuration $\mathcal{B}$. Note, even if the micro-continuum $\mathcal{S}$ is defined by the macroscopic basis vectors $\mathbf{g}_\alpha$, it is important to realize that the dimension of the micro-space does not have to coincide with the dimension of the macro-space, but must not be larger than three. Moreover, the applicable directors of the microspace can be freely associated with the basis vectors of the macrospace $\mathbf{g}_\alpha$. A variable orientation of the microspace with respect to the macrospace can be modelled by defining Eq. (314) as follows

$$\tilde{\mathbf{x}} = \mathbf{x}\left(\vartheta^k, t\right) + \zeta^\alpha\, Q_{\beta\alpha}\, \mathbf{g}_\beta\,, \tag{316}$$

where in general $Q_{\beta\alpha}$ denotes a constant rotation matrix of dimension (3×3). The spatial derivatives of the position vector in the current configuration with respect to the macro-coordinates ϑ^i are given by

$$\tilde{\mathbf{x}}_{,i} = \frac{\partial \tilde{\mathbf{x}}}{\partial \vartheta^i} = \mathbf{x}_{,i}\left(\vartheta^k, t\right) + \zeta^\alpha\, \mathbf{g}_{\alpha,i}\left(\vartheta^k, t\right)\,, \tag{317}$$

and with respect to the micro-coordinates ζ^α by

$$\tilde{\mathbf{x}}_{,\alpha} = \frac{\partial \tilde{\mathbf{x}}}{\partial \zeta^\alpha} = \mathbf{g}_\alpha\left(\vartheta^k, t\right)\,. \tag{318}$$

The generalized deformation gradient tensor is then expressed as follows

$$\tilde{\mathbf{F}} = \left(\mathbf{x}_{,i} + \zeta^\alpha\, \mathbf{g}_{\alpha,i}\right) \otimes \tilde{\mathbf{G}}^i + \mathbf{g}_\alpha \otimes \mathbf{I}^\alpha\,. \tag{319}$$

Note that second order derivatives have entered the formulation. This is a direct consequence of restricting the degrees of freedom to those of a classical continuum, while still keeping to a generalized deformation.

Analog to Sec. 7.5 a generalized *Cauchy-Green* deformation tensor is formulated based on the generalized deformation gradient as follows

$$\tilde{\mathbf{C}} = \tilde{\mathbf{F}}^T \tilde{\mathbf{F}}\,. \tag{320}$$

Neglecting higher order terms in ζ^α we can identify three strain measures

$$\tilde{\mathbf{C}}^I = \left(\mathbf{x}_{,k} \cdot \mathbf{x}_{,l} + \zeta^\alpha\, \left(\mathbf{x}_{,k} \cdot \mathbf{g}_{\alpha,l} + \mathbf{g}_{\alpha,k} \cdot \mathbf{x}_{,l}\right)\right)\; \tilde{\mathbf{G}}^k \otimes \tilde{\mathbf{G}}^l \tag{321}$$

$$\tilde{\mathbf{C}}^{II} = \left(\mathbf{x}_{,k} \cdot \mathbf{g}_\beta + \zeta^\alpha\, \mathbf{g}_{\alpha,k} \cdot \mathbf{g}_\beta\right)\; \left(\tilde{\mathbf{G}}^k \otimes \mathbf{I}^\beta + \mathbf{I}^\beta \otimes \tilde{\mathbf{G}}^k\right) \tag{322}$$

$$\tilde{\mathbf{C}}^{III} = \left(\mathbf{g}_\alpha \cdot \mathbf{g}_\beta\right)\; \mathbf{I}^\alpha \otimes \mathbf{I}^\beta\,. \tag{323}$$

For the sake of mathematical simplicity and computational performance we want to focus on $\tilde{\mathbf{C}}^I$ and neglect $\tilde{\mathbf{C}}^{II}$ and $\tilde{\mathbf{C}}^{III}$ in following. This is reasonable, because $\tilde{\mathbf{C}}^I$ contains the conventional as well as the higher-order strains, where the latter are linked to non-local deformation behaviour. Consequently, we define

$$\tilde{\mathbf{C}} := \tilde{\mathbf{C}}^I = \mathbf{C} + \zeta^\alpha \mathbf{K}_\alpha \,, \tag{324}$$

where $\mathbf{C}$ represents the conventional right *Cauchy-Green* deformation tensor and $\mathbf{K}_\alpha$ the higher order contributions of Eq. (321).

In summary, due to deformation of the micro-directors (Eq. 313) the micro-continuum undergoes rotation, stretch and shear and we find an analogy to the so-called *micromorph continuum* (Eringen, 1999). However, in this approach the micro-directors are hidden within the generalized deformation formulation (Eq. 314) and therefore, no additional degrees of freedom are involved. Any generalized strain measure derived from Eq. (314) can only be linear with respect to the micro-coordinate ζ^α, as the generalized deformation is linear in ζ^α. The micro-deformation $\boldsymbol{\xi}$ is obviously dependent on the macro deformation gradient $\mathrm{Grad}\,\mathbf{x}$, that is, the generalized deformation $\tilde{\mathbf{x}}$ is completely described by the unknown macro-displacement field $\mathbf{u}$ and its first derivatives.

In the following we want to confine ourselves to the quasi-static case and consider no time dependency.

Generalized variational formulation and the equilibrium equations

In a similar fashion as in Sec. 7.5, a generalized variational principle is established based on the generalized strain tensor $\tilde{\mathbf{C}}$ (Eq. 324). For this consider a non-linear boundary value problem in the domain $\mathcal{B} \times \mathcal{S}$ with the boundary $\partial\mathcal{B} \times \mathcal{S}$, where $\mathbf{n}$ defines the normal vector on $\partial\mathcal{B}$. For the static case and considering only mechanical processes, the *first law of thermodynamics* provides the following variational statement

$$\delta\Psi - \mathcal{W}_{ext} = 0 \,. \tag{325}$$

The external virtual work in the Lagrangian form is defined as follows

$$\mathcal{W}_{ext} = \int_{\mathcal{B}} \mathbf{b} \cdot \delta\mathbf{u} \; dV + \int_{\partial\mathcal{B}_N} \mathbf{t}^{(\mathbf{n})} \cdot \delta\mathbf{u} \; dA \tag{326}$$

where the external body force $\mathbf{b}$ acting in $\mathcal{B}$ and the external traction $\mathbf{t}^{(\mathbf{n})}$ acting on the Neumann boundary $\partial\mathcal{B}_N \subset \partial\mathcal{B}$ are obtained by integrating over the micro-space $\mathcal{S}$:

$$\mathbf{b}\left(\vartheta^k\right) = \frac{1}{V_{\mathcal{S}}} \int_{\mathcal{S}} \tilde{\rho}_0 \, \tilde{\mathbf{b}}\left(\vartheta^k, \zeta^\beta\right) \, dS \tag{327}$$

$$\mathbf{t}^{(\mathbf{n})}\left(\eta^r\right) = \frac{1}{V_{\mathcal{S}}} \int_{\mathcal{S}} \tilde{\rho}_0 \, \tilde{\mathbf{t}}^{(\mathbf{n})}\left(\eta^r, \zeta^\beta\right) \, dS \tag{328}$$

considering only micro-body forces $\tilde{\mathbf{b}}$ and $\tilde{\mathbf{t}}^{(\mathbf{n})}$, respectively, and η^r, $r = 1, 2$ represents the coordinate chart on $\partial\mathcal{B}$. The density of the micro-space in the reference configuration is denoted by $\tilde{\rho}_0$.

Furthermore, we assume now that the body under consideration $\mathcal{B} \times \mathcal{S}$ is hyperelastic and possesses an elastic potential Ψ represented by the stored strain energy per unit mass ψ. If the material is in addition homogeneous, Ψ depends upon position in $\mathcal{B} \times \mathcal{S}$ exclusively through the generalized strain measure $\tilde{\mathbf{C}}$. Thus, the internal virtual power in the Lagrangian form reads as follows

$$\delta\Psi = \int_{\mathcal{B}} \int_{\mathcal{S}} \tilde{\rho}_0 \, \frac{\partial \psi(\tilde{\mathbf{C}})}{\partial \tilde{\mathbf{C}}} : \delta\tilde{\mathbf{C}} \, dS \, dV \,, \tag{329}$$

where $\tilde{\mathbf{S}}$ is a generalised *second Piola-Kirchhoff*-like stress tensor. Integrating the internal virtual power (Eq. 329) over the micro-space leads to

$$\delta\Psi = \int_{\mathcal{B}} \frac{1}{2} \left\{ \mathbf{S} : \delta\mathbf{C} + \mathbf{M}^\alpha : \delta\mathbf{K}_\alpha \right\} dV \,, \tag{330}$$

with the force stress

$$\mathbf{S}\left(\vartheta^k\right) = \frac{1}{V_{\mathcal{S}}} \int_{\mathcal{S}} 2\, \tilde{\rho}_0 \, \frac{\partial \psi(\tilde{\mathbf{C}})}{\partial \tilde{\mathbf{C}}} \, dS \tag{331}$$

and higher-order and size-scale relevant stress

$$\mathbf{M}\left(\vartheta^k\right) = \frac{1}{V_{\mathcal{S}}} \int_{\mathcal{S}} 2\, \zeta^\alpha \, \tilde{\rho}_0 \, \frac{\partial \psi(\tilde{\mathbf{C}})}{\partial \tilde{\mathbf{C}}} \, dS \,. \tag{332}$$

and $\delta\tilde{\mathbf{C}} = \delta\mathbf{C} + \zeta^\alpha \, \delta\mathbf{K}_\alpha$ refers to Eq. (324). Note that the variational forms $\delta\mathbf{C}$ and $\delta\mathbf{K}_\alpha$ are constant over $\mathcal{S}$. It can be seen that the integration over the micro-continuum $\mathcal{S}$ leads to two different stress measures. Tensor $\mathbf{S}$ is energy conjugate to a first-order strain tensor $\mathbf{C}$ and tensors $\mathbf{M}_\alpha$ are energy conjugate to the second-order strain measures $\mathbf{K}_\alpha$. Hence, only tensors $\mathbf{M}^\alpha$ contain the internal length scales and therefore contribute to the non-local size-scale effects. Note that the integration of Eq. (329) over micro-continuum $\mathcal{S}$ within the intervals $[-\frac{l_\alpha}{2}, \frac{l_\alpha}{2}]$ provides the implicit incorporation of internal length scale parameters l_α, $\alpha = 1, n$ associated with the different dimensions of $\mathcal{S}$ and representing the assumed non-locality of the material constituents. Finally, substituting Eqs. (330) and (326) into

Eq. (325) we end up with a generalized variational formulation including strain gradients

$$\int_{\mathcal{B}} \left\{ \mathbf{S} : \delta\mathbf{C} + \mathbf{M}^{\alpha} : \delta\mathbf{K}_{\alpha} \right\} dV - \mathcal{W}_{ext} = 0\,. \tag{333}$$

The variational principle is supplemented by essential boundary conditions, the so-called Dirichlet boundary conditions

$$\mathbf{u} = \mathbf{h}_u \quad \text{on } \partial\mathcal{B}_D \tag{334}$$

$$\frac{\partial \mathbf{u}}{\partial \vartheta^{\alpha}}\, n_{\alpha} = \mathbf{h}_{\gamma} \quad \text{on } \partial\mathcal{B}_D\,, \tag{335}$$

which represent the microscopic average of the generalized displacement field $\tilde{\mathbf{u}}$ at points $\mathbf{X} \in \partial\mathcal{B}_D = \partial\mathcal{B} \setminus \partial\mathcal{B}_N$.

The governing equations of this functional with respect to macro-space $\mathcal{B}$ are found by integration by parts and applying *Gauss's theorem*

$$\begin{aligned} &-\frac{1}{\sqrt{G}}\frac{\partial}{\partial\vartheta^i}\left(\sqrt{G}\,\mathbf{F}\,\mathbf{S}\mathbf{G}^i\right) - \frac{1}{\sqrt{G}}\frac{\partial}{\partial\vartheta^i}\left(\sqrt{G}\left(\frac{\partial \mathbf{x}_{,k}}{\partial\vartheta^{\alpha}}\otimes\mathbf{G}^k\right)\boldsymbol{m}^{\alpha}\mathbf{G}^i\right) - \mathbf{b} \\ &+\frac{1}{\sqrt{G}}\frac{\partial}{\partial\vartheta^{\alpha}}\frac{\partial}{\partial\vartheta^i}\left(\sqrt{G}\,\mathbf{F}\,\boldsymbol{m}^{\alpha}\mathbf{G}^i\right) + \frac{\partial}{\partial\vartheta^{\alpha}}\mathbf{l}^{\alpha} = \mathbf{0} \quad \text{on } \mathcal{B} \end{aligned} \tag{336}$$

$$\mathbf{F}\,\mathbf{S}\mathbf{n} + \left(\frac{\partial \mathbf{x}_{,k}}{\partial\vartheta^{\alpha}}\otimes\mathbf{G}^k\right)\boldsymbol{m}^{\alpha}\mathbf{n} - \frac{\partial}{\partial\vartheta^i}\left(\mathbf{F}\,\boldsymbol{m}^{\alpha}\mathbf{G}^i\right) n_{\alpha} = \mathbf{t}^{(\mathbf{n})} \quad \text{on } \partial\mathcal{B}_N \tag{337}$$

$$\mathbf{F}\,\boldsymbol{m}^{\alpha}\mathbf{n} = \mathbf{q}^{(\mathbf{n})\,\alpha} \quad \text{on } \partial\mathcal{B}_N \tag{338}$$

where G denotes the determinate of the *Riemannian* metric coefficients $G_{ij} = \mathbf{G}_i \cdot \mathbf{G}_j$ and $\mathbf{F} = \mathrm{Grad}\,\mathbf{x}$ is the classical deformation gradient tensor. For simplicity we define the projection map $\pi_0^*(\tilde{\mathbf{G}}_i)$ (Eq. 223) to be the identity, thus, we have $\tilde{\mathbf{G}}_i := \mathbf{G}_i$.

Numerical examples For modelling purposes a *moving least square*-based meshfree method (Skatulla and Sansour, 2008) is employed, as it accommodates $C^1(\Omega)$ continuity and second order consistency approximation requirements of the the strain gradient approach outlined in the previous section. The numerical integration over the micro-continuum $\mathcal{S}$ is carried out with the help of the *Gauss quadrature*, the order of which is chosen to be second according to the used second-order basis polynomial of the meshfree approximation.

The numerical applications in this section can be distinguished in four sets. The first one focuses on size-scale effects in bending of small-scale structures, the second one investigates torsion of very thin beams, the third one is an extensive study of excavation openings in intact rock and the fourth one demonstrates oriented material behaviour for the bending of a thin sheet. The first three sets make use of the hyperelastic *Saint-Venant-Kirchhoff* constitutive model which involves as material parameters *Young's modulus* E and *Poission's ratio* ν, whereas the last one utilizes a non-linear hyperelastic material law.

In some of the examples a classical reference model a *Green* strain tensor based approach is utilized (Skatulla and Sansour, 2008).

Polymer micro-cantilever beam

The following example is a set of micro-cantilever beams, the problem definition of which has been already introduced in Sec. 7.4. The variational principle presented in the previous section allows to choose between one and three dimension for the micro-continuum so that the size of the micro-continuum is controlled by one to three internal length scale parameters l_1, l_2 and l_3. Here we assume $l_1 = l_2 = l_3 = l$. Corresponding to these parameters we have the directors or the basis vectors of the micro-continuum that are given by Eq. (313). In the following two cases are distinguished. These are firstly, the one-dimensional micro-continuum having only one micro-director expressed by

$$\mathbf{a}_3 = \frac{\partial \mathbf{x}}{\partial \vartheta^3} \tag{339}$$

and secondly, the three-dimensional micro-continuum incorporating three micro-directors defined by

$$\mathbf{a}_\alpha = \frac{\partial \mathbf{x}}{\partial \vartheta^\alpha} \qquad \text{with}\, \alpha = 1, 2, 3\,. \tag{340}$$

The graphs illustrated in Fig. 23 - Fig.26 combine the load-deflection curves of the tip of four different micro-cantilever beams of lengths $L = 1150\,\mu m$, $750\,\mu m$, $380\,\mu m$ and $200\,\mu m$. The width for all four beams is $b = 235\,\mu m$ and the ratio of length to height $L/h = 10$ is kept constant so that these four beams represent four different scaling levels of one and the same beam. The internal length scale parameter is assumed as $l = 42\,\mu m$. As depicted in the load-deflection graph (Fig. 23) the one-dimensional micro-space provides a close match to the experimental results indicated by corresponding dotted curves. Note that the experimental data for the beams with $L = 1150\,\mu m$

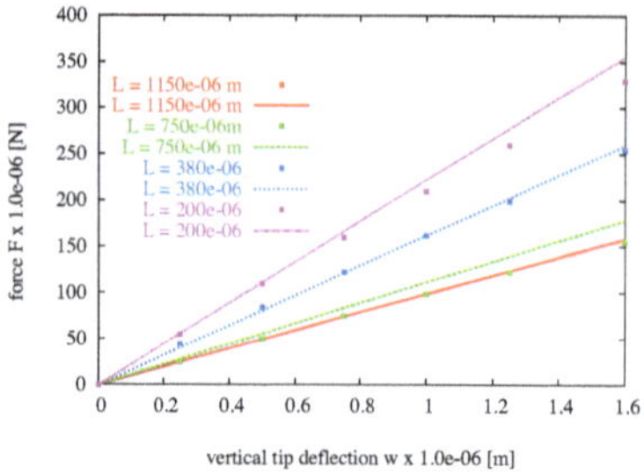

Figure 23. *load deflection diagram with* $l_3 = 42\,\mu m$ *and* $\nu = 0.3$

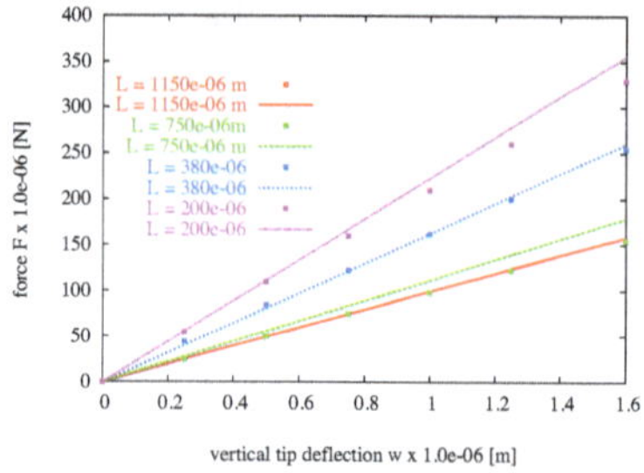

Figure 24. *load deflection diagram with* $l_1 = l_2 = l_3 = 42\,\mu m$ *and* $\nu = 0.3$

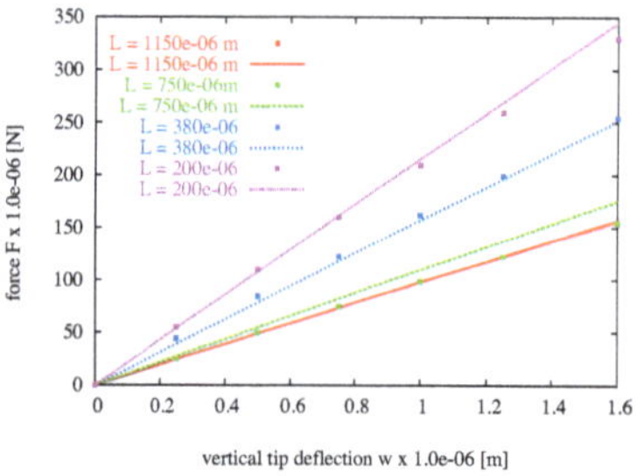

Figure 25. *load deflection diagram with* $l_3 = 42\,\mu m$ *and* $\nu = 0$

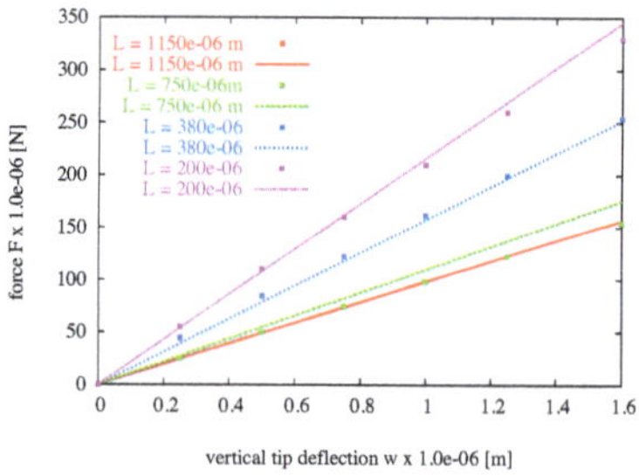

Figure 26. *load deflection diagram with* $l_1 = l_2 = l_3 = 42\,\mu m$ *and* $\nu = 0$

and 750 μm match each other. There is not a recognizable difference, if the microspace is chosen to be three-dimensional, shown in Fig. 24.

Neglecting the transversal contraction does not substantially change the results as illustrated in Figs. 25 and 26. This is in opposition to the use of the *Cosserat continuum*-based model (Eq. 306) in Sec. 7.4, where the choice of *Poisson's ratio* is of a great significance. This indicates that using the generalized deformation description (Eq. 314) the size-scale effects induced by the higher gradients outweigh the influence of the transversal contraction.

The modelling of small-scale structures such as the above micro-cantilever beam is discovered to be numerically critical, if the numerical integration over the micro-continuum involves very small numbers which fall below the minimum limit of double precision numbers. The solution leads then to a material behaviour which is significantly too soft. In order to achieve correct results it is necessary to change the geometrical units from meter to milli- or micrometer.

Clamped micro-sheet

The second example is a square sheet which is clamped at all its four edges. The problem definition is the same already studied in Sec. 7.4 and can be found there depicted in Fig. 1. Due to the symmetry only one quarter of the sheet is modelled by introducing the appropriate symmetry conditions for the displacement field. The material is chosen to be the same as for the micro-cantilever beam, but with a constant *Poisson ratio* $\nu = 0.3$. Again, four different scaling levels have been simulated with $L = 6.0\,mm$, $L = 3.0\,mm$, $L = 1.2\,mm$ and $L = 0.6\,mm$. The ratio of the sheet's length to its height $L/h = 300$ is fixed for all four sheets so that the results are comparable.

Preliminary studies showed that the difference between one-dimensional micro-continuum with micro-director $\mathbf{a}_3$ (Eq. 339) and the three-dimensional micro-continuum involving three micro-directors (Eq. 340) is negligible. This is clear as the bending of the sheet results in a curvature which is almost exclusively described by vector $\mathbf{a}_3$. Therefore, only the impact of a one-dimensional micro-continuum within the generalized variational formulation (Eq. 314) is investigated for scaling effects. The corresponding internal length scale parameter is assumed to be $l = 42\,\mu m$. The load-deflection

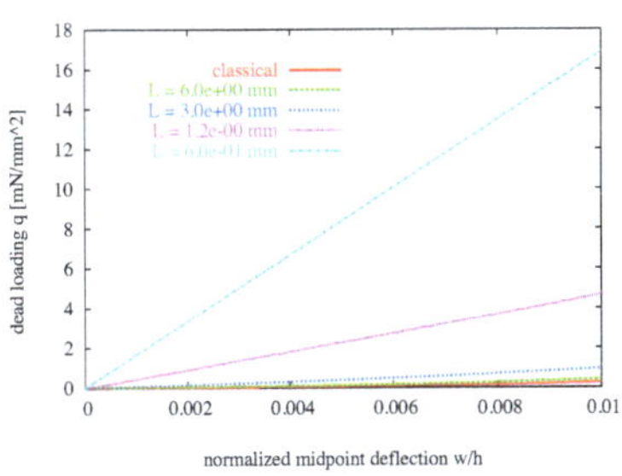

Figure 27. *load-deflection diagram*

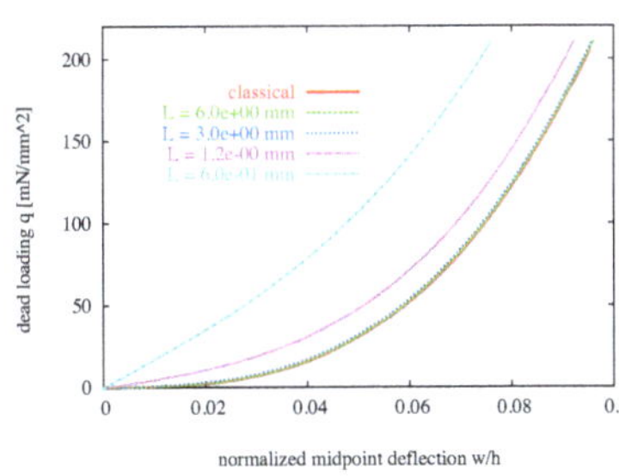

Figure 28. *load-deflection diagram*

curves for the vertical midpoint displacement in Fig. 27 and Fig. 28 illustrate distinct size-scale effects as the smaller sheets behave increasingly stiffer compared to the classical solution denoted by the red line. In Fig. 27

the relation of loading to deflection is for all sheets linear. Fig. 28 displays non-linear deformation behaviour for increasing loading. Interestingly, the curves which represent non-classical solutions do not show the same tendency to converge to the classical solution as observed for the modelling of this problem using the *Cosserat continuum* depicted in Fig. 7.

Torsion experiments on human bone specimens

This examples has already been presented in Sec. 7.4, where the reader can find further details. The problem is a set of thin beams with a square cross-section with side lengths $a = 5.0\,mm$, $2.0\,mm$, $1.0\,mm$ and $0.5\,mm$ depicted in Fig. 29. The ratio of the beam's length to side length $L/a = 100$ is kept constant for all four beams so that the results for twist θ and corresponding torsion moment M_T are comparable. The torsion moment is applied on the tip of the beams. The opposite end of the beams is clamped, that is all displacement degrees of freedom are held over the entire cross-section. The material is assumed to be hyperelastic and modelled by the *Saint-Venant-Kirchhoff* constitutive model which involves as material parameters *Young's modulus* E and *Poission's ratio* ν.

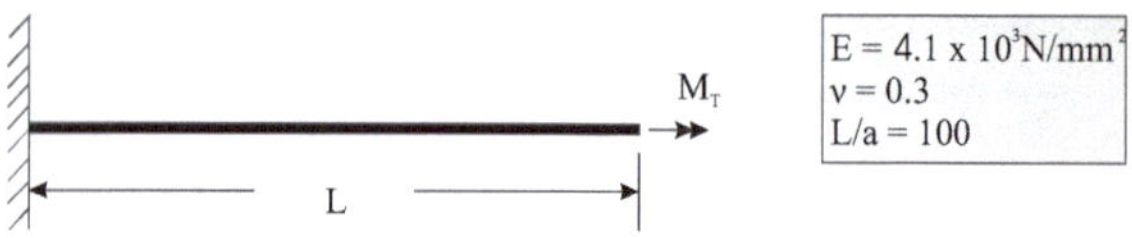

Figure 29. *problem configuration*

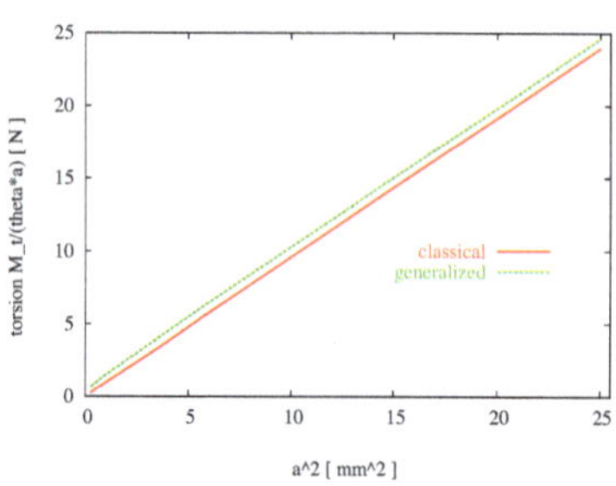

Figure 30. *diagram: normalized torsion vs. cross-section size*

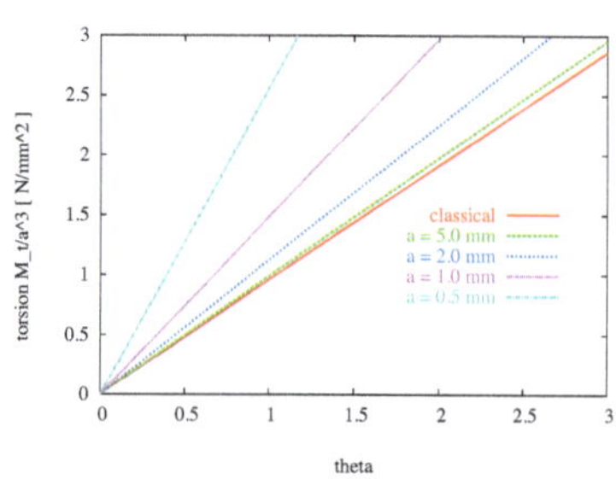

Figure 31. *diagram: normalized torsion vs. twist*

The micro-continuum $\mathcal{S}$ is chosen to be two-dimensional defined by two basis vectors or micro-directors which are given according to Eq. (313) by

$$\mathbf{a}_2 = \frac{\partial \mathbf{x}}{\partial \vartheta^2} \qquad \text{and} \qquad \mathbf{a}_3 = \frac{\partial \mathbf{x}}{\partial \vartheta^3}, \tag{341}$$

respectively. Associated with these two micro-directors are the so-called internal length scale parameters l_2 and l_3 which control the the size of the micro-continuum. Here, we assume $l_2 = l_3 = l = 0.5\,mm$ which is twice the value as used in the *Cosserat continuum*-based formulation (Eq. 306). The comparison of the simulation results of the micro-cantilever beams provided in this section with the results illustrated in Sec. 7.4 indicate that the internal length scale parameters in Eq. 306 and Eq. 333 can be related to each other by a factor 2.

If the twist is considered to be small, then the analytical relation of the applied torsion moment to the resulting twist θ is given by

$$M_T = C\,G\,a^4\,\frac{\theta}{L}, \tag{342}$$

where C is a constant and $L = 100\,a$.

The difference between the classical *Green strain tensor*-based and the generalized formulation (Eq. 333) is clearly recognizable in Fig. 30 which illustrates a linear computation. This scaling effect is also more distinctive than shown in Fig. 11 which is achieved using a variational principle based on the *Cosserat theory* (Eq. 306). The same also applies for non-linear modelling, where the decreasing cross-section leads to increasing rigidity displayed in Fig. 31, the magnitude of which is, however, much higher than for the *Cosserat continuum*-based model depicted in Fig. 12. The classical theory predicts that curves should fall for all side lengths a into one line which is the red curve.

Underground excavations

In the following underground excavations in intact rock conglomerate with two different geometrical configurations are modelled applying the classical *Green strain tensor*-based and the non-classical generalized formulation (Eq. 333). The micro-continuum is chosen to be two-dimensional and the two micro-directors are given, according to Eq. (313), by the spatial derivatives of the macroscopic placement vector $\mathbf{x}$ with respect to the macroscopic, the problem domain defining coordinates ϑ^1 and ϑ^2 as follows

$$\mathbf{a}_1 = \frac{\partial \mathbf{x}}{\partial \vartheta^1}, \quad \mathbf{a}_2 = \frac{\partial \mathbf{x}}{\partial \vartheta^2}. \tag{343}$$

The dimensions of the micro-continuum are controlled by the internal length scale parameters l_1 and l_2. These two parameters could be related to physical properties of the rock mass such as a certain lattice structure which provides that the deformation and the stress can not localize as much, or a sedimentary layering which leads to a oriented deformability. Both characteristics can be addressed by choosing the appropriate dimensionality of the microspace, where each dimension can be independently controlled by the internal length scale parameters. The latter will be demonstrated at the end. Note that sufficiently small values for these parameters provide the transition from the generalized continuum formulation to the classical one. The constitutive parameters of the used *Saint-Venant-Kirchhoff model* are Young's modulus $E = 20.0\,GPa$ and Poisson's ratio $\nu = 0$. As the meshfree code implemented in this work is three-dimensional the simulated results can be directly related to those achieved using the two-dimensional boundary element method (Hoek and Brown, 1982), if the transversal contraction is neglected. Both excavation configurations are simulated under plane strain conditions assuming the displacement vertical to the plane $u_3 = 0$. Since both problems are rotationally symmetric, only one quarter is modelled applying the appropriate symmetry conditions. The difference between a linear and non-linear simulation is found negligible.

The first example has a circular excavation opening with diameter $d = 4\,m$ depicted in Fig. 32. The diameter of the circular simulation domain $D = 10\,m$ is chosen to be large enough so tha't the pressure loading $p = 54.0189\,MPa$ which is applied on the boundary of the domain does not distort the stress distribution close to the opening. The problem is discretized with 26×31 particles in three in-plane layers.

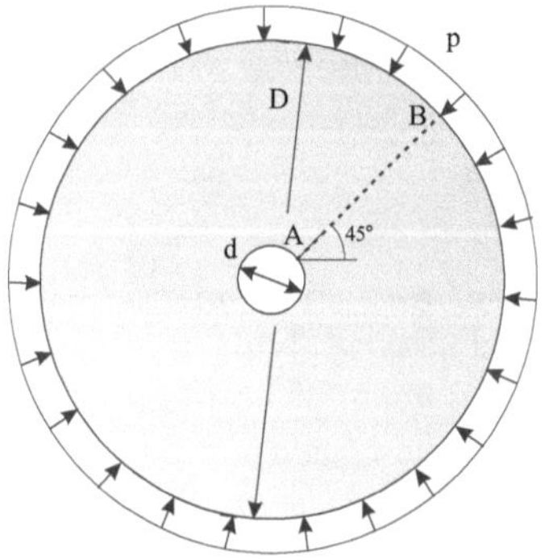

Figure 32. *problem definition*

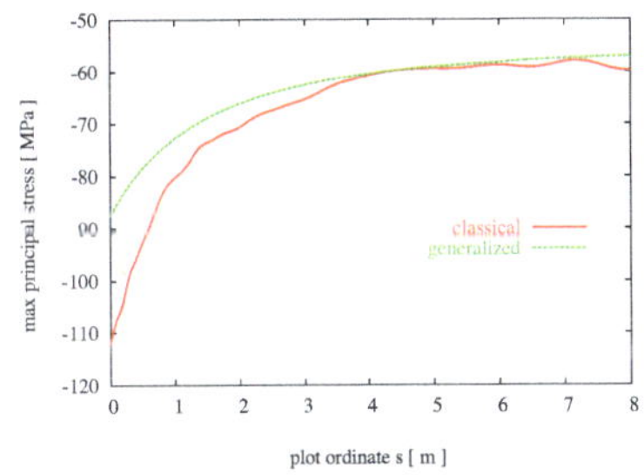

Figure 33. *maximum principal stress plotted along a line between point A and point B*

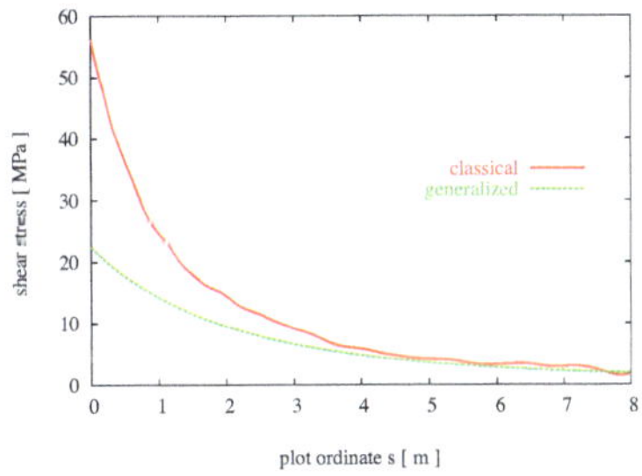

Figure 34. *shear stress plotted along a line between point A and point B*

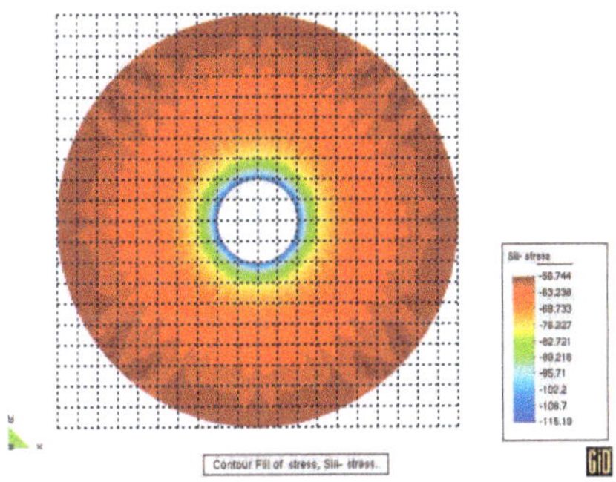

Figure 35. *max principal stress - classical solution* [MPa]

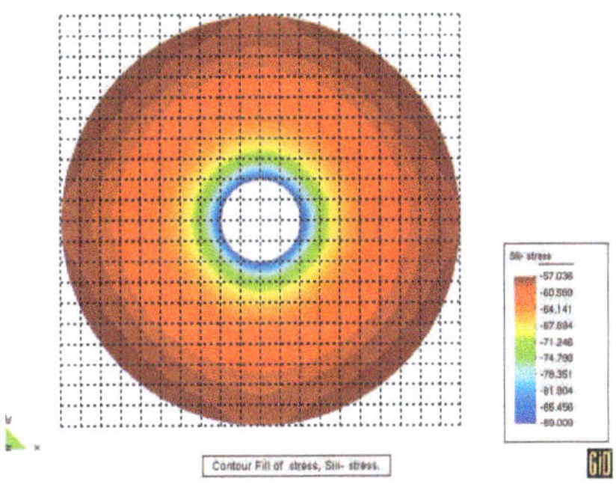

Figure 36. *max principal stress - generalized solution with a two-dimensional micro-continuum* [MPa]

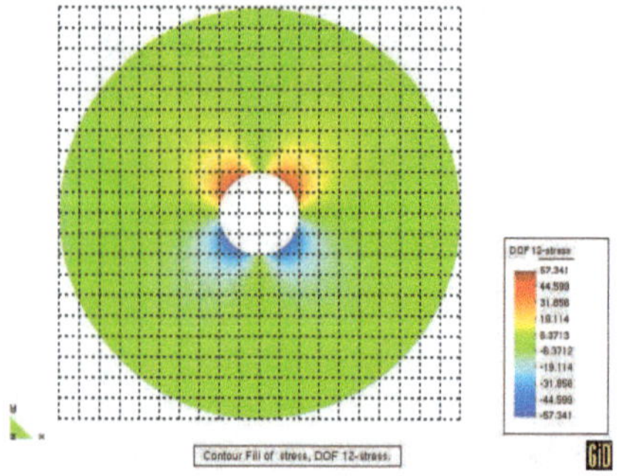

Figure 37. *plane shear stress - classical solution* [MPa]

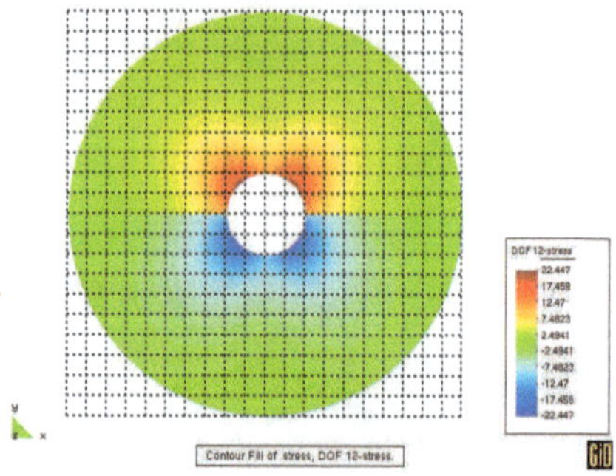

Figure 38. *plane shear stress - generalized solution with a two-dimensional micro-continuum* [MPa]

The classical solutions for maximum principal stress and shear stress are illustrated in Figs. 35 and 37, whereas the generalized continuum solution for internal length scale parameters $l_1 = l_2 = 4$ in Figs. 36 and 38. Note that the grid lines in all contour plots hfigures have a distance of $1.0\,m$ from each other. The illustrations of maximum principal stress and shear stress are supplemented each by plots along a line between point A and point B depicted in Figs. 33 and 34, where the solutions obtained with classical and generalized theory, respectively, are compared with each other. The generalized solutions for principal and shear stress distributions are significantly smoother than the classical ones, which is due to a regularizing effect of the second order deformation gradients in the formulation. The maximum principal stress at the excavation opening $\sigma = 115.19\,MPa$ illustrated in Fig. 35 agrees well with the boundary element solution (Hoek and Brown, 1982) which predicts $\sigma = 120\,MPa$. The magnitude of the in-plane shear stress near to the excavation for the non-classical model illustrated in Fig. 38 is less than one half of the *Green strain*-based approach depicted in Fig. 37. This clearly indicates that the energy stored in the problem domain splits into first order and second order strain energy parts and correspondingly, into first and second order stresses. Obviously this discrepancy depends on the choice of the micro-continuum.

The second examples has a square excavation opening of length and height $d = 4\,m$ depicted in Fig. refsquarePlateProbDef. The problem domain is rectangular with $D = 10\,m$ and the outer edges of the domain are subjected to a pressure loading $p = 60\,MPa$. The configuration is discretized with 4473 particles distributed in three in-plane layers. The classi-

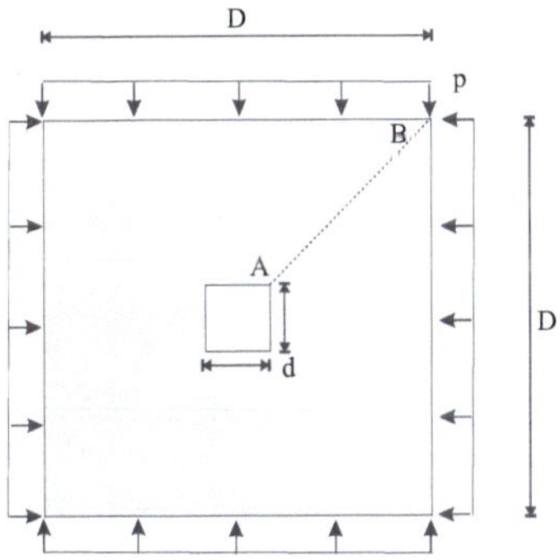

Figure 39. *problem definition*

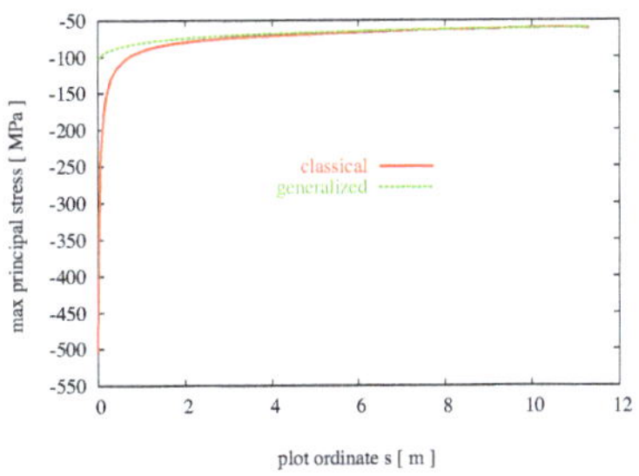

Figure 40. *maximum principal stress plotted along a line between point A and point B*

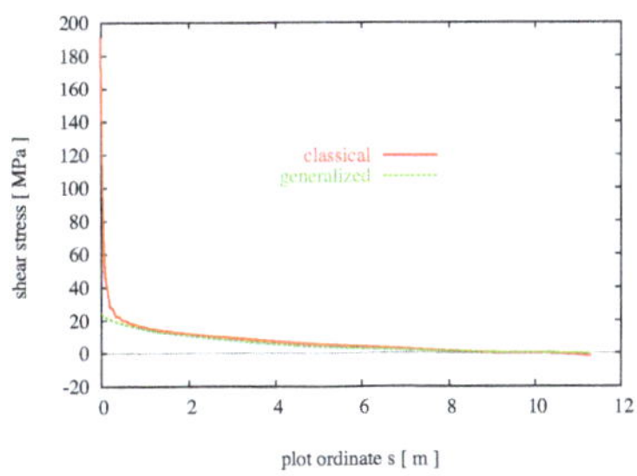

Figure 41. *shear stress plotted along a line between point A and point B*

cal solutions for maximum principal stress and and shear stress are depicted in Figs. 42 and 44. The corresponding plots for the generalized continuum solutions assuming the internal length scale parameters as $l_1 = l_2 = 4$ are shown in Figs. 43 and 45. Again, the grid lines in all contour plots are drawn with a distance of $1.0\,m$ from each other. Additionally, plots along a line between point A and point B of maximum principal stress and shear stress as depicted in Figs. 40 and 41 display the comparison of classical and non-classical results. The classical approach provides a peak value of the maximum principal stress in the corners of the excavation opening $\sigma = 504.4\,MPa$ illustrated in Fig. 42 which is only insignificantly lower than

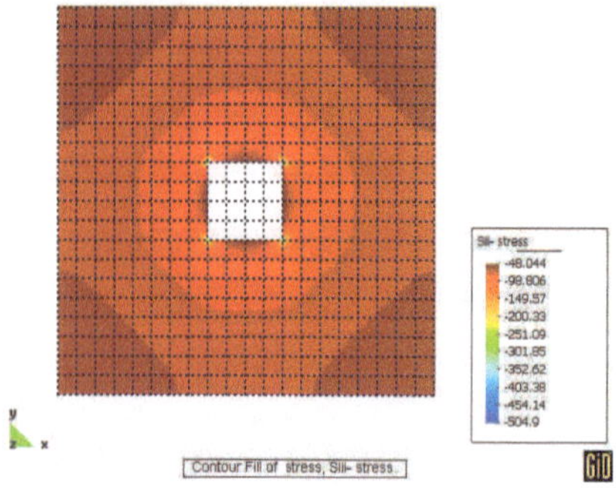

Figure 42. *max principal stress - classical solution* [MPa]

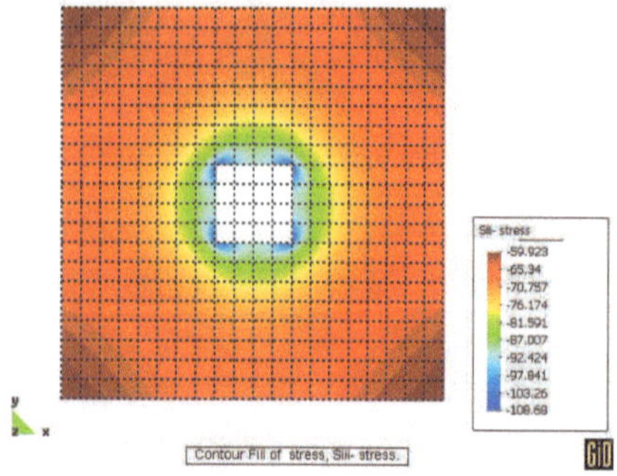

Figure 43. *max principal stress - generalized solution with a two-dimensional micro-continuum* [MPa]

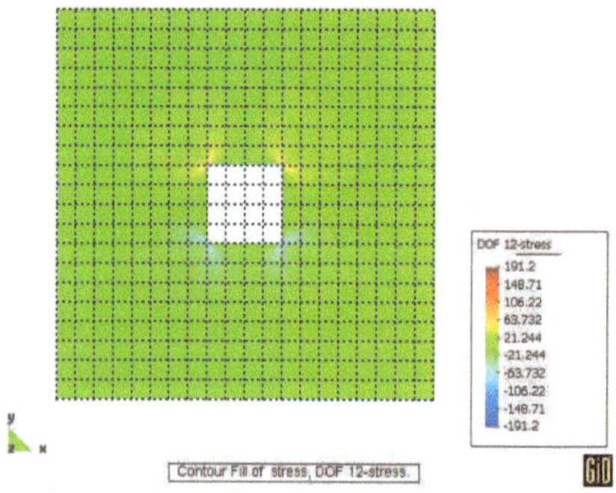

Figure 44. *plane shear stress - classical solution* [MPa]

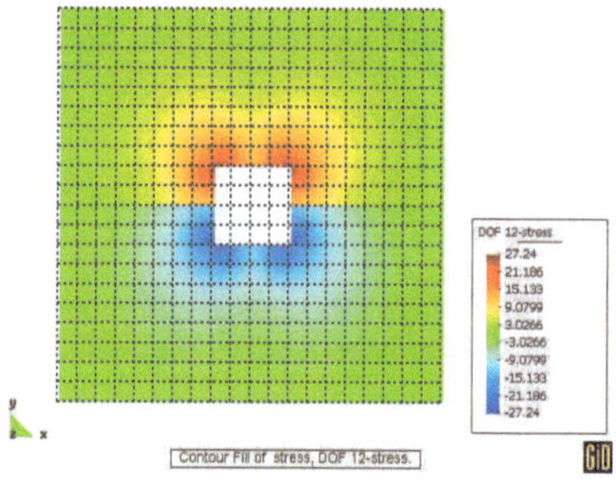

Figure 45. *plane shear stress - generalized solution with a two-dimensional micro-continuum* [MPa]

predicted in (Hoek and Brown, 1982) to $\sigma = 510\,MPa$. Similar to the application with circular opening, the maximum principal stress distribution is smoother for the non-classical approach.

Using the classical Green strain tensor based formulation the localized stress zone is confined between the first two layers of particles in radial direction. This is also the case for higher discretization levels than 4473 particles. This means that the size of the stress singularity is dependent on the particle distribution. This kind of behavior is known in FEM as mesh dependency. Contrarily, the use of the generalized formulation provides a solution for the stress fields that do not show such a dependency. The non-classical formulation has obviously a regularizing effect preventing the solution for stress and displacement fields to be dependent on the density of the particle distribution.

The modelling using generalized formulation results in significantly lower principle and shear stresses in the corner of the excavation opening than predicted by the classical theory shown in Figs. 40 and 41. The shear stress of the non-classical model is less than one seventh of the *Green strain*- based approach. However, the stresses are kept at significant values around the entire excavation boundary.

The difference in the solution between the classical model and non-classical one becomes negligible for $l_1 = l_2 = 0.01$ and smaller.

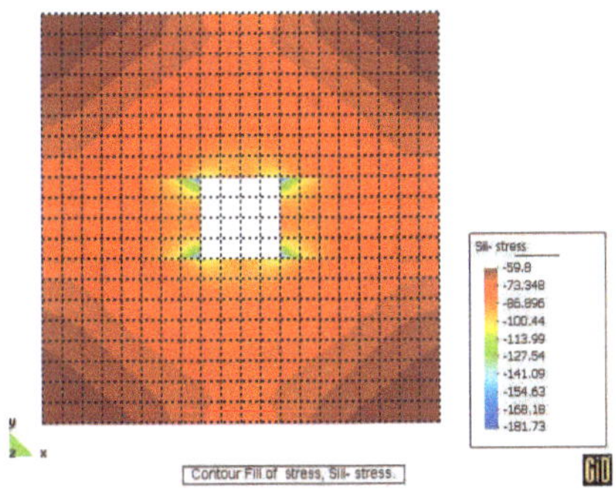

Figure 46. *max principal stress - generalized solution with a one-dimensional micro-continuum* [MPa]

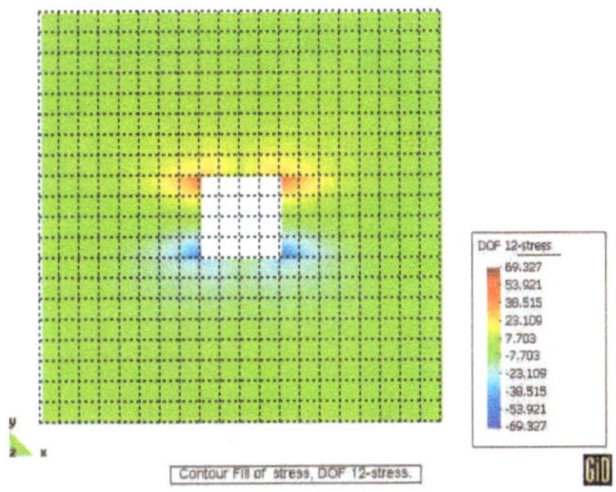

Figure 47. *plane shear stress - generalized solution with a one-dimensional micro-continuum* [MPa]

If the internal length scale parameters are set as $l_1 = 4$ and $l_2 < 0.01$, only the micro-structural deformation associated with the director $\mathbf{a}_1$ is considered. Accordingly, the stresses and the displacement fields depicted in Figs. 46 and 45 are obviously different and distorted compared to the results illustrated in Fig. 46 and 47.

Summarizing, these examples in rock mechanics demonstrate for the case of plane elasticity that the incorporation of elastic strain gradients can significantly change the distribution of stresses and displacement around excavation constructions. Furthermore, the generalized formulation Eqs. (333) proves to be independent of the particle distribution density unlike the classical Green strain tensor based formulation.

Simply supported micro-sheet

Finally, orientated material behaviour is modelled making use of the non-

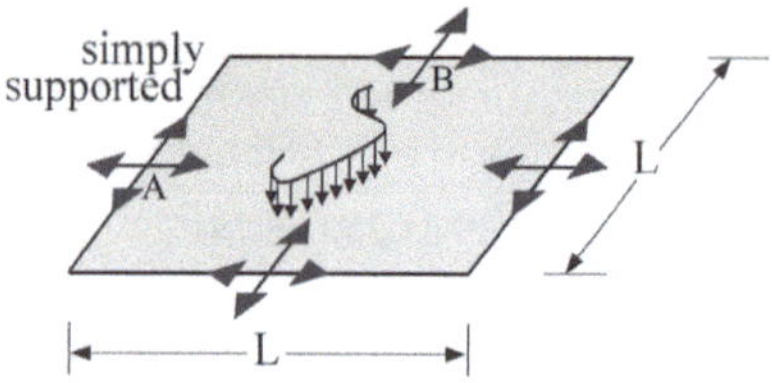

Figure 48. *problem definition*

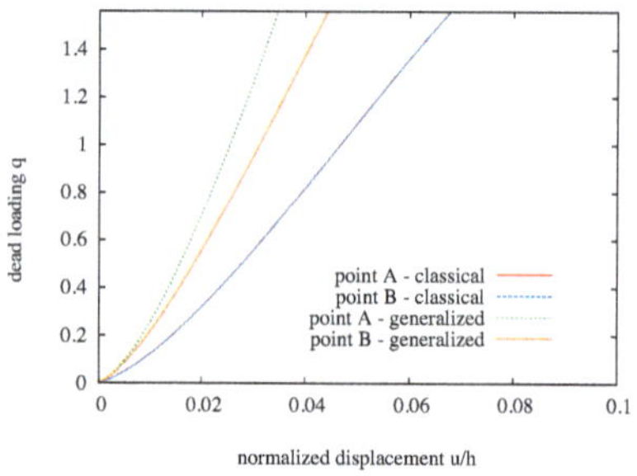

Figure 49. *load displacement diagram with* $l_1 = 4.2 \times 10^{-2}$

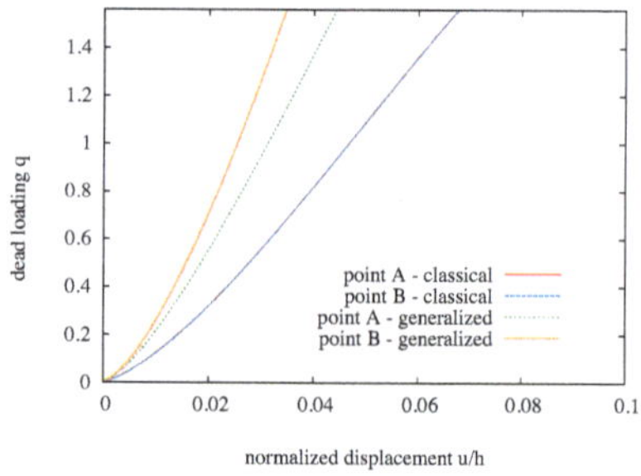

Figure 50. *load displacement diagram with* $l_2 = 4.2 \times 10^{-2}$

linear statistically based constitutive law of hyperelasticity (Arruda and Boyce, 1998; Sansour et al., 2003) which which involves as constitutive parameters the shear modulus $C_R = 1.0 \times 10^4$, the bulk modulus $\kappa = 1.0 \times 10^7$ and parameter $N = 8$. This example also aims to demonstrate that the

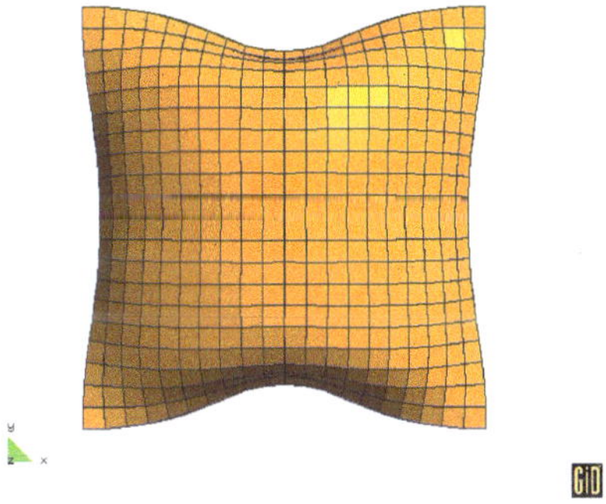

Figure 51. *deformed configuration at loading parameter* $q = 3.3$ *with* $l_1 = 4.2 \times 10^{-2}$

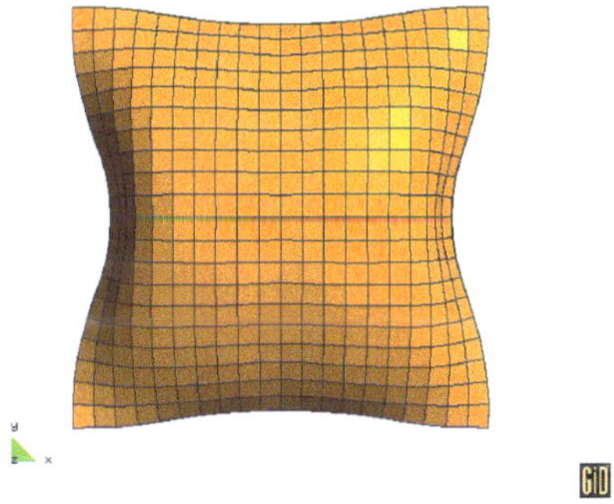

Figure 52. *deformed configuration at loading parameter* $q = 3.3$ *with* $l_2 = 4.2 \times 10^{-2}$

use of conventional material models within the proposed non-classical variational principle is very straightforward. The problem illustrated in the following is a square sheet with $L = 0.6$ and $h = 2.0 \times 10^{-3}$ which is simply supported at its outer edges in vertical direction and subjected to a dead loading depicted in Fig. 48. Only one quarter of the sheet is modelled with $11 \times 11 \times 3$ particles in longitudinal and thickness direction, respectively, and appropriate symmetry conditions are applied.

The oriented material response is achieved by a specific definition of the micro-continuum $\mathcal{S}$. That is, we consider a one-dimensional micro-continuum, the director or the basis vector of which is given by Eq. (313).

Firstly, we define the basis vector to be $\mathbf{g}_1$ and its corresponding internal length scale parameter is determined as $l_1 = 4.2 \times 10^{-2}$. In Fig. 49 it is illustrated that the magnitude of the displacement of point A in x-direction is larger than the one of point B in y-direction which the classical *Green strain tensor*-based model predicts to be equal as both curves fall into one line. The final deformed configuration of the sheet using the generalized formulation is illustrated in Fig. 51.

Secondly, the basis vector of the micro-continuum is now chosen to be $\mathbf{g}_2$ and the corresponding internal length scale-parameter is $l_2 = 4.2 \times 10^{-2}$. Again, a clear displacement difference of point A in x-direction and of point B in y-direction is recognizable displayed in Fig. 50. Now however, the displacement magnitude of point B is larger than the one of point A. The final deformed configuration of the sheet modelled with the help of the generalized formulation is depicted in Fig. 52.

7.6 Generalized continuum based on a triade of normal vectors $\mathbf{n}_\alpha$

Generalized deformation and strain measures In the previous section, Sec. 7.1, the basic principles to formulate a generalized deformation field are outlined. In this section a generalized formulation is considered as to consist of the macroscopic deformation field and, relative to it, a microscopic deformation field which is described by a triad of unit vectors $\mathbf{n}_\alpha$ attached to each material particle and defined by the following expression

$$\tilde{\mathbf{n}}_\alpha\left(\vartheta^k,t\right) = \frac{1}{2}\,\epsilon_{\alpha\beta\gamma}\,\frac{(\mathbf{g}_\beta\times\mathbf{g}_\gamma)}{|\mathbf{g}_\beta\times\mathbf{g}_\gamma|}\,, \tag{344}$$

where $\epsilon_{\alpha\beta\gamma}$ are the components of the *Levi-Civita tensor* (Eq. 92), $\mathbf{g}_\alpha\left(\vartheta^k,t\right)$ denote the tangent vectors of the macrospace $\mathcal{B}\subset\mathbb{E}(3)$ and $|\tilde{\mathbf{n}}_\alpha|=1$. Hence, the deformation of the micro-continuum is characterized by a rigid body rotation relative to the macroscopic placement as well as shear. This is true, because, firstly, the triad of the unit vectors $\tilde{\mathbf{n}}_\alpha$ rotates corresponding to the changing orientation of the planes $\mathbf{g}_\beta\times\mathbf{g}_\gamma$ to which these vectors are perpetual, and, secondly, because the orientation of those planes to each other is altered according to the deformation of the tangent vectors $\mathbf{g}_\beta$, $\mathbf{g}_\gamma$.

In this approach the micro-directors $\mathbf{a}_\alpha\left(\vartheta^k,t\right)$ in Eq. (226) are replaced by the expression in Eq. (344) and we arrive at

$$\tilde{\mathbf{x}}\left(\vartheta^k,\zeta^\beta,t\right) = \mathbf{x}\left(\vartheta^k,t\right) + \zeta^\alpha\,\tilde{\mathbf{n}}_\alpha\left(\vartheta^k,t\right)\,. \tag{345}$$

The advantage of this approach is that non-local material behaviour, e.g. rotation and shear of a lattice or a particle, can be described without involving additional degrees of freedom.

The tangent space $\mathcal{TG}$ in the reference configuration is defined by the pair $(\tilde{\mathbf{G}}_i\times\mathbf{I}_\alpha)$ given by

$$\tilde{\mathbf{G}}_i\left(\vartheta^k,\zeta^\beta\right) = \frac{\partial\tilde{\mathbf{X}}}{\partial\vartheta^i}\,,\qquad \mathbf{I}_\alpha\left(\vartheta^k\right) = \frac{\partial\tilde{\mathbf{X}}}{\partial\zeta^\alpha} = \tilde{\mathbf{N}}_\alpha\left(\vartheta^k\right) \tag{346}$$

and for the current configuration by

$$\tilde{\mathbf{g}}_i\left(\vartheta^k,\zeta^\beta,t\right) = \frac{\partial\tilde{\mathbf{x}}}{\partial\vartheta^i}\,,\qquad \mathbf{i}_\alpha\left(\vartheta^k,t\right) = \frac{\partial\tilde{\mathbf{x}}}{\partial\zeta^\alpha} = \tilde{\mathbf{n}}_\alpha\left(\vartheta^k,t\right)\,. \tag{347}$$

Furthermore, the spatial derivatives of the position vector in the current configuration with respect to the curvilinear coordinates of the macro-space ϑ^i are given by

$$\tilde{\mathbf{x}}_{,i}\left(\vartheta^k,\zeta^\beta,t\right) = \frac{\partial\tilde{\mathbf{x}}}{\partial\vartheta^i} = \mathbf{x}_{,i}\left(\vartheta^k,t\right) + \zeta^\alpha\tilde{\mathbf{n}}_{\alpha,i}\left(\vartheta^k,t\right)\,, \tag{348}$$

and with respect to the coordinates of the micro-space ζ^α by

$$\tilde{\mathbf{x}}_{,\alpha}\left(\vartheta^k, \zeta^\beta, t\right) = \frac{\partial \tilde{\mathbf{x}}}{\partial \zeta^\alpha} = \tilde{\mathbf{n}}_\alpha\left(\vartheta^k, t\right) . \tag{349}$$

With Eq. (348) and Eq. (349) a generalized deformation gradient is formulated as follows

$$\tilde{\mathbf{F}}\left(\vartheta^k, \zeta^\beta, t\right) = \left(\mathbf{x}_{,i} + \zeta^\alpha \tilde{\mathbf{n}}_{\alpha,i}\right) \otimes \tilde{\mathbf{G}}^i + \tilde{\mathbf{n}}_\alpha \otimes \tilde{\mathbf{N}}^\alpha . \tag{350}$$

Based on Eq. (350) the generalized right *Cauchy-Green* deformation tensor can be formulated in analogy to classical definition as

$$\tilde{\mathbf{C}}\left(\vartheta^k, \zeta^\beta, t\right) = \tilde{\mathbf{F}}^T \tilde{\mathbf{F}} . \tag{351}$$

Neglecting higher order terms in ζ^α and extracting only the dominant contributions to $\tilde{\mathbf{C}}$ (constant and linear in ζ^α) we can identity the following three parts of $\tilde{\mathbf{C}}$

$$\tilde{\mathbf{C}}^I\left(\vartheta^k, \zeta^\beta, t\right) = \mathbf{x}_{,k} \cdot \mathbf{x}_{,l} + \zeta^\alpha\left(\mathbf{x}_{,k} \cdot \tilde{\mathbf{n}}_{\alpha,l} + \tilde{\mathbf{n}}_{\alpha,k} \cdot \mathbf{x}_{,l}\right) \tilde{\mathbf{G}}^k \otimes \tilde{\mathbf{G}}^l \tag{352}$$

$$\tilde{\mathbf{C}}^{II}\left(\vartheta^k, \zeta^\beta, t\right) = \left(\mathbf{x}_{,k} \cdot \tilde{\mathbf{n}}_\beta + \zeta^\alpha\, \tilde{\mathbf{n}}_{\alpha,k} \cdot \tilde{\mathbf{n}}_\beta\right)\left(\tilde{\mathbf{G}}^k \otimes \tilde{\mathbf{N}}^\beta + \tilde{\mathbf{N}}^\beta \otimes \tilde{\mathbf{G}}^k\right) \tag{353}$$

$$\tilde{\mathbf{C}}^{III}\left(\vartheta^k, \zeta^\beta, t\right) = \left(\tilde{\mathbf{n}}_\alpha \cdot \tilde{\mathbf{n}}_\beta\right) \tilde{\mathbf{N}}^\alpha \otimes \tilde{\mathbf{N}}^\beta . \tag{354}$$

However, it is well possible to take $\tilde{\mathbf{C}} := \tilde{\mathbf{C}}^I$ by discarding $\tilde{\mathbf{C}}^{II}$ as well as $\tilde{\mathbf{C}}^{III}$ which is indeed adequate, if only the purpose of describing size-scale effects needs to be served.

For the quasi-static case, leaning on a classical variational formulation based on the classical *Cauchy-Green* deformation tensor, a generalized variational principle based on $\tilde{\mathbf{C}}$ can be established along the lines of

$$\int_{\mathcal{B}} \int_{\mathcal{S}} \tilde{\rho}_0 \frac{\partial \psi\left(\tilde{\mathbf{C}}\right)}{\partial \tilde{\mathbf{C}}} : \delta \tilde{\mathbf{C}}\, dS\, dV - \mathcal{W}_{ext} = 0 , \tag{355}$$

where $\psi(\tilde{\mathbf{C}})$ denotes the stored strain energy per unit mass and $\tilde{\rho}_0$ the density of the generalized space in the reference configuration.

Bibliography

E.C. Aifantis. Strain gradient interpretation of size effects. *International Journal for Fracture*, 95:299–314, 1999.

S. Akarapu and H.M. Zbib. Numerical analysis of plane cracks in strain-gradient elastic materials. *International Journal of Fracture*, 141:403–430, 2006.

R.K. Abu Al-Rub and G.Z. Voyiadjis. Analytical and experimental determination of the material intrinsic length scale of strain gradient plasticity theory from micro- and nano-indentation experiments. *International Journal of Plasticity*, 20:11391182, 2004.

E.M. Arruda and M.C. Boyce. A three-dimensional constitutive model for the large stretch behaviour of rubber elastic materials. *Journal of the Mechanics and Physics of Solids*, 41:389–412, 1998.

T. Bennett, I.M. Gitman, and H. Askes. Elasticity theories with higher-order gradients of inertia and stiffness for the modelling of wave dispersion in laminates. *Int. J. Fracture*, 148:185–193, 2007.

J.F. Besseling. *Irreversible Aspects of Continuum Mechanics*, chapter A thermodynamic approach to rheology. Springer, Wien, 1968.

S.R. Bodner and Y. Partom. Constitutive equations for elastic-viscoplastic strain-hardening materials. *ASME, J. Appl. Mech.*, 42:385–389, 1975.

F. Brickell and R.S. Clark. *Differentiable Manifolds.* Van Nostrand Reinhold Company, London, 1970.

H. Bufler. The biot stresses in nonlinear elasticity and the associated generalized variational principles. *Ing. Archiv*, 55:450–463, 1985.

R. Chambon, D. Cailleriea, and C. Tamagnini. A strain space gradient plasticity theory for finite strain. *Computer Methods in Applied Mechanics and Engineering*, 193:2797–2826, 2004.

H. Chen, G. Hu, and Z. Huang. Effective moduli for micropolar composite with interface effect. *International Journal of Solids and Structures*, 44: 8106–8118, 2007.

Y. Choquet-Bruhat, C. DeWitt-Morette, and M. Dillard-Bleick. *Analysis, Manifolds and Physics Part I.* North-Holland, Amsterdam, 1982.

E. Cosserat and F. Cosserat. *Théorie des corps déformables.* A. Hermann & Fils, Paris, 1909.

R. de Borst. A generalization of j_2-flow theory for polar continua. *Computer Methods in Applied Mechanics and Engineering*, 103:347–362, 1993.

T. Dillard, S. Forest, and P. Iennyb. Micromorphic continuum modelling of the deformation and fracture behaviour of nickel foams. *European Journal of Mechanics A/Solids*, 25:526–549, 2006.

B.A. Dubrovin, A.T. Fomenko, and S.P. Novikov. *Modern Geometry- Methods and Applications I.* Springer-Verlag, New York, 1984.

M. Epstein and M. de Leon. Geometrical theory of uniform cosserat media. *Journal of Geometry and Physics*, 26:127–170, 1998.

J.L. Ericksen and C. Truesdell. Exact theory of stress and strain in rods and shells. *Archive for Rational Mechanics and Analysis*, 1:295–323, 1957.

A.C. Eringen. Linear theory of micropolar viscoelasticity. *International Journal of Engineering Science*, 5:191–204, 1967.

A.C. Eringen. Micropolar fluids with stretch. *International Journal of Engineering Science*, 7:115–127, 1969.

A.C. Eringen. Theory of micromorphic materials with memory. *International Journal of Engineering Science*, 10:623–641, 1972.

A.C. Eringen. *Microcontinuum field theories I: Foundations and Solids*. Springer, New York, 1999.

A.C. Eringen and C.B. Kafadar. *Continuum Physics IV*. Academic Press, New York, 1976.

N.A. Fleck and J.R. Willis. A mathematical basis for strain-gradient plasticity theory. part ii: Tensorial plastic multiplier. *Journal of the Mechanics and Physics of Solids*, 57:1045–1057, 2009.

N.A. Fleck, G.M. Mller, M.F. Ashby, and J.W. Hutchinson. Strain gradient plasticity: theory and experiment. *Acta Metallurgica et Materialia*, 42: 475–487, 1994.

S. Forest. Micromorphic approach for gradient elasticity, viscoplasticity, and damage. *Journal of Engineering Mechanics*, 135:117–131, 2009.

S. Forest and R. Sievert. Nonlinear microstrain theories. *International Journal of Solids and Structures*, 43:7224–7245, 2006.

S. Forest, G. Cailletaud, and R. Sievert. A cosserat theory for elastoviscoplastic single crystals at finite deformation. *Archives of Mechanics*, 49:705–736, 1997.

S. Forest, F. Barbe, and G. Cailletaud. Cosserat modelling of size effects in the mechanical behaviour of polycrystals and multi-phase materials. *International Journal of Solids and Structures*, 37:7105–7126, 2000.

X.-L. Gao and H.M. Ma. Solution of eshelby's inclusion problem with a bounded domain and eshelby's tensor for a spherical inclusion in a finite spherical matrix based on a simplified strain gradient elasticity theory. *Journal of the Mechanics and Physics of Solids*, 58:779–797, 2010.

P. Grammenoudis and C. Tsakmakis. Isotropic hardening in micropolar plasticity. *Archive of Applied Mechanics*, 79:323–334, 2009.

P. Grammenoudis, C. Tsakmakis, and D. Hofer. Micromorphic continuum. part ii finite deformation plasticity coupled with damage. *International Journal of Non-Linear Mechanics*, 44:957–974, 2009.

P. Gudmundson. A unified treatment of strain gradient plasticity. *Journal of the Mechanics and Physics of Solids*, 52:1379–1406, 2004.

M.E. Gurtin and N. Ohno. A gradient theory of small-deformation, single-crystal plasticity that accounts for gnd-induced interactions between slip systems. *Journal of the Mechanics and Physics of Solids*, 59:320–343, 2011.

C.-S. Han, H. Gao, Y. Huang, and W.D. Nix. Mechanism-based strain gradient crystal plasticity - i. theory. *Journal of the Mechanics and Physics of Solids*, 53:1188–1203, 2005.

D. Hestens. *New Foundations for Classical Mechanics*. Kluwer Academic Publishers, Dordrecht, 1990.

C.B. Hirschberger, E. Kuhl, and P. Steinmann. On deformational and configurational mechanics of micromorphic hyperelasticity - theory and computation. *Computer Methods in Applied Mechanics and Engineering*, 196:4027–4044, 2007.

E. Hoek and E. T. Brown. *Underground excavations in rock*. The Institution of Mining and Metallurgy, London, 1982.

D. Iesan. On the microstretch piezoelectricity. *International Journal of Engineering Science*, 44:819–829, 2006.

R. Jänicke, S. Diebels, H.-G. Sehlhorst, and A. Düster. Two-scale modelling of micromorphic continua. a numerical homogenization scheme. *Continuum Mechanics and Thermodynamics*, 21:297–315, 2009.

G. Jelenic and M.A. Crisfield. Geometrically exact 3d beam theory: implementation of a strain-invariant finite element for static and dynamics. *Computer Methods in Applied Mechanics and Engineering*, 171:141–171, 1999.

D.J. Bammann J.R. Mayeur, D.L. McDowell. Dislocation-based micropolar single crystal plasticity: Comparison of multi- and single criterion theories. *Journal of the Mechanics and Physics of Solids*, 59:398–422, 2011.

J.K. Knowles and E. Sternberg. On the failure of ellipticity and the emergence of discontinuous deformation gradients in plane finite elastostatics. *Journal of Elasticity*, 8:329–379, 1978.

V. Kouznetsova, M.G.D. Geers, and W.A.M. Brekelmans. Multi-scale constitutive modelling of hetergeneous materials with a gradient-enhanced computational homogenization scheme. *International Journal for Numerical Methods in Engineering*, 54:1235–1260, 2002.

E. Kröner. Allgemeine kontinuumstheorie der versetzungen und eigenspannungen. *Archive for Rational Mechanics and Analysis*, 4:273–334, 1960.

E. Kröner. On the physical reality of torque stresses in continuum mechanics. *International Journal of Engineering Science*, 1:261–278, 1963.

R.S. Kumar and D.L. McDowell. Generalized continuum modeling of 2-d periodic cellular solids. *International Journal of Solids and Structures*, 41:7399–7422, 2004.

D.C.C. Lam, F. Yang, A.C.M Chong, J. Wang, and P. Tong. Experiments and theory in strain gradient elasticity. *Journal of the Mechanics and Physics of Solids*, 51:1477–1508, 2003.

R. Larsson and S. Diebels. A second-order homogenization procedure for multi-scale analysis based on micropolar kinematics. *International Journal for Numerical Methods in Engineering*, 69:2485–2512, 2006.

M. Lazar, G. Maugin, and E.C. Aifantis. Dislocations in second strain gradient elasticity. *International Journal of Solids and Structures*, 43: 1787–1817, 2006.

E.H. Lee. Elastic-plastic deformation at finite strains. *ASME J. Appl. Mech.*, 36:1–6, 1969.

M.T. Manzaria and R.A. Regueiro. Gradient plasticity modeling of geomaterials in a meshfree environment. part i: Theory and variational formulation. *Mechanics Research Communications*, 32:536–546, 2005.

F. A. McRobie and J. Lasenby. Simo-vu quoc rods using clifford algebra. *International Journal for Numerical Methods in Engineering*, 45, 1999.

C. McVeigha and W.K. Liu. Linking microstructure and properties through a predictive multiresolution continuum. *Computer Methods in Applied Mechanics and Engineering*, 197:3268–3290, 2008.

R.D. Mindlin. Micro-structure in linear elasticity. *Archive for Rational Mechanics and Analysis*, 16:51–78, 1964.

H-B. Mühlhaus. Application of cosserat theory in numerical solution of limit load problems. *Ingenieur-Archiv*, 59:124–137, 1989.

H.-B. Mühlhaus and E.C. Aifantis. A variational principle for gradient plasticity. *International Journal of Solids and Structures*, 28:845–857, 1991.

C.F. Niordson and B.N. Legarth. Strain gradient effects on cyclic plasticity. *Journal of the Mechanics and Physics of Solids*, 58:542–557, 2010.

T. Ohashi, M. Kawamukai, and H.M. Zbib. Multiscale modeling of size effects in fcc crystals: Discrete dislocation dynamics and density based gradient crystal plasticity. *Philosophical Magazine*, 87:1307–1326, 2007.

W. Pietraszkiewicz and J. Badur. Finite rotations in the description of continuum deformation. *International Journal of Engineering Science*, 21:1097–1115, 1983.

H-B Mhlhaus R. De Borst. Gradient-dependent plasticity: Formulation and algorithmic aspects. *International Journal for Numerical Methods in Engineering*, 35:521–539, 1992.

R.A. Regueiro. Finite strain micromorphic pressure-sensitive plasticity. *Journal of Engineering Mechanics*, 135:178–191, 2009.

C.W. Richards. Effects of size on the yielding of mild steel beams. In *Proceedings of the american society for testing and materials*, volume 58, pages 955–970, 1958.

C. Sansour. A unified concept of elastic-viscoplastic cosserat and micromorphic continua. *Journal de Physique IV Proceedings*, 8:341–348, 1998a.

C. Sansour. A theory of the elastic-viscoplastic cosserat continuum. *Archive of Mechanics*, 50:577–597, 1998b.

C. Sansour and H. Bednarczyk. The cosserat surface as shell model, theory and finite-element formulation. *Computer Methods in Applied Mechanics and Engineering*, 120:1–32, 1995.

C. Sansour and F. G. Kollmann. On theory and numerics of large viscoplastic deformation. *Comp. Meth. Appl. Mech. Engrg.*, 146:351–369, 1997.

C. Sansour and F.G. Kollmann. Large viscoplastic deformations of shells. theory and finite element formulation. *Computational Mechanics*, 21: 512–525, 1998.

C. Sansour and S. Skatulla. A micromorphic continuum-based formulation for inelastic deformations at finite strains. application to shear band formation. *International Journal of Solids and Structures*, 47:1546–1554, 2010.

C. Sansour and W. Wagner. A model of finite strain viscoplasticity based on unified constitutive equations. Theoretical and computational considerations with applications to shells. *Comp. Meth. Appl. Mech. Engrg.*, 191:423–450, 2001.

C. Sansour, S. Feih, and W. Wagner. On the performance of enhanced strain finite elements in large strain deformations of elastic shells. *International Journal for Computer-Aided Engineering and Software*, 20: 875–895, 2003.

J.C. Simo. The (symmetric) hessian for geometrically nonlinear models in solid mechanics: Intrinsic definition and geometric interpretation. *Computer Methods in Applied Mechanics and Engineering*, 96:189–200, 1992.

J.C. Simo and L. Vu-Quoc. A three-dimensional finite-strain rod model. part ii: Geometric and computational aspects. *Computer Methods in Applied Mechanics and Engineering*, 58, 1986.

J.C. Simo, N. Tarnow, and M. Doblare. Non-linear dynamics of three-dimensional rods: exact energy and momentum conservation algorithms. *International Journal for Numerical Methods in Engineering*, 38:1431–1473, 1995.

S. Skatulla and C. Sansour. Essential boundary conditions in meshfree methods via a modified variational principle. applications to shell computations. *Computer Assisted Mechanics and Engineering Sciences*, 15: 123–142, 2008.

L.J. Sluys. *Wave Propagation, Localization and Dispersion in Softening Solids*. PhD thesis, Delft University of Technology, Netherlands, 1992.

G.F. Smith. On isotropic functions of symmetric tensors, skew-symmetric tensors and vectors. *Int. J. Engng. Sci.*, 9:899–916, 1971.

A.J.M Spencer. *Theory of invariants in Continuum Physics I*, pages 239–353. Academic Press, New York, 1971.

P. Steinmann. A micropolar theory of finite deformation and finite rotation multiplicative elastoplasticity. *International Journal of Solids and Structures*, 31:1063–1084, 1994.

L.J. Sudak. Column buckling of multiwalled carbon nanotubes using non-local continuum mechanics. *Journal of Applied Physics*, 94:7281–7287, 2003.

D. Tian-min. New principles of work and energy as well as power and energy rate for continuum field theories. *Applied Mathematics and Mechanics*, 22:1231–1239, 2001.

N. Triantafyllidis and E.C. Aifantis. A gradient approach to localization of deformation. *International Journal of Elasticity*, 16:225–237, 1986.

V.S. Varadarajan. *Lie Groups, Lie Algebras, and their Representation.* Springer-Verlag, 1984.

I. Vardoulakis, G. Exakaktylos, and S.K. Kourkoulis. Bending of marble with intrinsic length scales: A gradient theory with surface energy and size effects. *Le Journal de Physique IV*, 8:399–406, 1998.

F. Vernerey, W.K. Liu, B. Moran, and G. Olson. Multi-length scale micromorphic process zone model. *Computational Mechanics*, 44:433–445, 2009.

T. Yalcinkaya, W.A.M. Brekelmans, and M.G.D. Geers. Deformation patterning driven by rate dependent non-convex strain gradient plasticity. *Journal of the Mechanics and Physics of Solids*, 59:1–17, 2011.

J.F.C Yang and R.S Lakes. Transient study of couple stress effects in compact bone: Torsion. *Journal of Biomechanical Engineering*, 103:275–279, 1981.

Y. Zhang and R. Larsson. Homogenization of delamination growth in an aca flip-chip joint based on micropolar theory. *European Journal of Mechanics A/Solids*, 28:433–444, 2009.

Z.H. Zhang, Z. Zhuang, Y. Gao, Z.L. Liu, and J.F. Nie. Cyclic plastic behavior analysis based on the micromorphic mixed hardening plasticity model. *Computational Materials Science*, 50:1136–1144, 2011.

H. To Zhu, H.M. Zbib, and E.C. Aifantis. Strain gradients and continuum modeling of size effect in metal matrix composites. *Acta Mechanica*, 121: 165–176, 1997.

H.T. Zhu, H.M. Zbib, and K.M. Khraisheh. Flow strength and size effect of an al-si-mg composite model system under multiaxial loadings. *Scripta Metall. Mater.*, 2:1895–1902, 1995.

Generalized Continuum Modelling of Crystal Plasticity

Samuel Forest
MINES ParisTech
Centre des Matériaux
CNRS UMR 7633
BP 87 91003 Evry Cedex, France

1 Cosserat crystal plasticity

The classical theory of continuum crystal plasticity is first recalled and then generalized to incorporate the effect of lattice curvature on material hardening. The specific notations used in this chapter are summarized in section 1.6.

1.1 Dislocation densities and classical continuum crystal plasticity

Elements from a statistical theory of dislocations The yielding and hardening behaviour of crystals mainly depends on the growth of the dislocation population and on the development of dislocation structures inside the volume element V of continuum mechanics. A precise account of the evolution of dislocation distribution in V still lies beyond current computing capacity, although promising results in that field are available Fivel et al. (1998). The incomplete information about the dislocation state permits probability predictions and suggests the use of statistical mechanics. Kröner proposed that the information be given in terms of n-point dislocation correlation tensor functions Kröner (1969, 1971). If the vectors $\underline{\boldsymbol{\xi}}(\underline{\boldsymbol{x}})$ and $\underline{\boldsymbol{b}}(\underline{\boldsymbol{x}})$ describe the line vector and Burgers vector of a dislocation located at $\underline{\boldsymbol{x}}$, the first correlation function reads :

$$\underset{\sim}{\alpha} = < \underline{\boldsymbol{b}} \otimes \underline{\boldsymbol{\xi}} > \tag{1}$$

where the brackets denote ensemble averaging. The next correlation function then is :

$$\underset{\approx}{\alpha}(\underline{\boldsymbol{x}}, \underline{\boldsymbol{x}}') = < (\underline{\boldsymbol{b}} \otimes \underline{\boldsymbol{\xi}})(\underline{\boldsymbol{x}}) \otimes (\underline{\boldsymbol{b}} \otimes \underline{\boldsymbol{\xi}})(\underline{\boldsymbol{x}}') > = \underset{\approx}{\alpha}(\underline{\boldsymbol{x}} - \underline{\boldsymbol{x}}') \tag{2}$$

if statistical uniformity is assumed. Second and fourth rank tensors $\underset{\sim}{\alpha}$ and $\underset{\approx}{\alpha}$ are indeed related to classical plastic state indicators used in classical crystal plasticity. For a large enough volume element V, it is resorted to the ergodic hypothesis so that ensemble averaging is replaced by volume averaging over V. In that case, $\underset{\sim}{\alpha}$ turns out to be identical to the so-called dislocation density tensor, or Nye's tensor, which is the basic variable of the continuum theory of dislocations Nye (1953). On the other hand, one invariant of the tensor $\underset{\approx}{\alpha}$ can be shown to be

$$\alpha_{ijij}(\underline{\mathbf{0}}) = L/V \tag{3}$$

where L is the length of dislocation lines inside V. The dislocation density ϱ well-known in the field of metallurgy.

Classical thermomechanics of single crystals Classical crystal plasticity theory is the paradigm of the model for anisotropic finite deformation plasticity. It has been founded in the original references Teodosiu (1970); Mandel (1971, 1973); Rice (1971); Teodosiu and Sidoroff (1976). More recent accounts of this continuum approach show the success of this approach based on multiplicative decomposition and the use of scalar dislocation densities ϱ Maugin (1992); Teodosiu (1997); Gumbsch and Pippan (2005); Bertram (2005). It is useful to recall the basis of this theory before generalizing it to include the effect of the dislocation density tensor.

The main field equations of continuum thermomechanics are:

- energy principle (local form)

$$\rho\dot{\epsilon} = \underset{\sim}{\boldsymbol{\sigma}} : \underset{\sim}{\boldsymbol{D}} + \rho r - div\, \underline{\boldsymbol{q}}$$

where $\underset{\sim}{\boldsymbol{D}}$ is the strain rate tensor, $\underset{\sim}{\boldsymbol{\sigma}}$ the Cauchy stress and $\underline{\boldsymbol{q}}$ the heat flux vector.

- entropy principle (local form)

$$\rho\dot{\eta} + div\left(\frac{\underline{\boldsymbol{q}}}{T}\right) - \frac{\rho r}{T} \geq 0$$

where η is the specific entropy function.

- Clausius-Duhem inequality

$$-\rho(\dot{e} - T\dot{\eta}) + \underset{\sim}{\boldsymbol{\sigma}} : \underset{\sim}{\boldsymbol{D}} - \frac{\underline{\boldsymbol{q}}}{T}.grad\, T \geq 0$$

The Helmholtz free energy density $\psi = \epsilon - Ts$ is introduced:

$$-\rho(\dot{\psi} + \eta\dot{T}) + \underset{\sim}{\boldsymbol{\sigma}} : \underset{\sim}{\boldsymbol{D}} - \frac{\underline{\boldsymbol{q}}}{T}.grad\, T \geq 0$$

- intrinsic and thermal dissipation

$$D^{th} = -\frac{\boldsymbol{q}}{T}.grad\,T$$

$$D^{i} = \rho(\dot{\psi} + s\dot{T}) + \underset{\sim}{\boldsymbol{\sigma}} : \underset{\sim}{\boldsymbol{D}}$$

The gradient of the transformation is called the deformation gradient $\underset{\sim}{\boldsymbol{F}}$. It is decomposed into elastic and plastic parts as shown in figure 1(a):

$$\underset{\sim}{\boldsymbol{F}} = \underset{\sim}{\boldsymbol{F}}^{e}.\underset{\sim}{\boldsymbol{F}}^{p} \tag{4}$$

The uniqueness of this decomposition is ensured in the case of the single crystal material element if the orientation of the intermediate configuration is such that the lattice vectors do not rotate from the initial to the intermediate configuration. This unique intermediate configuration is called isoclinic by Mandel (1973). The elastic strain and a stress tensor are defined with respect to the isoclinic configuration:

$$\underset{\sim}{\boldsymbol{E}}^{e} = \frac{1}{2}(\underset{\sim}{\boldsymbol{F}}^{e}.\underset{\sim}{\boldsymbol{F}}^{eT} - \underset{\sim}{\boldsymbol{1}}), \quad \underset{\sim}{\boldsymbol{\Pi}}^{e} = \frac{\rho_i}{\rho}\underset{\sim}{\boldsymbol{F}}^{e-1}.\underset{\sim}{\boldsymbol{\sigma}}.\underset{\sim}{\boldsymbol{F}}^{e-T} \tag{5}$$

The power of internal forces makes the link between the different stress tensors:

$$\frac{1}{\rho}\underset{\sim}{\boldsymbol{\sigma}} : (\dot{\underset{\sim}{\boldsymbol{F}}}.\underset{\sim}{\boldsymbol{F}}^{-1}) = \frac{1}{\rho_i}(\underset{\sim}{\boldsymbol{\Pi}}^{e} : \dot{\underset{\sim}{\boldsymbol{E}}}^{e} + (\underset{\sim}{\boldsymbol{F}}^{eT}.\underset{\sim}{\boldsymbol{F}}^{e}.\underset{\sim}{\boldsymbol{\Pi}}^{e}) : (\dot{\underset{\sim}{\boldsymbol{F}}}^{p}.\underset{\sim}{\boldsymbol{F}}^{p-1})) \tag{6}$$

The Clausius–Duhem inequality takes now the form

$$\rho(\frac{\underset{\sim}{\boldsymbol{\Pi}}^{e}}{\rho_i} - \frac{\partial\Psi}{\partial\underset{\sim}{\boldsymbol{E}}^{e}}) : \dot{\underset{\sim}{\boldsymbol{E}}}^{e} - \rho(s + \frac{\partial\Psi}{\partial T})\dot{T} - \rho\frac{\partial\Psi}{\partial\alpha} : \dot{\alpha} + D^{res} \geq 0 \tag{7}$$

The state variables are $(\underset{\sim}{\boldsymbol{E}}^{e}, T, \alpha)$ with additional internal variables α. The state functions are the internal energy density $e(\underset{\sim}{\boldsymbol{E}}^{e}, \eta, \alpha)$ and the Helmholtz free energy density $\psi(\underset{\sim}{\boldsymbol{E}}^{e}, T, \alpha)$. The exploitation of the second principle à la Coleman–Noll provides the state laws

$$\underset{\sim}{\boldsymbol{\Pi}}^{e} = \rho_i\frac{\partial\Psi}{\partial\underset{\sim}{\boldsymbol{E}}^{e}}, \quad X = \rho_i\frac{\partial\Psi}{\partial\alpha}, \quad s = -\frac{\partial\Psi}{\partial T} \tag{8}$$

The residual dissipation then reads

$$D^{res} = \underset{\sim}{\boldsymbol{M}} : (\dot{\underset{\sim}{\boldsymbol{F}}}^{p}.\underset{\sim}{\boldsymbol{F}}^{p-1}) - X\dot{\alpha} \geq 0 \tag{9}$$

This leads to the definition of the so–called Mandel stress tensor:

$$\underset{\sim}{\boldsymbol{M}} := \underset{\sim}{\boldsymbol{F}}^{eT}.\underset{\sim}{\boldsymbol{F}}^{e}.\underset{\sim}{\boldsymbol{\Pi}}^{e} = \underset{\sim}{\boldsymbol{C}}^{e}.\underset{\sim}{\boldsymbol{\Pi}}^{e} \tag{10}$$

We assume the existence of a convex dissipation potential $\Omega(\underset{\sim}{\boldsymbol{M}}, X)$ from which the flow rule and evolution equations for internal variables are derived:

$$\dot{\underset{\sim}{\boldsymbol{F}}}^p.\underset{\sim}{\boldsymbol{F}}^{p-1} = \frac{\partial\Omega}{\partial\underset{\sim}{\boldsymbol{M}}}, \qquad \dot{\alpha} = -\frac{\partial\Omega}{\partial X} \tag{11}$$

We present now a template model for single crystal behaviour introducing the specific internal variables: "dislocation density" ϱ^s, internal structure α^s. Take the free energy function as

$$\rho^i\psi(\underset{\sim}{\boldsymbol{E}}^e, \varrho^s, \alpha^s) = \frac{1}{2}\underset{\sim}{\boldsymbol{E}}^e : \underset{\approx}{\boldsymbol{c}} : \underset{\sim}{\boldsymbol{E}}^e + r_0\sum_{s=1}^{N}\varrho^s + \frac{1}{2}q\sum_{r,s=1}^{N}h^{rs}\varrho^r\varrho^s + \frac{1}{2}\sum_{s=1}^{N}\alpha^{s2} \tag{12}$$

The state laws then read

$$\begin{aligned} \underset{\sim}{\boldsymbol{\Pi}}^e &= \rho^i\frac{\partial\psi}{\partial\underset{\sim}{\boldsymbol{E}}^e} = \underset{\approx}{\boldsymbol{c}} : \underset{\sim}{\boldsymbol{E}}^e \\ r^s &= \rho^i\frac{\partial\psi}{\partial\varrho^s} = r_0 + q\sum_{r=1}^{N}h^{sr}\varrho^s \\ x^s &= \rho^i\frac{\partial\psi}{\partial\alpha^s} = c\alpha^s \end{aligned}$$

The driving force for plastic slip is the resolved shear stress on slip system s

$$\tau^s = (\underset{\sim}{\boldsymbol{M}}.\underline{\boldsymbol{n}}^s).\underline{\boldsymbol{m}}^s = \underset{\sim}{\boldsymbol{M}} : (\underline{\boldsymbol{m}}^s \otimes \underline{\boldsymbol{n}}^s) \tag{13}$$

The multimechanism crystal plasticity yield criterion for single crystal is Schmid's law:

$$f^s = |\tau^s - x^s| - r^s \tag{14}$$

The yield threshold is r^s. An internal stress (back–stress) x^s is introduced for each system s. Take the dissipation potential

$$\Omega(\underset{\sim}{\boldsymbol{M}}, r^s, x^s) = \frac{K}{n+1}\sum_{s=1}^{N}\langle\frac{f^s}{K}\rangle^{n+1} \tag{15}$$

from which flow and hardening rules are derived

$$\dot{\underset{\sim}{\boldsymbol{\varepsilon}}}^p = \frac{\partial\Omega}{\partial\underset{\sim}{\boldsymbol{\sigma}}} = \sum_{s=1}^{N}\dot{\gamma}^s\underline{\boldsymbol{m}}^s \overset{sym}{\otimes} \underline{\boldsymbol{n}}^s, \quad \dot{\varrho}^s = -\frac{\partial\Omega}{\partial r^s} = \dot{v}^s, \quad \dot{\alpha}^s = -\frac{\partial\Omega}{\partial x^s} = \dot{\gamma}^s \tag{16}$$

with

$$\dot{v}^s = \langle\frac{f^s}{K}\rangle^n, \quad \dot{\gamma}^s = \dot{v}^s\, sign(\tau^s - r^s) \tag{17}$$

Nonlinear hardening rules are more realistic for actual materials in the form:

- nonlinear isotropic hardening

$$\varrho^s = 1 - \exp(-bv^s), \quad r^s = r_0 + q \sum_{r=1} h^{sr}(1 - \exp(-bv^s)) \tag{18}$$

- nonlinear kinematic hardening

$$\dot{\alpha}^s = \dot{\gamma}^s - d\dot{v}^s \alpha^s \tag{19}$$

For monotonic loading, this equation integrates in

$$x^s = \frac{c}{d}(\pm 1 - \exp(-dv^s)) \tag{20}$$

The isotropic hardening involves an interaction matrix between slip systems. This corresponds to the notion of latent hardening Besson et al. (2001); Fivel and Forest (2004). The general form of this matrix for FCC crystals is

$$\left[\begin{array}{ccccccccccccc}
 & B4 & B2 & B5 & D4 & D1 & D6 & A2 & A6 & A3 & C5 & C3 & C1 \\
B4 & h_1 & h_2 & h_2 & h_4 & h_5 & h_5 & h_5 & h_6 & h_3 & h_5 & h_3 & h_6 \\
B2 & & h_1 & h_2 & h_5 & h_3 & h_6 & h_4 & h_5 & h_5 & h_5 & h_6 & h_3 \\
B5 & & & h_1 & h_5 & h_6 & h_3 & h_5 & h_3 & h_6 & h_4 & h_5 & h_5 \\
D4 & & & & h_1 & h_2 & h_2 & h_6 & h_5 & h_3 & h_6 & h_3 & h_5 \\
D1 & & & & & h_1 & h_2 & h_3 & h_5 & h_6 & h_5 & h_5 & h_4 \\
D6 & & & & & & h_1 & h_5 & h_4 & h_5 & h_3 & h_6 & h_5 \\
A2 & & & & & & & h_1 & h_2 & h_2 & h_6 & h_5 & h_3 \\
A6 & & & & & & & & h_1 & h_2 & h_3 & h_5 & h_6 \\
A3 & & & & & & & & & h_1 & h_5 & h_4 & h_5 \\
C5 & & & & & & & & & & h_1 & h_2 & h_2 \\
C3 & & & & & & & & & & & h_1 & h_2 \\
C1 & & & & & & & & & & & & h_1
\end{array}\right]$$

A simplified form of the matrix is $h_1 = 1, \quad h_2 = h_3 = h_4 = h_5 = h_6 = h$. If additionally $h = 1$, we obtain Taylor hardening.

Improved evolution equations for continuum crystal plasticity are obtained by comparing continuum and discrete dislocation dynamics methods Šiška et al. (2009).

1.2 A Cosserat theory for elastoviscoplastic single crystals at finite deformation

Mandel Mandel (1973) introduced the notion of oriented microelements characterized by some hidden directors into the theory of elastoviscoplasticity. The epoch-making expression "trièdre directeur" is directly taken from the Cosserat brothers' well-known work Cosserat and Cosserat (1909). The relative rotation of neighbouring microelements may induce local couple stresses. To a first approximation Mandel neglects them and regards the single crystal and the polycrystal as a classical continuum. We propose here the strict treatment of the single crystal as a Cosserat continuum.

Nye noticed that after bending or torsion a crystal contains excess dislocations of a definite sign that give rise to lattice curvature Nye (1953). In a modelling of single crystals with more reference to dislocations, this additional deformation possibility should be taken into account. Kröner claimed that the macroscopic response of a medium to lattice curvature is the existence of actual Cosserat couple stresses Kröner (1963). The couple-stresses may be of the same order of magnitude as force-stresses under some circumstances Hehl and Kröner (1958); McClintock et al. (1958). In these early works Kröner regards the dislocated crystal as a Cosserat medium. However, his theory deals with symmetric force-stresses and he suggests later that there may be fundamental differences between dislocation theory and Cosserat theory Kröner, E. (1967).

The reason for such a misunderstanding stems from the frequent use in literature of the Cosserat continuum as the medium in which single dislocations may be embedded. Kessel computes the force and couple stress fields around a screw and an edge dislocation in a Cosserat continuum Kessel (1970). In this chapter, we claim that the continuum containing a large number of dislocations in the sense of the continuum theory of dislocations, can be modelled as a Cosserat continuum. Kröner argues that the rotation of the crystal lattice with dislocations is not the eigenrotation of physical particles but the rotation of a structure. This pleads against the constrained Cosserat theory that is usually used in the continuum theory of dislocations. As a result, in the Cosserat theory presented here, the rotational degrees of freedom are independent of the displacement field. Whereas the definition of the Cosserat directors involved in the continuum theory of dislocation is left unspecified, the "trièdre directeur" in this work is clearly made of three orthogonal lattice vectors attached to each volume element. As for them, Claus and Eringen also erect a lattice triad at every point of the continuum Eringen and Claus (1970). A most interesting point in their work is that they resort to a micromorphic continuum. They also propose a phenomenological treatment of micromorphic elastoplasticity but they do not derive the crystallographic expressions of plastic slip and curvature nor the necessary constitutive equations.

The characteristic size of the volume element must be such that it contains a large number of dislocations and that a mean crystal orientation can be unambiguously defined at each time. The elements of the presented theory have been proposed in Forest et al. (1997). They have been recently extended and implemented in finite element code by Clayton et al. (2006); Mayeur et al. (2011).

Kinematics of the Cosserat continuum A material point $M \in V$ at time t_0 is described by its position $\underline{X}$ and its inner state, for an arbitrary initial placement, chosen as the reference configuration. At time t, its position is $\underline{x}(\underline{X}, t)$ and its inner state $\underset{\sim}{R}(\underline{X}, t)$, in a given reference frame E. If $(\underline{d}_i)_{i=1,3}$ are three orthogonal lattice vectors in a released state at t and $(\underline{d}_i^0)_{i=1,3}$ their initial position in E, then the rotation $\underset{\sim}{R}$ is defined through

$$\underline{d}_i = \underset{\sim}{R}.\underline{d}_i^0, \quad with \quad \underset{\sim}{R}.\underset{\sim}{R}^T = \underset{\sim}{1}, \quad \underset{\sim}{R}(\underline{X}, t_0) = \underset{\sim}{1} \ and \ Det \ \underset{\sim}{R} = 1 \tag{21}$$

A rotating frame $E^\sharp(M)$ is attached to the lattice structure at each point $M \in V$ and each tensor variable y considered with respect to $E^\sharp$ will be denoted ${}^\sharp y$.

The rotation field $\underset{\sim}{R}(\underline{X}, t)$ can be replaced by the vector field $\underline{\Phi}(\underline{X}, t)$ given by equation (118). The three components of $\underline{\Phi}$ are three degrees of freedom of the continuum in addition to the three components of the displacement field

$$\underline{u}(\underline{X}, t) = \underline{x}(\underline{X}, t) - \underline{X} \tag{22}$$

Here, $\underline{u}$ and $\underline{\Phi}$ are regarded as independent kinematic variables which can be connected only on the balance or constitutive level or by some constraint.

The deformation gradient classically links a current infinitesimal material segment $d\underline{x}$ with its initial position $d\underline{X}$

$$d\underline{x} = \underset{\sim}{F}.d\underline{X} \tag{23}$$

so that

$$\underset{\sim}{F} = \underline{u} \otimes \nabla = u_{i,j} \ \underline{e}_i \otimes \underline{e}_j \tag{24}$$

(in the absence of other indication, partial derivatives are taken with respect to the X_j).

Similarly, we compute the variation $d\underset{\sim}{R}$ of microrotation along a material segment $d\underline{X}$. Defining $\delta\underline{\Phi}$ by

$$(d\underset{\sim}{R}).\underset{\sim}{R}^T = \underset{\sim}{1} \times \delta\underline{\Phi} = \underset{\approx}{\epsilon}.\delta\underline{\Phi} \tag{25}$$

we derive

$$\delta\underline{\Phi} = -\frac{1}{2} \underset{\approx}{\epsilon} : (d\underset{\sim}{R}.\underset{\sim}{R}^T) = \underset{\sim}{\Gamma} \ d\underline{X} \tag{26}$$

with

$$\underset{\sim}{\Gamma} = \frac{1}{2} \underset{\approx}{\epsilon} : (\underset{\sim}{R}.(\underset{\sim}{R}^T \otimes \nabla)) \tag{27}$$

The notation $\delta\underline{\Phi}$ means that $\delta\underline{\Phi}$ is not a total differential, as can be seen from (26). Contrary to $\underset{\sim}{F}$, $\underset{\sim}{\Gamma}$ generally is not invertible. With respect to the local space frame $E^\sharp$,

$$^\sharp d\underline{x} = {}^\sharp\underset{\sim}{F}.d\underline{X} \quad and \quad {}^\sharp\delta\underline{\Phi} = {}^\sharp\underset{\sim}{\Gamma}.d\underline{X} \tag{28}$$

where $^\sharp d\underline{x} = \underset{\sim}{R}^T.d\underline{x}$ and $^\sharp\delta\underline{\Phi} = \underset{\sim}{R}^T.\delta\underline{\Phi}$ and

$$^\sharp\underset{\sim}{F} = \underset{\sim}{R}^T.\underset{\sim}{F}, \quad {}^\sharp\underset{\sim}{\Gamma} = \underset{\sim}{R}^T.\underset{\sim}{\Gamma} \tag{29}$$

It can be seen that the relative measures $^\sharp\underset{\sim}{F}$ and $^\sharp\underset{\sim}{\Gamma}$ are invariant under any Euclidean transformation Kafadar and Eringen (1971). Accordingly they are natural Cosserat strains for the development of constitutive equations. They are called respectively the Cosserat deformation tensor and the wryness (or bend-twist, or torsion-curvature) tensor. An alternative expression of the wryness tensor is then

$$^\sharp\underset{\sim}{\Gamma} = -\frac{1}{2}\,\underset{\approx}{\epsilon} : (\underset{\sim}{R}^T.(\underset{\sim}{R}\otimes\nabla)) \tag{30}$$

One defines next the velocity and the gyration tensor Stojanović (1970); Eringen (1976)

$$\underline{v} = \dot{\underline{u}} = \dot{u}_i\,\underline{e}_i \quad and \quad \underset{\sim}{\nu} = \dot{\underset{\sim}{R}}.\underset{\sim}{R}^T \tag{31}$$

which can be replaced by the associated gyration vector

$$\overset{\times}{\underline{\nu}} = -\frac{1}{2}\,\underset{\approx}{\epsilon} : \underset{\sim}{\nu} \tag{32}$$

since it is antisymmetric. The time derivative of the Cosserat strains can be related to the gradient of the latter quantities:

$$^\sharp\dot{\underset{\sim}{F}}.{}^\sharp\underset{\sim}{F}^{-1} = \underset{\sim}{R}^T.(\underline{v}\otimes\nabla^c - \underset{\sim}{1}\times\overset{\times}{\underline{\nu}}).\underset{\sim}{R} \tag{33}$$

$$^\sharp\dot{\underset{\sim}{\Gamma}}.{}^\sharp\underset{\sim}{F}^{-1} = \underset{\sim}{R}^T.(\overset{\times}{\underline{\nu}}\otimes\nabla^c).\underset{\sim}{R} \tag{34}$$

where $\nabla^c = \dfrac{\partial}{\partial x_i}\,\underline{e}_i = \underset{\sim}{F}^{-T}\,\nabla$ (Euclidean representation, c stands for current).

$\underline{v}\otimes\nabla^c - \underset{\sim}{\nu}$ is the relative velocity gradient and describes the local motion of the material element with respect to the microstructure.

Sthenics In order to introduce forces and stresses and to deduce the equilibrium equations, we resort to the method of virtual power developed by Germain in the case of micromorphic media Germain (1973b). The virtual motions are the velocity $\underline{v}$ and the gyration $\overset{\times}{\underline{v}}$ (or microrotation rate vector). The next step is to choose the form of the virtual power of a system of forces. Within the framework of a first gradient theory, the virtual power of the internal forces is a linear form of the virtual motions and their gradient. The principle of material frame indifference requires that this linear form should be invariant under any Euclidean transformation. That is why we will work with the objective quantities $\underline{v} \otimes \boldsymbol{\nabla}^c - \underset{\sim}{v}$ and $\overset{\times}{\underline{v}} \otimes \boldsymbol{\nabla}^c$. The dual quantities involved in the linear form of the virtual power of the internal forces are denoted $\underset{\sim}{\boldsymbol{\sigma}}$ and $\underset{\sim}{\boldsymbol{\mu}}$ respectively and are assumed to be objective tensors. For objectivity reasons the dual variable associated with $\underline{v}$ is zero. For any subdomain $D \subset V$

$$
\begin{aligned}
P^{(i)} &= -\int_D \left(\underset{\sim}{\boldsymbol{\sigma}} : (\underline{v} \otimes \boldsymbol{\nabla}^c - \underset{\sim}{v}) + \underset{\sim}{\boldsymbol{\mu}} : (\overset{\times}{\underline{v}} \otimes \boldsymbol{\nabla}^c)\right) dV \\
&= -\int_D \left(\sigma_{ij}\, v_{i,j} + \mu_{ij}\, \overset{\times}{v}_{i,j} - \sigma_{ij}\, v_{ij}\right) dV \\
&= -\int_D \left(\sigma_{ij}\, v_i + \mu_{ij} \overset{\times}{v}_i\right)_{,j} dV && (35) \\
&+ \int_D \left(\sigma_{ij,j}\, v_i + (\mu_{ij,j} - \epsilon_{ikl}\, \sigma_{kl})\, \overset{\times}{v}_i\right) dV && (36) \\
&= -\int_{\partial D} \left(\underline{v}\, \underset{\sim}{\boldsymbol{\sigma}} + \overset{\times}{\underline{v}}\, \underset{\sim}{\boldsymbol{\mu}}\right) .\underline{n}\, dS \\
&+ \int_D \left(\underline{v} .div\, \underset{\sim}{\boldsymbol{\sigma}} + \overset{\times}{\underline{v}} .(div\, \underset{\sim}{\boldsymbol{\mu}} + 2 \overset{\times}{\underline{\boldsymbol{\sigma}}})\right) dV && (37)
\end{aligned}
$$

(in this section the partial derivatives are taken with respect to the current configuration). The virtual power of external forces reads

$$
P^{(e)} = \int_D \left(\underline{f} .\underline{v} + \underline{c} .\overset{\times}{\underline{v}}\right) dS \qquad (38)
$$

The virtual power of contact forces must then be defined

$$
P^{(c)} = \int_{\partial D} \left(\underline{t} .\underline{v} + \underline{m} .\overset{\times}{\underline{v}}\right) dS \qquad (39)
$$

The dual quantities of the velocity and microrotation rate in $P^{(e)}$ and $P^{(i)}$ have the dimensions of force and moment respectively. The principle of virtual power then states that

$$\forall D \subset V, \forall(\underline{v}, \underline{\overset{\times}{v}}) \quad P^{(i)} + P^{(e)} + P^{(c)} = 0 \tag{40}$$

In particular

$$\forall D \subset V, \forall(\underline{v}, \underline{\overset{\times}{v}}) \; / \; \underline{v} = \underline{\overset{\times}{v}} = 0 \; on \; \partial D, \tag{41}$$

$$\int_D \left(\underline{v} .(div \; \underset{\sim}{\sigma} + \underline{f}) + \underline{\overset{\times}{v}} .(div \; \underset{\sim}{\mu} + 2\underline{\overset{\times}{\sigma}} + \underline{c}) \right) dV = 0 \tag{42}$$

Assuming that the quantities are continuous on V, the local equilibrium equations follow from (42)

$$\begin{cases} div \; \underset{\sim}{\sigma} + \underline{f} = 0 \\ div \; \underset{\sim}{\mu} + 2\underline{\overset{\times}{\sigma}} + \underline{c} = 0 \end{cases} \tag{43}$$

As a result, the principle of virtual power becomes

$$\forall D \subset V, \forall(\underline{v}, \underline{\overset{\times}{v}}) \quad \int_{\partial D} \left((\underset{\sim}{\sigma} .\underline{n} - \underline{t}).\underline{v} + (\underset{\sim}{\mu} .\underline{n} - \underline{m}).\underline{\overset{\times}{v}} \right) dV = 0 \tag{44}$$

from which the boundary conditions are deduced

$$\begin{cases} \underset{\sim}{\sigma} .\underline{n} = \underline{t} \\ \underset{\sim}{\mu} .\underline{n} = \underline{m} \end{cases} \tag{45}$$

$\underset{\sim}{\sigma}$ is called the Cauchy force stress tensor and $\underset{\sim}{\mu}$ the couple-stress tensor. They are generally not symmetric.

Hyperelasticity

Energy balance Let ϵ be the internal energy per unit mass, $\underline{q}$ the heat flux vector, ρ the current density. The energy balance equation reads then

$$\rho \, \dot{\epsilon} = \underset{\sim}{\sigma} : (\underline{v} \otimes \boldsymbol{\nabla}^c - \underset{\sim}{v}) + \underset{\sim}{\mu} : (\underline{\overset{\times}{v}} \otimes \boldsymbol{\nabla}^c) - div \; \underline{q} \tag{46}$$

(any other inner heat supply is excluded).

According to the thermodynamics of irreversible processes, the entropy principle is written

$$\rho \, \dot{\eta} + div \; (\frac{\underline{q}}{T}) \geq 0 \tag{47}$$

where T denotes the temperature and η the entropy per unit mass.

Introducing the free energy $\psi = \epsilon - \eta T$ and combining the energy and entropy equations, one derives the Clausius-Duhem inequality

$$-\rho\,(\dot{\psi} + \eta\dot{T}) + {}^\sharp\underset{\sim}{\boldsymbol{\sigma}} : ({}^\sharp\dot{\underset{\sim}{\boldsymbol{F}}}.{}^\sharp\underset{\sim}{\boldsymbol{F}}^{-1}) + {}^\sharp\underset{\sim}{\boldsymbol{\mu}} : ({}^\sharp\dot{\underset{\sim}{\boldsymbol{\Gamma}}}.{}^\sharp\underset{\sim}{\boldsymbol{F}}^{-1}) - \frac{1}{T}\,\underline{\boldsymbol{q}}.T\boldsymbol{\nabla}^c \geq 0 \quad (48)$$

where

$$\left\{\begin{array}{lcl} {}^\sharp\underset{\sim}{\boldsymbol{\sigma}} & = & \underset{\sim}{\boldsymbol{R}}^T.\underset{\sim}{\boldsymbol{\sigma}}.\underset{\sim}{\boldsymbol{R}} \\ {}^\sharp\underset{\sim}{\boldsymbol{\mu}} & = & \underset{\sim}{\boldsymbol{R}}^T.\underset{\sim}{\boldsymbol{\mu}}.\underset{\sim}{\boldsymbol{R}} \end{array}\right. \quad (49)$$

are rotated stress tensors with respect to the space frame $E^\sharp$ attached to the microstructure.

A material is said to be hyperelastic if its free energy and entropy are functions of ${}^\sharp\underset{\sim}{\boldsymbol{F}}$, ${}^\sharp\underset{\sim}{\boldsymbol{\Gamma}}$ and temperature only. The Clausius-Duhem inequality (48) becomes

$$-\left(\rho\,\frac{\partial\psi}{\partial{}^\sharp\underset{\sim}{\boldsymbol{F}}} - {}^\sharp\underset{\sim}{\boldsymbol{\sigma}}.{}^\sharp\underset{\sim}{\boldsymbol{F}}^{-T}\right) : {}^\sharp\dot{\underset{\sim}{\boldsymbol{F}}} \qquad - \left(\rho\,\frac{\partial\psi}{\partial{}^\sharp\underset{\sim}{\boldsymbol{\Gamma}}} - {}^\sharp\underset{\sim}{\boldsymbol{\mu}}.{}^\sharp\underset{\sim}{\boldsymbol{F}}^{-T}\right) : {}^\sharp\dot{\underset{\sim}{\boldsymbol{\Gamma}}}$$
$$- \left(\rho\,\eta + \rho\,\frac{\partial\psi}{\partial T}\right)\dot{T} - \frac{1}{T}\,\underline{\boldsymbol{q}}.T\boldsymbol{\nabla}^c \geq 0$$

Since this expression is linear in ${}^\sharp\dot{\underset{\sim}{\boldsymbol{F}}}$, ${}^\sharp\dot{\underset{\sim}{\boldsymbol{\Gamma}}}$ and $\dot{T}$, the last inequality implies

$$\eta = -\frac{\partial\psi}{\partial T} \quad (50)$$

and

$$\left\{\begin{array}{lcl} {}^\sharp\underset{\sim}{\boldsymbol{\sigma}} & = & \rho\,\dfrac{\partial\psi}{\partial{}^\sharp\underset{\sim}{\boldsymbol{F}}}.{}^\sharp\underset{\sim}{\boldsymbol{F}}^T \\ {}^\sharp\underset{\sim}{\boldsymbol{\mu}} & = & \rho\,\dfrac{\partial\psi}{\partial{}^\sharp\underset{\sim}{\boldsymbol{\Gamma}}}.{}^\sharp\underset{\sim}{\boldsymbol{F}}^T \end{array}\right. \quad (51)$$

Linear case; isotropic elasticity Strain and torsion-curvature are small if $\|{}^\sharp\underset{\sim}{\boldsymbol{F}} - \underset{\sim}{\boldsymbol{1}}\| \ll 1$ and $\|{}^\sharp\underset{\sim}{\boldsymbol{\Gamma}}\|\, l \ll 1$, where l is a characteristic length. If, in addition, microrotations remain small, i.e. if $\|\underline{\boldsymbol{\Phi}}\| \ll 1$, then

$$\left\{\begin{array}{lcl} \underset{\sim}{\boldsymbol{R}} & \simeq & \underset{\sim}{\boldsymbol{1}} + \underset{\sim}{\boldsymbol{1}} \times \underline{\boldsymbol{\Phi}} = \underset{\sim}{\boldsymbol{1}} - \underset{\approx}{\boldsymbol{\epsilon}}.\underline{\boldsymbol{\Phi}} \\ {}^\sharp\underset{\sim}{\boldsymbol{F}} & \simeq & \underset{\sim}{\boldsymbol{1}} + \underline{\boldsymbol{u}} \otimes \boldsymbol{\nabla} + \underset{\approx}{\boldsymbol{\epsilon}}.\underline{\boldsymbol{\Phi}} = \underset{\sim}{\boldsymbol{1}} + \underset{\sim}{\boldsymbol{e}} \\ {}^\sharp\underset{\sim}{\boldsymbol{\Gamma}} & \simeq & \underline{\boldsymbol{\Phi}} \otimes \boldsymbol{\nabla} = \underset{\sim}{\boldsymbol{\kappa}} \end{array}\right. \quad (52)$$

Furthermore, ${}^\sharp\underset{\sim}{\boldsymbol{\sigma}} \simeq \underset{\sim}{\boldsymbol{\sigma}}$ and ${}^\sharp\underset{\sim}{\boldsymbol{\mu}} \simeq \underset{\sim}{\boldsymbol{\mu}}$. Accordingly, for linear elasticity, two four-rank elasticity tensors are introduced

$$\left\{\begin{array}{lcl} \underset{\sim}{\boldsymbol{\sigma}} & = & \underset{\approx}{\boldsymbol{E}} : \underset{\sim}{\boldsymbol{e}} \\ \underset{\sim}{\boldsymbol{\mu}} & = & \underset{\approx}{\boldsymbol{C}} : \underset{\sim}{\boldsymbol{\kappa}} \end{array}\right. \quad (53)$$

(No coupling between strain and torsion-curvature is possible as soon as point symmetry is assumed even for the less symmetric solid Kessel (1964)). Some symmetry properties of these tensors are derived from the hyperelasticity conditions (51)

$$E_{ijkl} = E_{klij} \; and \; C_{ijkl} = C_{klij} \tag{54}$$

Further symmetry conditions can be gained if material symmetries are taken into account. The form of the Cosserat elasticity tensors for all symmetry classes has been established by Kessel Kessel (1964). For a triclinic solid, 90 independent constants are necessary instead of 21 in the classical case (for a solid without point symmetry, Kessel found 171 constants!). We now consider the example of isotropic elasticity. The two classical Lamé constants λ, μ are complemented by 4 additional parameters according to

$$\left\{ \begin{array}{l} \underset{\sim}{\sigma} = \lambda \, \underset{\sim}{1} \, Tr \, \underset{\sim}{e} + 2\mu \, ^{\{}\underset{\sim}{e}^{\}} + 2\mu_c \, ^{\}}\underset{\sim}{e}^{\{} \\ \underset{\sim}{\mu} = \alpha \, \underset{\sim}{1} \, Tr \, \underset{\sim}{\kappa} + 2\beta \, ^{\{}\underset{\sim}{\kappa}^{\}} + 2\gamma \, ^{\}}\underset{\sim}{\kappa}^{\{} \end{array} \right. \tag{55}$$

Elastoplastic Cosserat single crystals The works of Sawczuk, Lippmann and Besdo are the first milestones in the plasticity theory of Cosserat continua at small strains Sawczuk (1967); Lippmann (1969); Besdo (1974). In the case of single crystals we resort to recent results in Cosserat theory at large strains Sievert (1992); Sansour (1998b).

Strain decomposition In single crystals, non-homogeneous plastic deformations induce non-homogeneous permanent lattice rotations, which are associated with plastic lattice curvature. That is why elastic and plastic Cosserat deformations and curvatures are introduced: $^\sharp\underset{\sim}{F}^e, {}^\sharp\underset{\sim}{F}^p, {}^\sharp\underset{\sim}{\Gamma}^e$ *and* $^\sharp\underset{\sim}{\Gamma}^p$. Strain partition rules must then be proposed.

The multiplicative decomposition proposed in Mandel (1971) is adopted here but only for the Cosserat deformation gradient

$$^\sharp\underset{\sim}{F} = {}^\sharp\underset{\sim}{F}^e . {}^\sharp\underset{\sim}{F}^p \tag{56}$$

The expression

$$^\sharp\dot{\underset{\sim}{F}} . {}^\sharp\underset{\sim}{F}^{-1} = {}^\sharp\dot{\underset{\sim}{F}}^e . {}^\sharp\underset{\sim}{F}^{e-1} + {}^\sharp\underset{\sim}{F}^e . {}^\sharp\dot{\underset{\sim}{F}}^p . {}^\sharp\underset{\sim}{F}^{p-1} . {}^\sharp\underset{\sim}{F}^{e-1} \tag{57}$$

has to be substituted in the Clausius-Duhem inequality (48). The most natural assumption is that the hyperelastic relations still have the form

$$\left\{ \begin{array}{l} ^\sharp\underset{\sim}{\sigma} = \rho \, \dfrac{\partial \psi}{\partial {}^\sharp\underset{\sim}{F}^e} . {}^\sharp\underset{\sim}{F}^{eT} \\ ^\sharp\underset{\sim}{\mu} = \rho \, \dfrac{\partial \tilde{\psi}}{\partial {}^\sharp\underset{\sim}{\Gamma}^e} . {}^\sharp\underset{\sim}{F}^{eT} \end{array} \right. \tag{58}$$

It has been proved in Sievert (1992) that the multiplicative decomposition (56) results from the hyperelastic relation (58)a. Similarly, for the relations (58) to hold the form of the decomposition of the overall curvature tensor cannot be arbitrary. One must have :

$$ {}^{\sharp}\underset{\sim}{\Gamma} = {}^{\sharp}\underset{\sim}{\Gamma}^{e} . {}^{\sharp}\underset{\sim}{F}^{p} + {}^{\sharp}\underset{\sim}{\Gamma}^{p} \tag{59} $$

An elastic-plastic decomposition of displacement or rotation like in Steinmann (1994) is not recommended, because these non-objective variables can not be connected with the quantities energy and dissipation. Such a connection is possible only on the level of strains. The form of the decomposition (59) is similar to that used by Eringen and Kafadar for the symmetry transformation of a Cosserat fluid. The decomposition (59) has been assumed in Dluzewski (1991). Then the elastic constitutive equations (58) follow necessarily.

The previous decompositions of total deformation and curvature, together with the hyperelastic relations (58) make it possible to define a unique intermediate isoclinic configuration for which force and couple stresses are simultaneously released. This is illustrated by the actualized picture of Cosserat crystal plasticity in figure 1(b).

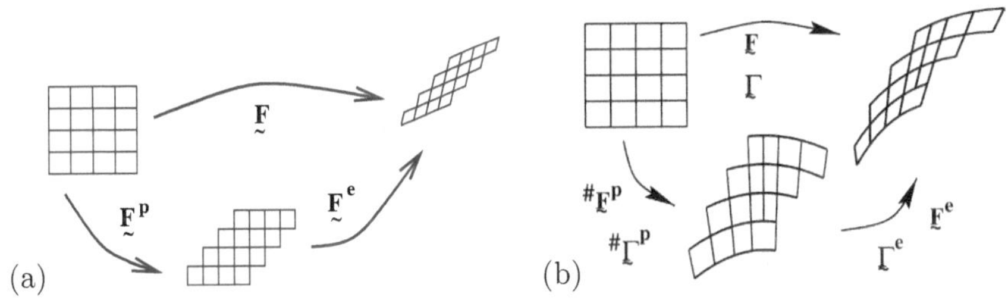

Figure 1. Kinematics of classical crystal plasticity (a) and Cosserat crystal crystal plasticity (b).

Kinematics of elastoplastic Cosserat single crystals The plastic deformation of single crystals is the result of slip processes on slip systems. For each slip system s, we define

$$ \underline{m}^{s} = \underline{b}^{s} / \|\underline{b}^{s}\| \tag{60} $$

where $\underline{\boldsymbol{b}}^{\,s}$ is the Burgers vector. $\underline{\boldsymbol{n}}^{\,s}$ is the unit vector normal to the slip plane. As a result, the plastic strain rate takes the form

$$ {}^{\sharp}\dot{\underset{\sim}{\boldsymbol{F}}}^{p}.{}^{\sharp}\underset{\sim}{\boldsymbol{F}}^{p-1} = \sum_{s\in S} \dot{\gamma}^{s}\, {}^{\sharp}\underset{\sim}{\boldsymbol{P}}^{s} \tag{61} $$

γ^s is the amount of slip for the system s. ${}^{\sharp}\underset{\sim}{\boldsymbol{P}}^{s}$ is given by the kinematics of slip

$$ {}^{\sharp}\underset{\sim}{\boldsymbol{P}}^{s} = {}^{\sharp}\underline{\boldsymbol{m}}^{\,s} \otimes {}^{\sharp}\underline{\boldsymbol{n}}^{\,s} \tag{62} $$

If we go back to the Eulerian representation

$$ \underline{\boldsymbol{v}} \otimes \boldsymbol{\nabla}^{c} = \dot{\underset{\sim}{\boldsymbol{F}}}.\underset{\sim}{\boldsymbol{F}}^{-1} = \dot{\underset{\sim}{\boldsymbol{R}}}.\underset{\sim}{\boldsymbol{R}}^{T} + \underset{\sim}{\boldsymbol{R}}.{}^{\sharp}\dot{\underset{\sim}{\boldsymbol{F}}}^{e}.{}^{\sharp}\underset{\sim}{\boldsymbol{F}}^{e-1}.\underset{\sim}{\boldsymbol{R}}^{T} \tag{63} $$

$$ + \; \underset{\sim}{\boldsymbol{R}}.{}^{\sharp}\underset{\sim}{\boldsymbol{F}}^{e}.{}^{\sharp}\dot{\underset{\sim}{\boldsymbol{F}}}^{p}.{}^{\sharp}\underset{\sim}{\boldsymbol{F}}^{p-1}.{}^{\sharp}\underset{\sim}{\boldsymbol{F}}^{e-1}.\underset{\sim}{\boldsymbol{R}}^{T}, \tag{64} $$

we can split the last expression into its symmetric and skew–symmetric parts:

$$ {}^{\{}\underline{\boldsymbol{v}} \otimes \boldsymbol{\nabla}^{c\}} = {}^{\{}\underset{\sim}{\boldsymbol{R}}.{}^{\sharp}\dot{\underset{\sim}{\boldsymbol{F}}}^{e}.{}^{\sharp}\underset{\sim}{\boldsymbol{F}}^{e-1}.\underset{\sim}{\boldsymbol{R}}^{T\}} + \sum_{s\in S} \dot{\gamma}^{s}\, \{\, {}^{\star}\underline{\boldsymbol{m}}^{\,s} \otimes {}^{\star}\underline{\boldsymbol{n}}^{\,s\}} \tag{65} $$

and

$$ {}^{\}}\underline{\boldsymbol{v}} \otimes \underline{\boldsymbol{\nabla}}^{\,c\{} - \dot{\underset{\sim}{\boldsymbol{R}}}.\underset{\sim}{\boldsymbol{R}}^{T} = {}^{\}}\underset{\sim}{\boldsymbol{R}}.{}^{\sharp}\dot{\underset{\sim}{\boldsymbol{F}}}^{e}.{}^{\sharp}\underset{\sim}{\boldsymbol{F}}^{e-1}.\underset{\sim}{\boldsymbol{R}}^{T\{} + \sum_{s\in S} \dot{\gamma}^{s}\, \}\, {}^{\star}\underline{\boldsymbol{m}}^{\,s} \otimes {}^{\star}\underline{\boldsymbol{n}}^{\,s\{} \tag{66} $$

where we have noted

$$ {}^{\star}\underline{\boldsymbol{m}}^{\,s} = \underset{\sim}{\boldsymbol{R}}.{}^{\sharp}\underset{\sim}{\boldsymbol{F}}^{e}.{}^{\sharp}\underline{\boldsymbol{m}}^{\,s} \;\; \textit{and} \;\; {}^{\star}\underline{\boldsymbol{n}}^{\,s} = \underset{\sim}{\boldsymbol{R}}.{}^{\sharp}\underset{\sim}{\boldsymbol{F}}^{e-T}.{}^{\sharp}\underline{\boldsymbol{n}}^{\,s} \tag{67} $$

Equation (66) clearly shows that the relative rotation rate of material lines with respect to the microstructure is due to the lattice rotation associated with slip processes, if elastic contributions are neglected.

We would like to compare the proposed formulation with Mandel's work. We are working with invariant tensors written in the microstructure space frame in order to get rid of undetermined rotations. An equivalent method is to deal with the so-called isoclinic configuration introduced by Teodosiu and Mandel Teodosiu (1970); Mandel (1971). Their description reads

$$ \underset{\sim}{\boldsymbol{F}} = \underset{\sim}{\boldsymbol{E}}.\underset{\sim}{\boldsymbol{P}} \tag{68} $$

where the rotation $\underset{\sim}{\boldsymbol{R}}^{isoclinic}$ appearing in the polar decomposition of $\underset{\sim}{\boldsymbol{E}}$ links the isoclinic reference frame to the working space frame. As a result comparing (56) and (68) one can think of the equivalence

$$ \underset{\sim}{\boldsymbol{E}} = \underset{\sim}{\boldsymbol{R}}\,{}^{\sharp}\underset{\sim}{\boldsymbol{F}}^{e} \tag{69} $$

However, considering the respective polar decompositions

$$\underset{\sim}{\boldsymbol{E}} = \underset{\sim}{\boldsymbol{R}}^{isoclinic}\,\underset{\sim}{\boldsymbol{U}}^{e} \quad and \quad {}^{\sharp}\underset{\sim}{\boldsymbol{F}}^{e} = \underset{\sim}{\boldsymbol{R}}^{e}\,\underset{\sim}{\boldsymbol{U}}^{e} \tag{70}$$

we should have then

$$\underset{\sim}{\boldsymbol{R}}^{isoclinic} = \underset{\sim}{\boldsymbol{R}}.\underset{\sim}{\boldsymbol{R}}^{e} \tag{71}$$

Regarding the elastic behaviour in the classical case, lattice vectors are material vectors with respect to the intermediate configuration. Within the proposed framework this is not exactly true any more. There is an additional rotation $\underset{\sim}{\boldsymbol{R}}^{e}$ of material fibres with respect to the microstructure, that could be attributed to the presence of heterogeneities. Nevertheless the constitutive theory must be such that $\underset{\sim}{\boldsymbol{R}}^{e}$ remains a corrective term. The equivalence of the two theories is established if ${}^{\sharp}\underset{\sim}{\boldsymbol{F}}^{e}$ is symmetric.

The plastic lattice curvature and torsion are due to the presence of dislocations with a non-vanishing resulting Burgers vector. The curvature planes and torsion axes are therefore related to crystallographic directions. They can be represented by the effect of continuous edge and screw dislocations for each slip system. That is why we propose the following kinematics for the plastic wryness

$${}^{\sharp}\underset{\sim}{\dot{\boldsymbol{\Gamma}}}^{p}.{}^{\sharp}\underset{\sim}{\boldsymbol{F}}^{p-1} = \sum_{s\in S} \frac{\dot{\theta}^{s}}{l}\,{}^{\sharp}\underset{\sim}{\boldsymbol{Q}}^{s} \tag{72}$$

The θ^{s} are angles that measure the plastic curvature and torsion over a characteristic length l. Explicit forms for ${}^{\sharp}\underset{\sim}{\boldsymbol{Q}}^{s}$ are given in section 4.1. In section 3.4.2 an alternative more simple treatment for the plastic curvature is proposed.

Dissipation In the Clausius-Duhem inequality (48), a contribution to the overall entropy production is due to the development of rotation gradients. If no hardening variables are introduced, the intrinsic dissipation rate is

$$\begin{aligned}\dot{D} &= \underset{\sim}{\boldsymbol{\sigma}} : \underset{\sim}{\boldsymbol{R}}.{}^{\sharp}\underset{\sim}{\boldsymbol{F}}^{e}.{}^{\sharp}\underset{\sim}{\dot{\boldsymbol{F}}}^{p}.{}^{\sharp}\underset{\sim}{\boldsymbol{F}}^{p-1}.{}^{\sharp}\underset{\sim}{\boldsymbol{F}}^{e-1}.\underset{\sim}{\boldsymbol{R}}^{T}\\ &+ \underset{\sim}{\boldsymbol{\mu}} : \underset{\sim}{\boldsymbol{R}}.{}^{\sharp}\underset{\sim}{\boldsymbol{\Gamma}}^{e}.{}^{\sharp}\underset{\sim}{\dot{\boldsymbol{F}}}^{p}.{}^{\sharp}\underset{\sim}{\boldsymbol{F}}^{p-1}.{}^{\sharp}\underset{\sim}{\boldsymbol{F}}^{e-1}\,\underset{\sim}{\boldsymbol{R}}^{T} + \underset{\sim}{\boldsymbol{\mu}} : \underset{\sim}{\boldsymbol{R}}.{}^{\sharp}\underset{\sim}{\dot{\boldsymbol{\Gamma}}}^{p}.{}^{\sharp}\underset{\sim}{\boldsymbol{F}}^{-1}.\underset{\sim}{\boldsymbol{R}}^{T}\end{aligned} \tag{73}$$

Taking (61) and (72) into account,

$$\dot{D} = \sum_{s\in S}\dot{\gamma}^{s}\,\underset{\sim}{\boldsymbol{\sigma}} : {}^{\star}\underset{\sim}{\boldsymbol{P}}^{s} + \sum_{s\in S}\dot{\theta}^{s}\,\underset{\sim}{\boldsymbol{\mu}} : {}^{\star}\underset{\sim}{\boldsymbol{Q}}^{s} + \underset{\sim}{\boldsymbol{\mu}} : \underset{\sim}{\boldsymbol{R}}.{}^{\sharp}\underset{\sim}{\boldsymbol{\Gamma}}^{e}.{}^{\sharp}\underset{\sim}{\dot{\boldsymbol{F}}}^{p}.{}^{\sharp}\underset{\sim}{\boldsymbol{F}}^{p-1}.{}^{\sharp}\underset{\sim}{\boldsymbol{F}}^{e-1}.\underset{\sim}{\boldsymbol{R}}^{T} \tag{74}$$

where

$$\left\{\begin{aligned}{}^{\star}\underset{\sim}{\boldsymbol{P}}^{s} &= \underset{\sim}{\boldsymbol{R}}.{}^{\sharp}\underset{\sim}{\boldsymbol{F}}^{e}.{}^{\sharp}\underset{\sim}{\boldsymbol{P}}^{s}.{}^{\sharp}\underset{\sim}{\boldsymbol{F}}^{e-1}.\underset{\sim}{\boldsymbol{R}}^{T}\\ {}^{\star}\underset{\sim}{\boldsymbol{Q}}^{s} &= \underset{\sim}{\boldsymbol{R}}.{}^{\sharp}\underset{\sim}{\boldsymbol{F}}^{e}.{}^{\sharp}\underset{\sim}{\boldsymbol{Q}}^{s}.{}^{\sharp}\underset{\sim}{\boldsymbol{F}}^{e-1}.\underset{\sim}{\boldsymbol{R}}^{T}\end{aligned}\right.$$

Three terms appear in the dissipation. The first one is the classical one: slip processes due to irreversible dislocation motion are dissipative. The second one is due to the evolution of plastic curvature and torsion. It is clear that homogeneous lattice rotation is definitely not a dissipative process but plastic curvature due to non-homogeneous lattice rotation is related to the existence of accommodation dislocations and therefore must be associated with dissipation. The last term reveals the independence of the elastic curvature-torsion measure from plastic changes of the material lines in the intermediate configuration. This is due to the lattice concept, which means that the elastic behaviour, established in (58), is primary not influenced by plastic straining. Thus, the elastic strain measures are related to lattice line-elements and the referring to material lines produces an additional term in the plastic wryness rate. However, at small elastic strains, this term vanishes.

1.3 Closure of the continuum theory of dislocations

Closure problem of the continuum theory of dislocations The origin of the continuum theory of dislocations goes back to Nye's epoch-making work Nye (1953). He introduced the dislocation density tensor $\underset{\sim}{\alpha}$ and he established a link between $\underset{\sim}{\alpha}$ and the lattice curvature. Kröner proposed a general presentation of the theory and gave the set of partial differential equations to be solved in the linear static case for a given distribution of dislocations and here for an infinite body Kröner (1958)

$$\begin{cases} \underset{\sim}{\beta} = \underset{\sim}{\beta}^e + \underset{\sim}{\beta}^p \\ \underset{\sim}{\sigma} = \underset{\approx}{E} : \{\underset{\sim}{\beta}^e\} \\ div\, \underset{\sim}{\sigma} = 0 \\ curl\, \underset{\sim}{\beta}^e = \underset{\sim}{\alpha} \end{cases}$$

where $\underset{\sim}{\beta} = \underline{u} \otimes \nabla = u_{i,j}\, \underline{e}_i \otimes \underline{e}_j$. In this part, we use Kröner's notations for historical reasons. It must be noted that strictly speaking the non-objective quantity $\underset{\sim}{\beta}$ cannot be decomposed entirely into an elastic and plastic part but the usual notations of the continuum theory of dislocations and of classical plasticity theory can be reconciled by the concept of isoclinic configuration. The continuum theory of dislocations is a way to think of dislocation theory as a physical field theory. Despite the fact that the system (75) enables us to find the stress strain field around dislocations in some given arrangement, such a theory cannot bridge the gap between dislocation theory and plasticity theory since it does not predict the motion of dislocations: the dislocation distribution must be known at each step. In the dynamic theory of continuous distributions of dislocations, Kröner and Mura, T. (1963) introduce the dislocation velocity tensor $\underset{\sim}{V}$ which is related to the plastic deformation rate $\dot{\underset{\sim}{\beta}}^p$ by

$$\dot{\underset{\sim}{\beta}}^p = -\left(\underset{\sim}{\epsilon} : \underset{\approx}{V}\right)^T \tag{75}$$

and we still have

$$\dot{\underset{\sim}{\alpha}} = -curl\, \dot{\underset{\sim}{\beta}}^p \tag{76}$$

For a single dislocation

$$\underline{\underline{\underline{V}}} = \underline{v} \otimes \underline{\xi} \otimes \underline{b} \tag{77}$$

$\underline{v}$ is the dislocation velocity vector, $\underline{\xi}$ is the dislocation line vector (notation of Hirth and Lothe (1982)) and $\underline{b}$ the Burgers vector. In this case stress and strain can be obtained provided that $\underset{\sim}{\alpha}$ and $\underset{\approx}{V}$ are given at each time, which is of no help to derive a plasticity theory. For, the continuum theory of dislocations does not provide constitutive equations. As pointed out by Hahn and Jaunzemis (1973), in a complete theory of dislocations, the density and motion of dislocations should be derivable from the knowledge of initial conditions (and boundary conditions) only. This is what we call the closure problem of the continuum theory of dislocations.

Two attempts to derive the missing constitutive equations must be mentioned. On the one hand Mura showed how the von Mises yield criterion and Prandtl-Reuss relations can be explained in terms of the dislocation velocity tensor and a so-called "gliding force" Mura, T. (1965). The underlying constitutive assumption is a linear relation between $\underset{\approx}{V}$ and the gliding force. According to Lardner (1969); Eisenberg (1970), constitutive equations are also necessary to link plasticity and dislocation theories. On the other hand Hahn and Jaunzemis distinguish mobile dislocations (M) from immobile ones (I) with common line and Burgers vectors. $A^{ab} = A_I^{ab} + A_M^{ab}$ is the number of dislocations of Burgers vector $\underline{b}^{\,a}$ and line vector $\underline{\xi}^{\,b}$. Using a large strain formulation, (75) combined with (77) yields

$$\dot{\underset{\sim}{F}}^p\, \underset{\sim}{F}^{p-1} = \sum_{a,b} \underline{b}^{\,a} \otimes (\underline{\xi}^{\,b} \times \underline{v}^{\,ab})\, A_M^{ab} = \sum_a A^a V^a\, \underline{b}^{\,a} \otimes \underline{n}^{\,a} \tag{78}$$

where $A^a V^a\, \underline{n}^{\,a} = \sum_b A_M^{ab}\, \underline{\xi}^{\,b} \times \underline{v}^{\,ab}$ is normal to the slip plane. Evolution equations are proposed for A_I^a and A_M^a. Isotropic and kinematic hardening and a viscous stress are also introduced in the modelling.

Recent advances have been reported in the development of a dynamic continuum theory of dislocations Acharya (2010). Applications to continuum modelling of pile-ups around particles can be found in Taupin et al.

(2010). A common feature of such generalized continuum dislocations theories is that they are associated with the existence of lattice curvature induced back-stress and corresponding kinematic hardening Steinmann (1996); Forest (2008); Cordero et al. (2010).

Statistical description of dislocation distribution The dislocation network and the distribution of dislocation sources within a volume element of single crystal often is or becomes so intricate that an exact description of all dislocation lines and Burgers vectors must be abandoned. Instead some overall and statistical information about the distribution may be sufficient for the modelling of the plastic behaviour of the element. The only known attempts to develop a complete statistical theory of dislocations goes back to Zorski (1968) and Kröner (1969). The systematic approach comes up against tremendous difficulties which are still not overcome. This explains why the concepts reviewed in this section are only rudimentary tools which do not exhaust the complexity of dislocation structures.

Dislocation density tensor and the continuum theory of dislocations Within the framework of the continuum theory of dislocations, the characteristic size l of the volume element is taken large enough for the effects of the dislocations within it to be averaged. The distribution of dislocations is made continuous by letting $b = \|\underline{\boldsymbol{b}}\|$ approach zero and increasing the number n of dislocations of each kind so as to keep nb constant Nye (1953). The definition of the Burgers vector can be extended to continuous distributions of dislocations. For that purpose one refers to the kinematic description proposed by Mandel making use of the isoclinic configuration and of the strain partition given by (68). In (68), $\underset{\sim}{\boldsymbol{E}}$ relates the infinitesimal vectors $d\underline{\boldsymbol{\zeta}}$ and $d\underline{\boldsymbol{x}}$, where $d\underline{\boldsymbol{\zeta}}$ results from the cutting and releasing operations from the infinitesimal current lattice vector $d\underline{\boldsymbol{x}}$

$$d\underline{\boldsymbol{\zeta}} = \underset{\sim}{\boldsymbol{E}}^{-1}\, d\underline{\boldsymbol{x}} \tag{79}$$

From this, it can be seen that the decomposition (68) actually goes back to Bilby et al. (1957).

Accordingly, if S is a smooth surface containing $\underline{\boldsymbol{x}}$ in the current configuration and bounded by the closed line c, the true Burgers vector is defined as

$$\underline{\boldsymbol{b}} = \oint_c \underset{\sim}{\boldsymbol{E}}^{-1}.d\underline{\boldsymbol{x}} \tag{80}$$

The application of Stokes' formula (125) leads to the definition of the so-called true dislocation density tensor

$$\underset{\sim}{\alpha} = -curl^c \, \underset{\sim}{E}^{-1} = \underset{\sim}{E}^{-1} \times \nabla^c = -\epsilon_{jkl} \, E^{-1}_{ik,l} \, \underline{e}_i \otimes \underline{e}_j \tag{81}$$

such that

$$\underline{b} = \int_S \underset{\sim}{\alpha}.\underline{n} \; dS \tag{82}$$

If the surface is infinitesimal of normal $\underline{n}$, $d\underline{b} = \underset{\sim}{\alpha}.\underline{n} \; dS$ is the resulting true Burgers vector of dislocations crossing the surface dS. It is convenient to associate each component α_{ij} of the dislocation density tensor with a (super)dislocation characterized by its line vector $\underline{e}_j$ and its Burgers vector $b_i \, \underline{e}_i$ (no summation). As a result, the diagonal components of $\underset{\sim}{\alpha}$ represent screw dislocations and the out-of-diagonal ones edge dislocations. For n dislocations per unit surface of Burgers vector $\underline{b}$ and line vector $\underline{\xi}$, we have

$$\underset{\sim}{\alpha} = n \, \underline{b} \otimes \underline{\xi} \tag{83}$$

Scalar dislocation densities and crystal plasticity In the classical continuum theory of dislocations, the description of the dislocation distribution is restricted to the dislocation density tensor. It enables one to compute stress-strain fields for special distributions and even discrete dislocations for which $\underset{\sim}{\alpha}$ becomes the sum of Dirac's functions. However the classical continuum theory of dislocations has failed to describe the elastoplastic behaviour of single crystals. The main reason is that the dislocation density tensor is not the relevant variable to explain the hardening processes. In Hahn and Jaunzemis (1973), the kinematics of plastic deformation are derived from the dislocation velocity tensor and corresponds exactly to the purely mechanical description of slip processes proposed by Mandel. The next step is the introduction of hardening variables as in the classical macroscopic plasticity theory. They are related to usual scalar dislocation densities that are commonly used by metal physicists and which represent the total length of dislocation lines within a volume element. The multiplication and interaction of dislocations is responsible for the hardening of single crystals and the scalar densities are reliable measures for it. This type of description culminates with the work of Mandel, Zarka and Teodosiu and Sidoroff Mandel (1971); Zarka (1972); Teodosiu and Sidoroff (1976). In these theories the dislocation density tensor is not even mentioned since it is not the relevant quantity any more. Constitutive equations for hardening variables are proposed in a more or less phenomenological way and several elementary dislocation interaction processes are taken into account.

The main successes of these theories are the modelling of the tensile behaviour of single crystals, the lattice rotations and the cyclic behaviour of single and polycrystals Cailletaud (1992).

Proposed description In this work we claim that both types of descriptions are required for the modelling of non-homogeneous deformation of single crystals. That is why the statistical description of dislocation distribution must contain at least:

• the dislocation density tensor $\underset{\sim}{\alpha}$ which accounts for the resulting Burgers vector across any infinitesimal surface,

and

• scalar dislocation densities ρ^s or the associated hardening variables, for instance r^s and x^s already used in Méric and Cailletaud (1991). The kinematic hardening variables x_s are a measure for microscopically non-homogeneous spatial dislocation distributions that give rise to a vanishing resulting Burgers vector (dislocation cells...). Additional variables (densities of mobile and immobile dislocations...) may also be necessary.

It must be noted that the dislocation density tensor and the scalar dislocation densities are related respectively to the one-point and two-point dislocation correlations introduced by Kröner Kröner (1969), as explained in section 1.1.

The scalar dislocation densities are necessary to account for the hardening or softening behaviour of the material whereas the dislocation density tensor may play a significant role when strong lattice incompatibilities are present.

Link between the dislocation density tensor and the lattice torsion-curvature tensor

Classical analysis at small strains and small rotations Nye introduces the rotation vector $\underline{\mathbf{\Phi}}$ of the lattice and the curvature tensor $\underset{\sim}{\kappa} = \underline{\mathbf{\Phi}} \otimes \mathbf{\nabla}$ At small strains and small rotations, the strain and rotation rate decomposition into elastic and plastic parts reads

$$\begin{aligned} \dot{\underset{\sim}{\beta}} &= \dot{\underset{\sim}{\beta}}^e + \dot{\underset{\sim}{\beta}}^p = \dot{\underset{\sim}{\varepsilon}} + \underset{\sim}{\omega} \\ &= \dot{\underset{\sim}{\varepsilon}}^e + \underset{\sim}{\omega}^e + \dot{\underset{\sim}{\varepsilon}}^p + \underset{\sim}{\omega}^p \end{aligned}$$

$\underset{\sim}{\omega}^p = \underset{\sim}{\omega} - \underset{\sim}{\omega}^e = \underset{\sim}{\omega} - \underset{\sim}{\mathbf{1}} \times \dot{\underline{\mathbf{\Phi}}}^e$ represents the relative rotation of material lines with respect to the lattice. As a result, relation (76) becomes

$$\dot{\underset{\sim}{\alpha}} = curl\, \dot{\underset{\sim}{\beta}}^e = curl\, \dot{\underset{\sim}{\varepsilon}}^e + curl\, \underset{\sim}{\omega}^e \tag{84}$$

In a way similar to Kröner (1958), we derive

$$\begin{aligned}
curl\, \underset{\sim}{\omega}^e &= \epsilon_{jkl}\, \omega_{ik,l}\, \underline{e}_i \otimes \underline{e}_j \\
&= -\epsilon_{jkl}\, \epsilon_{ikm}\, \dot{\Phi}_{m,l}\, \underline{e}_i \otimes \underline{e}_j \\
&= -\epsilon_{klj}\epsilon_{kmi}\, \dot{\kappa}_{ml}\, \underline{e}_i \otimes \underline{e}_j \\
&= -(\delta_{ml}\delta_{ij} - \delta_{il}\delta_{mj})\, \dot{\kappa}_{ml} \underline{e}_i \otimes \underline{e}_j \\
&= \dot{\underset{\sim}{\kappa}}^T - Tr\, \dot{\underset{\sim}{\kappa}}\, \underset{\sim}{1}
\end{aligned}$$

Neglecting the elastic strain, one obtains the expression proposed by Nye and its inverse form

$$\underset{\sim}{\alpha} = \underset{\sim}{\kappa}^T - Tr\, \underset{\sim}{\kappa}\, \underset{\sim}{1}, \quad \underset{\sim}{\kappa} = \underset{\sim}{\alpha}^T - \frac{1}{2}\, Tr\, \underset{\sim}{\alpha}\, \underset{\sim}{1} \tag{85}$$

Keeping the elastic term

$$\underset{\sim}{\alpha} = curl\, \underset{\sim}{\varepsilon}^e + \underset{\sim}{\kappa}^T - Tr\, \underset{\sim}{\kappa}\, \underset{\sim}{1} \tag{86}$$

Analysis for the Cosserat theory Within the framework of the Cosserat theory for single crystals presented in part 2, we propose the following definition for the true dislocation density tensor

$$\underset{\sim}{\alpha} = -curl^c\, ({}^\sharp\underset{\sim}{\boldsymbol{F}}^{e-1}.\underset{\sim}{\boldsymbol{R}}^T) \tag{87}$$

We try now to link the dislocation density tensor and the wryness tensor. Equation (87) becomes

$$\underset{\sim}{\alpha} = -\left(\epsilon_{jkl}\, \frac{\partial^\sharp F^{e-1}_{im}}{\partial x_l}\, R_{km} + {}^\sharp F^{e-1}_{im}\, \epsilon_{jkl}\, R^T_{mk,L}\, F^{-1}_{Ll} \right) \underline{e}_i \otimes \underline{e}_j \tag{88}$$

(the comma denotes again a derivative with respect to the reference configuration). Note that,

$$\underset{\sim}{\boldsymbol{\Gamma}} = \frac{1}{2}\, \underset{\approx}{\epsilon} : (\underset{\sim}{\boldsymbol{R}}.(\underset{\sim}{\boldsymbol{R}}^T \otimes \boldsymbol{\nabla})) \implies \underset{\sim}{\boldsymbol{R}}.(\underset{\sim}{\boldsymbol{R}}^T \otimes \boldsymbol{\nabla}) = \underset{\approx}{\epsilon}.\underset{\sim}{\boldsymbol{\Gamma}} \tag{89}$$

or, in components,

$$R^T_{mk,l} = -R^T_{mu}\, \epsilon_{ukv}\, \Gamma_{vl} \tag{90}$$

As a result (87) can now be written

$$\begin{aligned}
\underset{\sim}{\alpha} &= \underset{\sim}{\boldsymbol{A}}^{el} - {}^\sharp F^{e-1}_{im}\, R^T_{mu}\, \epsilon_{klj}\, \epsilon_{kvu}\, \Gamma_{vL}\, F^{-1}_{Ll}\, \underline{e}_i \otimes \underline{e}_j \\
&= \underset{\sim}{\boldsymbol{A}}^{el} + {}^\sharp\underset{\sim}{\boldsymbol{F}}^{e-1}.\underset{\sim}{\boldsymbol{R}}^T.\left((\underset{\sim}{\boldsymbol{\Gamma}}.\underset{\sim}{\boldsymbol{F}}^{-1})^T - Tr\,(\underset{\sim}{\boldsymbol{\Gamma}}.\underset{\sim}{\boldsymbol{F}}^{-1})\, \underset{\sim}{1}\right) \\
&= \underset{\sim}{\boldsymbol{A}}^{el} + {}^\sharp\underset{\sim}{\boldsymbol{F}}^{e-1}.\left(({}^\sharp\underset{\sim}{\boldsymbol{\Gamma}}.{}^\sharp\underset{\sim}{\boldsymbol{F}}^{-1})^T - Tr\,({}^\sharp\underset{\sim}{\boldsymbol{\Gamma}}.{}^\sharp\underset{\sim}{\boldsymbol{F}}^{-1})\, \underset{\sim}{1}\right)\, \underset{\sim}{\boldsymbol{R}}^T
\end{aligned}$$

where $\underset{\sim}{A}^{el} = \epsilon_{jkl} \dfrac{\partial^\sharp F^{e-1}_{im}}{\partial x_l} R_{km}\, \underline{e}_i \otimes \underline{e}_j$. It can be checked that equation (86) is retrieved for small strains and rotations. We define

$$^\sharp\underset{\sim}{\alpha} = \underset{\sim}{\alpha}.\underset{\sim}{R} \tag{91}$$

and $\underset{\sim}{R}.^\sharp\underset{\sim}{F}^e\ \underset{\sim}{\alpha}$ can be interpreted as the Cosserat counterpart of the local dislocation density tensor introduced in the classical continuum theory of dislocations.

Geometrically necessary dislocations and statistically stored dislocations According to Ashby (1971), dislocations become stored in a plastically non-homogeneous solid for two reasons: dislocations are either required for the compatible deformation of various parts of the specimen or they accumulate by trapping each other in a random way. This gives rise on the one hand to the density ρ_G of so-called geometrically necessary dislocations and on the other hand to the density ρ_S of statistically stored dislocations. The density ρ_G can be computed approximately in some situations like plastic bending or punching. This variable comes directly from the continuum theory of dislocations and corresponds to the components of the dislocation density tensor $\underset{\sim}{\alpha}$. In contrast, the density ρ_S belongs to the second group of variables that have been listed and called hardening variables. However, as shown by Ashby in the case of two-phase alloys, geometrically necessary dislocations may lead to additional hardening. In the following, we will try to model this coupling effect between the two types of variables that describe the dislocation distribution. The relative importance of ρ_G and ρ_S depends on the amount of overall plastic deformation, and on the type of solicitation. Clearly ρ_G can dominate in the case of strong deformation gradients.

1.4 Explicit constitutive equations

We propose a complete set of constitutive equations for the elastoviscoplastic deformation and intrinsic curvature of metal single crystals.

Kinematics of plastic deformation and curvature The plastic flow due to slip on various slip systems is given by equation (61). Similarly an expression of the plastic curvature evolution has been proposed (equation (72)). An expression of $^\sharp\underset{\sim}{Q}^s$ is now derived from the analysis of the dislocation density tensor. The scalar γ^s represent the amount of slip due to the passage of dislocations of type s through the volume element. As for them the scalars θ^s represent the plastic curvature due to dislocations trapped in the volume element of characteristic length l. We will assume that the curvature axes or say the expression $^\sharp\underset{\sim}{F}^{e-1}.\left((^\sharp\underset{\sim}{\Gamma}^p.^\sharp\underset{\sim}{F}^{-1})^T - Tr\,(^\sharp\underset{\sim}{\Gamma}^p.^\sharp\underset{\sim}{F}^{-1})\,\underset{\sim}{1}\right)$ are given by the dislocation density tensor $^\sharp\underset{\sim}{\alpha}$. Conversely, the curvature axes $^\sharp\underset{\sim}{\Gamma}^p.^\sharp\underset{\sim}{F}^{p-1}$ are then given by $\left((^\sharp\underset{\sim}{F}^e.^\sharp\underset{\sim}{\alpha})^T - \dfrac{1}{2}\,Tr\,(^\sharp\underset{\sim}{F}^e.^\sharp\underset{\sim}{\alpha})\,\underset{\sim}{1}\right).^\sharp\underset{\sim}{F}^e$ for each type of dislocation. The amount of curvature θ^s will be computed using constitutive equations proposed in the next section. Furthermore we drop the factors $^\sharp\underset{\sim}{F}^e$ for simplicity (small elastic strains). We must now distinguish :

Curvature due to edge dislocations ($\perp$)

For edge dislocations, $^\sharp\underset{\sim}{\alpha} = b\,^\sharp\underline{m} \otimes {}^\sharp\underline{\xi}$.
$\underline{\xi}$ is the dislocation line vector and the normal to the glide plane is defined as

$$\underline{n} = \underline{m} \times \underline{\xi} \tag{92}$$

The associated curvature is then $b\,^\sharp\underline{\xi} \otimes {}^\sharp\underline{m}$, so that we take

$$^\sharp\underset{\sim}{Q}_\perp = {}^\sharp\underline{\xi} \otimes {}^\sharp\underline{m} \tag{93}$$

Arrays of edge dislocations of the same type give rise to lattice curvature in the plane $(\underline{m}\,,\underline{b})$. The rotation vector $\underline{\Phi}$ has the same direction as the dislocation line vector.

Torsion due to screw dislocations ($\odot$)

For a screw dislocation, $^\sharp\underset{\sim}{\alpha} = b\,^\sharp\underline{m} \otimes {}^\sharp\underline{m}$
The associated curvature is then $b\,(^\sharp\underline{m} \otimes {}^\sharp\underline{m} - \frac{1}{2}\,\underset{\sim}{1})$, so that we take

$$^\sharp\underset{\sim}{Q}_\odot = {}^\sharp\underline{m} \otimes {}^\sharp\underline{m} - \frac{1}{2}\,\underset{\sim}{1} \tag{94}$$

As a result screw dislocations cause lattice torsion about the three reference axes. Kröner Kröner (1958) noticed that a planar array of crossed screw dislocations with perpendicular Burgers vectors produces a twist of the lattice about the third direction. This is equivalent to a grain boundary of the second kind. Grain boundaries of the first kind are generated by an array of edge dislocations with parallel Burgers and line vectors. Note that the result (94) is different from that proposed in Dluzewski (1992),

$$\underset{\sim}{\Gamma}^{Dluzewski} = b\,(^\sharp\underline{m} \otimes {}^\sharp\underline{m} - \underset{\sim}{1}) \tag{95}$$

which gives no torsion with respect to the dislocation line axis. This seems to hold only when the couple that can be derived from the classical stress field around a screw dislocation is not released Friedel (1964); Hirth and Lothe (1982).

At last we give the proposed kinematics of the plastic lattice torsion-curvature

$$ {}^\sharp\dot{\underset{\sim}{\boldsymbol{\Gamma}}}^p . {}^\sharp\underset{\sim}{\boldsymbol{F}}^{p-1} = \sum_{s\in S} \left(\frac{\dot\theta^s_\perp}{l}\, {}^\sharp\underline{\boldsymbol{\xi}}^{\,s} \otimes {}^\sharp\underline{\boldsymbol{m}}^{\,s} + \frac{\dot\theta^s_\odot}{l} \left({}^\sharp\underline{\boldsymbol{m}}^{\,s} \otimes {}^\sharp\underline{\boldsymbol{m}}^{\,s} - \frac{1}{2}\, \underset{\sim}{\boldsymbol{1}}\right)\right) \tag{96}$$

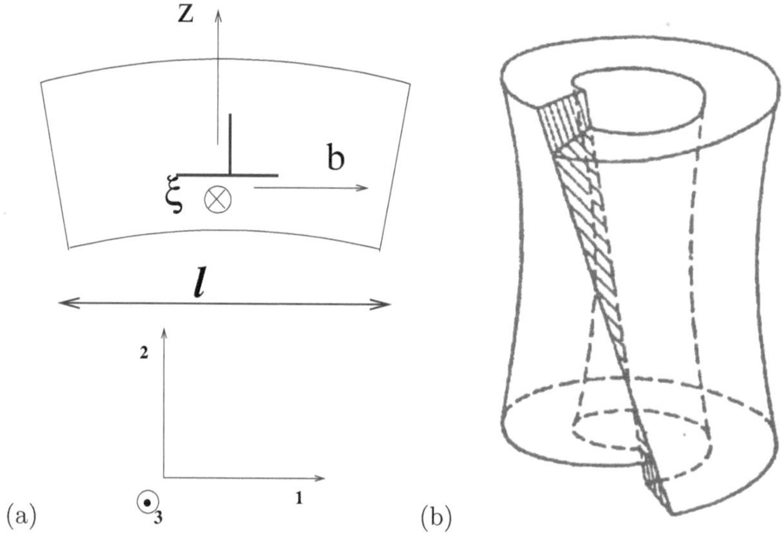

Figure 2. Lattice curvature induced by edge dislocations (a); lattice torsion induced by screw dislocations (b).

Generalized Schmid's law

Peach and Koehler's force Kröner shows that Peach and Koehler's formula giving the force on a dislocation $(\underline{\boldsymbol{b}}, \underline{\boldsymbol{\xi}})$ due to a stress field $\underset{\sim}{\boldsymbol{\sigma}}$ applies also for a non-symmetric stress tensor Kröner (1956). But it is important to derive again the formula taking care of any transposition. The force $\underline{\boldsymbol{f}}$ per unit length of dislocation is defined through

$$ \underline{\boldsymbol{f}} . d\underline{\boldsymbol{x}} = \underline{\boldsymbol{b}} . (\underset{\sim}{\boldsymbol{\sigma}}\, \underline{\boldsymbol{n}})\, dS = ((\underline{\boldsymbol{b}}\, \underset{\sim}{\boldsymbol{\sigma}}) \times \underline{\boldsymbol{\xi}}) . d\underline{\boldsymbol{x}} \tag{97}$$

where

$$ \underline{\boldsymbol{n}}\, dS = \underline{\boldsymbol{\xi}} \times d\underline{\boldsymbol{x}} \tag{98}$$

The dislocation can move in its plane only if the component of the force in the glide plane

$$b\tau = \underline{f}.(\underline{n} \times \underline{\xi}) = \underset{\sim}{\sigma} : (\underline{b} \otimes \underline{n}) \quad (99)$$

reaches a threshold. This is the physical meaning of Schmid's criterion. We will use this criterion to compute the slip rate on slip system s

$$\dot{\gamma}^s = \left(\frac{Max(0, \mid \tau^s - x^s \mid - r^s)}{k^s} \right)^{n^s} sign\ (\tau^s - x^s) \quad (100)$$

where $\tau^s = {}^{\sharp}\underset{\sim}{\sigma} : {}^{\sharp}\underset{\sim}{P}^s$ according to (99), and x^s and r^s are internal kinematic and isotropic hardening variables. x^s and r^s represent respectively a back-stress and the yield threshold, which are supposed to describe with sufficient accuracy the dislocation structure with a view to modelling the hardening behaviour.

Evolution law for the viscoplastic torsion-curvature variables We consider an array of edge dislocations with $\underline{b} = b\,\underline{e}_1$, the normal to the glide plane $\underline{n} = \underline{e}_2$ *and* $\underline{\xi} = -\underline{e}_3$ (see figure 2). At small strains they produce a curvature

$$\underset{\sim}{\kappa}^p = -nb\,\underline{e}_3 \otimes \underline{e}_1 = \kappa^p_{31}\,\underline{e}_3 \otimes \underline{e}_1 \quad (101)$$

We will assume that such geometrically necessary dislocations are produced by local dislocation sources if the local moment $\underset{\sim}{\mu} = m\,\underline{e}_3 \otimes \underline{e}_1$ ($m < 0$ *here*) is so high that the imposed curvature cannot be accommodated elastically any longer.

Generalizing the previous example, we propose the following expression of the viscoplastic curvature rate

$$\dot{\theta}^s = \left(\frac{Max(0, \mid {}^{\sharp}\underset{\sim}{\mu} : {}^{\sharp}\underset{\sim}{Q}^s \mid - l\, r^s_c)}{l\, k_c} \right)^{n_c} sign\ ({}^{\sharp}\underset{\sim}{\mu} : {}^{\sharp}\underset{\sim}{Q}^s) \quad (102)$$

where r^s_c denotes the threshold and k_c and n_c is a viscosity parameter. The formula is to be applied successively for edge and screw dislocations belonging to the same system. Equations (100) and (102) and the hardening rules of the next section close the theory based on multicriteria and associative flow rules. Accordingly this theory is part of associative generalized plasticity.

Expression of the free energy and hardening rules The key-point of the thermodynamical analysis of a constitutive model for a dissipative system is the choice of the relevant internal variables on which the free energy may depend. We propose such a formulation of the previous model in the linear case for simplicity. In addition to the observable variables deformation, curvature and temperature ($\underset{\sim}{e}$, $\underset{\sim}{\boldsymbol{\kappa}}$, T) or equivalently ($\underset{\sim}{e}^e$, $\underset{\sim}{\boldsymbol{\kappa}}^e$, T), the free energy is assumed to depend on the following internal variables:

- the variables ϱ_S^s, which are similar to the densities of statistically stored dislocations, and which are defined by

$$\dot{\varrho}_S^s = \mid \dot{\gamma}^s \mid \tag{103}$$

- the variables ϱ_G^s, which are similar to the densities of geometrically necessary dislocations, and which are defined by

$$\varrho_G^s = \mid \frac{b\theta^s}{l} \mid \tag{104}$$

- the kinematic hardening variables α^s.

We postulate then that the free energy is a quadratic form of these variables according to

$$\begin{aligned}
\rho\psi(\underset{\sim}{e}^e, \underset{\sim}{\boldsymbol{\kappa}}^e, T, \alpha^s, \varrho_S^s, \varrho_G^s) &= \tfrac{1}{2}\, \underset{\sim}{e}^e : \underset{\approx}{\boldsymbol{E}} : \underset{\sim}{e}^e + \tfrac{1}{2}\, \underset{\sim}{\boldsymbol{\kappa}}^e : \underset{\approx}{\boldsymbol{C}} : \underset{\sim}{\boldsymbol{\kappa}}^e \;+\; \tfrac{1}{2} \textstyle\sum_{s\in S}\; c\alpha^{s2} \\
&+\;\; r_0 \textstyle\sum_{s\in S} \varrho_S^s \;+\; \tfrac{1}{2} \textstyle\sum_{r,s\in S}\; h^{sr} \varrho_S^s \varrho_S^r \\
&+\;\; r_{c0} \textstyle\sum_{s\in S} \varrho_G^s \;+\; \tfrac{1}{2} \textstyle\sum_{r,s\in S}\; h_c^{sr} \varrho_G^s \varrho_G^r \\
&+\;\; \textstyle\sum_{r,s\in S}\; h_I^{sr} \varrho_S^s \varrho_G^r \;+\; f(T)
\end{aligned}$$

Hardening matrices h^{rs} and h_c^{rs} have been introduced for each population of dislocations following Mandel (1965), but a coupling term associated with the matrix h_I^{sr} must be added.

Assuming then that the thermodynamical forces corresponding to the variables ϱ_S^s, ϱ_G^s and α^s respectively are r^s, r_c^s and x^s, the following hardening rules are derived :

Isotropic hardening

$$r^s = \rho \frac{\partial \psi}{\partial \varrho_S^s} = r_0 + \sum_{r\in S} h^{sr} \varrho_S^r + \sum_{r\in S} h_I^{sr} \varrho_G^r \tag{105}$$

$$r_c^s = \rho \frac{\partial \psi}{\partial \varrho_G^s} = r_{c0} + \sum_{r\in S} h_c^{sr} \varrho_G^r + \sum_{r\in S} h_I^{sr} \varrho_S^r \tag{106}$$

Note that, for simplicity, we have omitted to split the terms $h_c^{sr}\varrho_G^r$ into $h_{c\perp}^{sr}\varrho_{G\perp}^r + h_{c\odot}^{sr}\varrho_{G\odot}^r$ in order to distinguish the contributions of edge and screw dislocations. The same holds for the terms involving matrix h_I^{sr}. Furthermore, a similar thermodynamical formulation can be worked out for non-linear isotropic hardening. It can be seen that a coupling between plastic deformation and curvature naturally arises from our choice of the free energy. The existence of additional hardening due to plastic curvature must be investigated experimentally.

Kinematic hardening

$$x^s = \rho\frac{\partial\psi}{\partial\alpha^s} = c^s\,\alpha^s \tag{107}$$

We refer to Méric and Cailletaud (1991) for the expression of the non-linear evolution law for kinematic hardening:

$$\dot{\alpha}^s = \dot{\gamma}^s - d\mid\dot{\gamma}^s\mid\alpha^s \tag{108}$$

Dissipation Introducing now the internal variables in the expression of the intrinsic dissipation rate derived in 2.5, one obtains

$$\begin{aligned}\dot{D} = \sum_{r\in S}\Big(&\tau^s\,\dot{\gamma}^s - x^s\dot{\alpha}^s - r^s\dot{\varrho}_S^s + \nu_\perp^s\,\dot{\theta}_\perp^s + \nu_\odot^s\,\dot{\theta}_\odot^s - r_{c\perp}^s\dot{\varrho}_{G\perp}^s \\ &- r_{c\odot}^s\dot{\varrho}_{G\odot}^s\Big)\end{aligned} \tag{109}$$

where

$$\nu_\perp^s = \frac{1}{l}\,{}^\sharp\underset{\sim}{\boldsymbol{\mu}}:{}^\sharp\underset{\sim}{\boldsymbol{Q}}_\perp^s \quad and \quad \nu_\odot^s = \frac{1}{l}\,{}^\sharp\underset{\sim}{\boldsymbol{\mu}}:{}^\sharp\underset{\sim}{\boldsymbol{Q}}_\odot^s \tag{110}$$

The multiplication and motion of dislocations are dissipative processes. The three first terms in (109) account for dissipation due to slip activity whereas the remaining terms account for multiplication of geometrically necessary dislocations. In many cases the last terms can be neglected. But when strong lattice rotation gradients develop, they may well be the leading terms.

Some conditions on the material parameters can then be derived from the entropy principle. Taking the flow rules (100) and (102) and the definitions (103) and (104) into account, equation (109) can be rewritten under the form

$$\begin{aligned}\dot{D} &= \textstyle\sum_{r\in S}\ \big(((\tau^s - x^s)\ sign(\dot{\gamma}^s) - r^s + cd\alpha^{s2})\mid\dot{\gamma}^s\mid \\ &+ \dot{\theta}_\perp^s\,(\nu_\perp^s - r_{c\perp}^s\ sign(\theta_{G\perp}^s)) + \dot{\theta}_\odot^s(\nu_\odot^s - r_{c\odot}^s\ sign(\theta_{G\odot}^s))\big)\end{aligned}$$

It can be checked that the positivity of the intrinsic dissipation rate is ensured if $c\ d\ >\ 0$ and if the marix $h^{rs}_{c\perp,\odot}$ is such that $r^{s}_{c\perp,\odot}$ is always positive.

1.5 Conclusions

Recent advances in the mechanics of generalized continua have been used to develop a Cosserat theory for single crystals at finite deformation and curvature. The decomposition of the relative deformation gradient into an elastic and plastic part is multiplicative as usual, whereas the wryness tensor admits a mixed additive-multiplicative decomposition. We have assumed that the plastic lattice curvature and torsion are accommodated respectively by edge and screw dislocations belonging to each slip system. The curvature and torsion angles over a characteristic length due to each type of dislocation are internal variables in addition to the cumulative amounts of slip for each slip system. Explicit constitutive equations have been proposed in the case of elastoviscoplasticity. An important consequence of the theory is that the plastic lattice curvature and torsion as well as the plastic spin are associated with dissipation. The production of geometrically necessary dislocations is clearly a dissipative process. There is an overwhelming tendency to include these microstructural features of dislocated crystals into the framework of generalized continua. Since the pioneering work of Günther, Claus and Eringen Eringen and Claus (1970) resorted to a micromorphic continuum. As for them, Smyshlyaev and Fleck (1996) prefer to develop a strain gradient theory of slip. However they replace this plasticity problem by a problem of non-linear elasticity at small strains. In contrast our theory has the tremendous advantage to provide a set of kinematic and constitutive equations in elastoviscoplasticity at finite deformation on a physical and thermodynamical basis. Finally Lachner et al. (1994) have shown that polycrystals also can be regarded as Cosserat media. Homogenization techniques should enable one to derive a polycrystal model from the present theory.

Only a precise enough description of dislocation distribution within a volume element can enable one to model the plastic behaviour of single crystals. For that purpose, the continuum theory of dislocations resorts to the dislocation density tensor. In contrast, macroscopic elastoplasticity theory involves hardening variables which are related to scalar dislocation densities. In both theories, the expression "dislocation density" is seen to have a very different meaning. The dislocation density tensor and the scalar dislocation densities are independent measures of the dislocation distribution. The most important advantage of the proposed theory is to combine both descriptions within a single constitutive framework.

It must be noticed that only slip processes have been taken into account in the present work. Further developments are necessary to include climb processes, which may play a significant role during creep.

A coupling between plastic curvature and plastic deformation has been introduced on the level of the hardening rule to represent the influence of slip plane curvature on further dislocation motion. Experimental evidence of such hardening effects have been provided for instance in Jaoul (1965).

It is clear that the difference between the classical theory and the Cosserat theory can appear only if deformation and more precisely lattice rotation is not homogeneous. The theory can therefore be applied to the prediction and the simulation of localized deformation modes like shear bands in single crystals. A theoretical analysis of such material instabilities is presented in Petryk (1992). The case of single crystals is investigated in Asaro and Rice (1977); Duszek-Perzyna and Perzyna (1993). An analysis and numerical simulations of localization phenomena in single crystals are presented in Forest and Cailletaud (1995) for the classical theory. In Forest (1998) we have performed a bifurcation analysis for single crystals undergoing single slip using the Cosserat theory. Some crucial differences with respect to the classical case have been pointed out. For instance, according to the classical theory, slip bands and kink bands can occur for the same critical hardening modulus. This is no longer true for the Cosserat theory, which is strongly supported by experimental evidence.

1.6 Appendix: Notations

In this work, $\underline{\boldsymbol{a}}$ denotes a vector of the Euclidean space E, $\underset{\sim}{\boldsymbol{A}}$ a second-rank Euclidean tensor, and $\underset{\approx}{\boldsymbol{A}}$ (resp. $\underset{\simeq}{\boldsymbol{A}}$) a third-rank tensor when operating on a vector (resp. a second-rank tensor). The same third-rank tensor is denoted $\underline{\underline{\underline{\boldsymbol{A}}}}$ when regarded as a 3-linear form. The tensor product of two vectors $\underline{\boldsymbol{a}}\,,\underline{\boldsymbol{b}}$ is such that, for all $\underline{\boldsymbol{x}} \in E$,

$$\begin{aligned} \underline{\boldsymbol{x}}\,.(\underline{\boldsymbol{a}} \otimes \underline{\boldsymbol{b}}) &= \underline{\boldsymbol{x}}\,.\underline{\boldsymbol{a}}\;\underline{\boldsymbol{b}} \\ (\underline{\boldsymbol{a}} \otimes \underline{\boldsymbol{b}}).\underline{\boldsymbol{x}} &= \underline{\boldsymbol{b}}\,.\underline{\boldsymbol{x}}\;\underline{\boldsymbol{a}} \end{aligned}$$

where the dot denotes the inner product on E.

Let $(\underline{\boldsymbol{e}}_1, \underline{\boldsymbol{e}}_2, \underline{\boldsymbol{e}}_3)$ be a positive oriented orthonormal basis of oriented E with dimension 3. When written in components, the double contraction of second-rank tensors reads

$$\underset{\sim}{\boldsymbol{A}} : \underset{\sim}{\boldsymbol{B}} = A_{ij} B_{ij} \tag{111}$$

We note $\underset{\sim}{\boldsymbol{\epsilon}}$ the Levi-Civita tensor

$$\underline{\underline{\underline{\boldsymbol{\epsilon}}}} = Det(\underline{\boldsymbol{e}}_j, \underline{\boldsymbol{e}}_k, \underline{\boldsymbol{e}}_l)\; \underline{\boldsymbol{e}}_j \otimes \underline{\boldsymbol{e}}_k \otimes \underline{\boldsymbol{e}}_l \tag{112}$$

Notice the useful identity

$$\epsilon_{ijk}\, \epsilon_{ilm} = \delta_{jl}\, \delta_{km} - \delta_{jm}\, \delta_{kl} \tag{113}$$

The following result concerning third-rank tensors is used,

$$\begin{array}{llll} If & \underset{\sim}{\boldsymbol{B}} & = \frac{1}{2}\,\underset{\approx}{\boldsymbol{\epsilon}} : \underset{\approx}{\boldsymbol{A}} & = \frac{1}{2}\,\epsilon_{ikl}\,A_{klj}\,\underline{\boldsymbol{e}}_i \otimes \underline{\boldsymbol{e}}_j \\ and & A_{ijk} & = -A_{jik} & \\ then & \underset{\approx}{\boldsymbol{A}} & = \underset{\approx}{\boldsymbol{\epsilon}} \cdot \underset{\sim}{\boldsymbol{B}} & = \epsilon_{ijm}\,B_{mk}\,\underline{\boldsymbol{e}}_i \otimes \underline{\boldsymbol{e}}_j \otimes \underline{\boldsymbol{e}}_k \end{array}$$

The cross product is defined by

$$\underline{\boldsymbol{a}} \times \underline{\boldsymbol{b}} \;=\; \underset{\approx}{\boldsymbol{\epsilon}} : (\underline{\boldsymbol{a}} \otimes \underline{\boldsymbol{b}}) \;=\; \epsilon_{ijk}\,a_j b_k\,\underline{\boldsymbol{e}}_i \tag{114}$$

The symmetric and antisymmetric parts of tensor $\underset{\sim}{\boldsymbol{A}}$ are respectively denoted $\{\underset{\sim}{\boldsymbol{A}}\}$ and $\}\underset{\sim}{\boldsymbol{A}}\{$. There is then one and only one vector $\overset{\times}{\underline{\boldsymbol{A}}}$ such that, for all $\underline{\boldsymbol{x}}$,

$$\}\underset{\sim}{\boldsymbol{A}}\{\,.\underline{\boldsymbol{x}} \;=\; \overset{\times}{\underline{\boldsymbol{A}}} \times \underline{\boldsymbol{x}}, \quad \overset{\times}{\underline{\boldsymbol{A}}} \;=\; -\frac{1}{2}\,\underset{\approx}{\boldsymbol{\epsilon}} : \underset{\sim}{\boldsymbol{A}} \;=\; -\frac{1}{2}\,\epsilon_{klm}\,A_{lm}\,\underline{\boldsymbol{e}}_k \tag{115}$$

Following Trostel [56], we define a cross product between a second-rank tensor and a vector

$$\begin{array}{rcl} (\underline{\boldsymbol{a}} \otimes \underline{\boldsymbol{b}}) \times \underline{\boldsymbol{c}} & = & \underline{\boldsymbol{a}} \otimes (\underline{\boldsymbol{b}} \times \underline{\boldsymbol{c}}) \\ \underline{\boldsymbol{a}} \times (\underline{\boldsymbol{b}} \otimes \underline{\boldsymbol{c}}) & = & (\underline{\boldsymbol{a}} \times \underline{\boldsymbol{b}}) \otimes \underline{\boldsymbol{c}} \end{array}$$

so that

$$\underset{\sim}{\boldsymbol{A}} \times \underline{\boldsymbol{c}} \;=\; -(\underline{\boldsymbol{c}} \times \underset{\sim}{\boldsymbol{A}}^T)^T \tag{116}$$

As a result

$$\}\underset{\sim}{\boldsymbol{A}}\{ \;=\; \underset{\sim}{\boldsymbol{1}} \times \overset{\times}{\underline{\boldsymbol{A}}} \;=\; -\underset{\approx}{\boldsymbol{\epsilon}}.\overset{\times}{\underline{\boldsymbol{A}}} \tag{117}$$

Any element $\underset{\sim}{\boldsymbol{R}}$ of the orthogonal group can be represented by the element $\underline{\boldsymbol{\Phi}}$ of the associated Lie group such that

$$\underset{\sim}{\boldsymbol{R}} \;=\; \exp(\underset{\sim}{\boldsymbol{1}} \times \underline{\boldsymbol{\Phi}}) \;=\; \exp(-\underset{\approx}{\boldsymbol{\epsilon}}.\underline{\boldsymbol{\Phi}}) \tag{118}$$

Concerning tensor analysis, our notations are,

nabla operator

$$\boldsymbol{\nabla} \;=\; {}_{,i}\,\underline{\boldsymbol{e}}_i \tag{119}$$

gradient operator

$$grad\ f \;=\; \boldsymbol{\nabla} f \;=\; f_{,i}\,\underline{\boldsymbol{e}}_i \tag{120}$$

$$grad\ \underline{\boldsymbol{u}} \;=\; \underline{\boldsymbol{u}} \otimes \boldsymbol{\nabla} \;=\; u_{i,j}\,\underline{\boldsymbol{e}}_i \otimes \underline{\boldsymbol{e}}_j \tag{121}$$

curl operator

$$curl\ \underline{\boldsymbol{u}} \;=\; \underline{\boldsymbol{u}} \times \boldsymbol{\nabla} \;=\; \epsilon_{ijk}\,u_{j,k}\,\underline{\boldsymbol{e}}_i \tag{122}$$

$$curl\ \underset{\sim}{\boldsymbol{A}} \;=\; \underset{\sim}{\boldsymbol{A}} \times \boldsymbol{\nabla} \;=\; \epsilon_{hjk}\,A_{ij,k}\,\underline{\boldsymbol{e}}_i \otimes \underline{\boldsymbol{e}}_h \tag{123}$$

Note that

$$\}\underline{\boldsymbol{u}} \otimes \boldsymbol{\nabla}\{ \;=\; \frac{1}{2}\,\underset{\approx}{\boldsymbol{\epsilon}}.(curl\ \underline{\boldsymbol{u}}) \tag{124}$$

We have made a wide use of theorem

$$\oint_L \underset{\sim}{\boldsymbol{A}}.\underline{\boldsymbol{l}}\ dl \;=\; -\int_S (curl\ \underset{\sim}{\boldsymbol{A}}).\underline{\boldsymbol{n}}\ dS \tag{125}$$

where the open surface S is bounded by the contour L.

2 The micromorphic approach to strain gradient crystal plasticity

2.1 The micromorphic approach

A unifying thermomechanical framework is presented by Forest (2009) that reconciles several classes of gradient elastoviscoplasticity and damage models proposed in the literature during the last 40 years. It is based on the introduction of the micromorphic counterpart $^{\chi}\phi$ of a selected state or internal variable ϕ in a standard constitutive model. Following the method of virtual power, the power of internal forces is extended by the contribution of micromorphic power. In addition to the classical balance of momentum equation, a balance of micromorphic momentum is derived that involves generalized stress tensors associated with invariant generalized strain measures based on $^{\chi}\phi$ and its first gradient $\nabla^{\chi}\phi$. The corresponding additional boundary conditions are also deduced from the procedure. The power of generalized forces is assumed to contribute to the energy balance equation. The free energy density function is then assumed to depend on invariant generalized strain measures involving in particular the relative generalized strain $\phi - {}^{\chi}\phi$ and its gradient $\nabla^{\chi}\phi$. The Coleman–Noll procedure is applied to derive the state laws and residual dissipation from the entropy principle.

When applied to the deformation gradient itself, $\phi \equiv \underset{\sim}{\boldsymbol{F}}$, the micromorphic theory of Eringen and Mindlin is recovered Eringen and Suhubi (1964); Mindlin (1964) together with its extension to finite deformation elastoviscoplasticity Sansour (1998a); Forest and Sievert (2003). If the selected variable is the cumulative plastic strain, the theory reduces to the so–called "nonlocal implicit gradient–enhanced elastoplasticity model" provided that simplified linear relationships are adopted between generalized stresses and strains Peerlings et al. (2001); Engelen et al. (2003). The same holds if the micromorphic variable coincides with a damage variable Peerlings et al. (2004).

There are two possible sources of coupling between the macro and micro variables in the proposed approach. If the relative generalized strain associated with $\phi - {}^{\chi}\phi$ explicitly appears in the extended power of internal forces, the associated relative stress acts as a coupling between the balance equation for standard momentum and for generalized momentum, as in Eringen's micromorphic continuum. The relative generalized strain can also be included as an argument of the free energy function which eventually leads to the wanted coupling between standard and generalized stresses. Both formulations have been illustrated recently for the so–called "microstrain" theory Forest and Sievert (2006).

It the internal constraint is introduced that the micromorphic variable

${}^{\chi}\phi$ remains as close as possible to the macroscopic variable ϕ, the micromorphic model reduces to the second gradient or gradient of internal variable approach as defined in Maugin (1990); Maugin and Muschik (1994); Papenfuss and Forest (2006). When the micromorphic variable is the deformation gradient itself, the second gradient theory according to Mindlin and Eshel (1968) is obtained. If the selected variable is the cumulative plastic strain, the constrained micromorphic theory delivers Aifantis strain gradient plasticity model according to Aifantis (1987); Fleck and Hutchinson (2001). When applied to the full plastic strain tensors, the strain gradient plasticity models initially proposed by Forest and Sievert (2003) and Gurtin (2003) are derived.

General procedure for introducing micromorphic variables

We start from an elastoviscoplasticity model formulation within the framework of the classical Cauchy continuum theory and classical continuum thermodynamics according to Germain et al. (1983); Maugin (1999). The material behaviour is characterized by the reference set of state variables

$$STATE0 = \{\underset{\sim}{\boldsymbol{F}}, \quad T, \quad \alpha\} \tag{126}$$

on which the free energy density function ψ may depend. The deformation gradient is denoted by $\underset{\sim}{\boldsymbol{F}}$ whereas α represents the whole set of internal variables of arbitrary tensorial order accounting for nonlinear processes at work inside the material volume element, like isotropic and kinematic hardening variables. The absolute temperature is T.

The proposed systematic method for the enhancement of the previous continuum and constitutive theory to incorporate generalized strain gradient effects proceeds as follows:

1. Select a variable ϕ from the set of state variables, which is supposed to carry the targeted gradient effects:

$$\phi \in \{\underset{\sim}{\boldsymbol{F}}, \quad \alpha, \quad T\} \tag{127}$$

It can be a tensor variable of arbitrary rank. For the illustration, it is treated as a scalar quantity in this section. The case $\phi = T$ is treated in Forest and Amestoy (2008).

2. Introduce the micromorphic variable ${}^{\chi}\phi$ associated with ϕ. It has the same tensor rank and same physical dimension as ϕ.
3. Extend the virtual power of internal forces to the power done by the micromorphic variable and its first gradient:

$$\mathcal{P}^{(i)}(\underline{\boldsymbol{v}}^{\star}, {}^{\chi}\dot{\phi}^{\star}) = -\int_{\mathcal{D}} p^{(i)}(\underline{\boldsymbol{v}}^{\star}, {}^{\chi}\dot{\phi}^{\star})\, dV \tag{128}$$

$$p^{(i)}(\underline{v}^{\star}, {}^{\chi}\dot{\phi}^{\star}) = \underset{\sim}{\sigma} : \nabla \underline{v}^{\star} + a\, {}^{\chi}\dot{\phi}^{\star} + \underline{b}\,.\nabla^{\chi}\dot{\phi}^{\star} \tag{129}$$

where $\mathcal{D}$ is a subdomain of the current configuration Ω of the body. The Cauchy stress is $\underset{\sim}{\sigma}$ and a and $\underline{b}$ are generalized stresses associated with the micromorphic variable and its first gradient.

4. Extend then the power of contact forces as follows:

$$\mathcal{P}^{(c)}(\underline{v}^{\star}, {}^{\chi}\dot{\phi}^{\star}) = \int_{\mathcal{D}} p^{(c)}(\underline{v}^{\star}, {}^{\chi}\dot{\phi}^{\star})\, dV, \quad p^{(c)}(\underline{v}^{\star}, {}^{\chi}\dot{\phi}^{\star}) = \underline{t}\,.\underline{v}^{\star} + a^{c}\, {}^{\chi}\dot{\phi}^{\star} \tag{130}$$

where $\underline{t}$ is the traction vector.

5. Extend the power of forces acting at a distance by introducing, if necessary, generalized body forces:

$$\mathcal{P}^{(e)}(\underline{v}^{\star}, {}^{\chi}\dot{\phi}^{\star}) = \int_{\mathcal{D}} p^{(e)}(\underline{v}^{\star}, {}^{\chi}\dot{\phi}^{\star})\, dV \tag{131}$$

$$p^{(e)}(\underline{v}^{\star}, {}^{\chi}\dot{\phi}^{\star}) = \rho \underline{f}\,.\underline{v}^{\star} + a^{e}\, {}^{\chi}\dot{\phi}^{\star} + \underline{b}^{e}.\nabla^{\chi}\dot{\phi}^{\star} \tag{132}$$

where $\rho\underline{f}\,, a^{e}, \underline{b}^{e}$ account for given simple and generalized body forces. Following Germain (1973a), given body couples and double forces working with the gradient of the velocity field, can also be introduced in this theory.

6. Formulate the generalized principle of virtual power with respect to the velocity and micromorphic variable fields, presented here in the static case only:

$$\mathcal{P}^{(i)}(\underline{v}^{\star}, {}^{\chi}\dot{\phi}) + \mathcal{P}^{(e)}(\underline{v}^{\star}, {}^{\chi}\dot{\phi}^{\star}) + \mathcal{P}^{(c)}(\underline{v}^{\star}, {}^{\chi}\dot{\phi}) = 0, \quad \forall \mathcal{D} \subset \Omega, \quad \forall \underline{v}^{\star}, {}^{\chi}\dot{\phi} \tag{133}$$

The method of virtual power according to Maugin (1980) is used then to

7. Derive the standard local balance of momentum equation:

$$div\, \underset{\sim}{\sigma} + \rho \underline{f} = 0, \quad \forall \underline{x} \in \Omega \tag{134}$$

and the generalized balance of micromorphic momentum equation:

$$div\, (\underline{b} - \underline{b}^{e}) - a + a^{e} = 0, \quad \forall \underline{x} \in \Omega \tag{135}$$

8. Derive the associated boundary conditions for the simple and generalized tractions:

$$\underline{t} = \underset{\sim}{\sigma}.\underline{n}\,, \quad \forall \underline{x} \in \partial\mathcal{D} \tag{136}$$

$$a^{c}_{\phi} = (\underline{b} - \underline{b}^{e}).\underline{n}\,, \quad \forall \underline{x} \in \partial\mathcal{D} \tag{137}$$

9. Enhance the local balance of energy by the generalized micromorphic power already including in the power of internal forces (129):

$$\rho\dot{\epsilon} = p^{(i)} - div\, \underline{q} + \rho r \tag{138}$$

where ϵ is the specific internal energy, $\underline{q}$ the heat flux vector and r denotes external heat sources.

10. Enlarge the state space to include the micromorphic variable and its first gradient:

$$STATE = \{\underset{\sim}{\boldsymbol{F}}, \quad T, \quad \alpha, \quad {}^{\chi}\phi, \quad \boldsymbol{\nabla}^{\chi}\phi\} \tag{139}$$

The free energy density function ψ will in general be a function of the generalized relative strain variable e defined as:

$$e = \phi - {}^{\chi}\phi \tag{140}$$

thus introducing a coupling between macro and micromorphic variables.

11. Formulate the entropy principle in its local form:

$$-\rho(\dot{\psi} + \eta\dot{T}) + p^{(i)} - \frac{\underline{q}}{T}.\boldsymbol{\nabla}T \geq 0 \tag{141}$$

where it was assumed that the entropy production vector is still equal to the heat vector divided by temperature, as in the classical thermomechanics according to Coleman and Noll (1963). Again, the enhancement of the theory goes through the enriched power density of internal forces (129).

12. Exploit the entropy principle according to classical continuum thermodynamics to derive the state laws. For that purpose, the following constitutive functions are introduced:

$$\psi = \hat{\psi}(\underset{\sim}{\boldsymbol{F}}^e, T, \alpha, {}^{\chi}\phi, \boldsymbol{\nabla}^{\chi}\phi), \underset{\sim}{\boldsymbol{\sigma}} = \hat{\underset{\sim}{\boldsymbol{\sigma}}}(\underset{\sim}{\boldsymbol{F}}^e, T, \alpha, {}^{\chi}\phi, \boldsymbol{\nabla}^{\chi}\phi) \tag{142}$$

$$\eta = \hat{\eta}(\underset{\sim}{\boldsymbol{F}}^e, T, \alpha, {}^{\chi}\phi, \boldsymbol{\nabla}^{\chi}\phi) \tag{143}$$

$$a = \hat{a}(\underset{\sim}{\boldsymbol{F}}^e, T, \alpha, {}^{\chi}\phi, \boldsymbol{\nabla}^{\chi}\phi), \underline{\boldsymbol{b}} = \hat{\underline{\boldsymbol{b}}}(\underset{\sim}{\boldsymbol{F}}^e, T, \alpha, {}^{\chi}\phi, \boldsymbol{\nabla}^{\chi}\phi) \tag{144}$$

$$\tag{145}$$

where $\underset{\sim}{\boldsymbol{F}}^e$ represents the elastic part of total deformation. Its precise definition depends however on the retained decomposition of total deformation into elastic and plastic contribution. The usual multiplicative decomposition is adopted below for the illustration. The state laws follow:

$$\underset{\sim}{\sigma} = \rho \frac{\partial \hat{\psi}}{\partial \underset{\sim}{F}^e} . \underset{\sim}{F}^{eT} \quad \eta = -\frac{\partial \hat{\psi}}{\partial T} \quad X = \rho \frac{\partial \hat{\psi}}{\partial \alpha} \tag{146}$$

$$a = \frac{\partial \hat{\psi}}{\partial^{\chi} \phi} \tag{147}$$

$$\underline{b} = \frac{\partial \hat{\psi}}{\partial \nabla^{\chi} \phi} \tag{148}$$

and the residual dissipation is

$$D^{res} = W^p - X\dot{\alpha} - \frac{\underline{q}}{T} . \nabla T \geq 0 \tag{149}$$

where W^p represents the (visco–)plastic power and X the thermodynamic force associated with the internal variable α. The existence of a convex dissipation potential depending on the thermodynamic forces can then be assumed from which the evolution rules for internal variables are derived, that identically fulfill the entropy inequality, as usually done in classical continuum thermomechanics Germain et al. (1983).

In the following, this methodology will be applied to existing theories of plasticity.

At this first stage, there will be no need for considering generalized external body forces so that $a^e = 0, \underline{b}^e = 0$. In the most simple model, assuming isotropic material behavior for brevity, the additional contributions to the free energy are taken as quadratic functions of e and $\nabla^{\chi}\phi$:

$$\psi(\underset{\sim}{F}, T, \alpha, {}^{\chi}\phi, \nabla^{\chi}\phi) = \psi^1(\underset{\sim}{F}, \alpha, T) + \psi^2(e = \phi - {}^{\chi}\phi, \nabla^{\chi}\phi, T), \quad \textit{with} \tag{150}$$

$$\rho\psi^2 = \frac{1}{2} H_{\chi} (\phi - {}^{\chi}\phi)^2 + \frac{1}{2} A \nabla^{\chi}\phi . \nabla^{\chi}\phi \tag{151}$$

After inserting the state laws (147) and (148)

$$a = \rho \frac{\partial \psi}{\partial^{\chi} \phi} = -H_{\chi} (\phi - {}^{\chi}\phi) \tag{152}$$

$$\underline{b} = \rho \frac{\partial \psi}{\partial \nabla^{\chi} \phi} = A \nabla^{\chi} \phi \tag{153}$$

into the additional balance equation (135),

$$a = div\, \underline{b} \tag{154}$$

the following partial differential equation is obtained:

$$\phi = {}^{\chi}\phi - \frac{A}{H_\chi}\Delta^{\chi}\phi \tag{155}$$

where Δ is the Laplace operator. This type of equation is encountered at several places in the mechanics of generalized continua especially in the linear micromorphic theory Mindlin (1964); Eringen (1999); Forest and Sievert (2003); Dillard et al. (2006) and in the so–called implicit gradient theory of plasticity and damage Peerlings et al. (2001); Engelen et al. (2003); Peerlings et al. (2004). Note however that this equation corresponds to a special quadratic potential and represents the simplest micromorphic extension of the classical theory. It involves a characteristic length scale defined by:

$$l_c^2 = \frac{A}{H_\chi} \tag{156}$$

This length is real for positive values of the ratio A/H_χ. The additional material parameters H_χ and A are assumed to be positive in this work. This does not exclude a softening material behaviour that can be induced bu the proper evolution of the internal variables (including $\phi \in \alpha$ itself).

Full micromorphic and microstrain theories

The micromorphic theory proposed in Eringen and Suhubi (1964) and Mindlin (1964) is retrieved by choosing

$$\phi \equiv \underset{\sim}{\boldsymbol{F}} \tag{157}$$

i.e. the selected variable ϕ is the full deformation gradient itself. The associated micromorphic variable is

$$^{\chi}\phi \equiv \underset{\sim}{\boldsymbol{\chi}} \tag{158}$$

where $\underset{\sim}{\boldsymbol{\chi}}(\underline{\boldsymbol{x}})$ is the generally nonsymmetric and non compatible field of microdeformation introduced by these authors. Following the approach sketched in the previous section, the power of internal forces is extended by the micromorphic power, written here in the small deformation framework for the sake of brevity:

$$p^{(i)} = \underset{\sim}{\boldsymbol{\sigma}} : \boldsymbol{\nabla}\dot{\underline{\boldsymbol{u}}} + \underset{\sim}{\boldsymbol{a}} : \dot{\underset{\sim}{\boldsymbol{\chi}}} + \underset{\approx}{\boldsymbol{B}} \,\vdots\, \boldsymbol{\nabla}\dot{\underset{\sim}{\boldsymbol{\chi}}} \tag{159}$$

where $\underline{\boldsymbol{u}}$ is the displacement field. An (infinitesimal) change of observer of rate $\underset{\sim}{\boldsymbol{w}}$ changes the gradient of the velocity field into $\boldsymbol{\nabla}\dot{\underline{\boldsymbol{u}}} + \underset{\sim}{\boldsymbol{w}}$ and the microdeformation rate into $\dot{\underset{\sim}{\chi}} + \underset{\sim}{w}$. The principle of (infinitesimal) material frame indifference requires the invariance of $p^{(i)}$ with respect to (infinitesimal) Euclidean changes of observers Gurtin (2003). As a result, the sum $\underset{\sim}{\sigma} + \underset{\sim}{a}$ must be a symmetric second–rank tensor. The power density of internal forces can therefore be rewritten in the following form

$$p^{(i)} = \underset{\sim}{\sigma} : \dot{\underset{\sim}{\varepsilon}} + \underset{\sim}{s} : (\nabla \dot{\underline{u}} - \dot{\underset{\sim}{\chi}}) + \underset{\approx}{S} \vdots \nabla \dot{\underset{\sim}{\chi}} \tag{160}$$

where $\underset{\sim}{\sigma}$ is symmetric, $\underset{\sim}{s}$ is the generally nonsymmetric relative stress tensor, and $\underset{\approx}{S}$ the higher order stress tensor introduced in the formulation of Eringen (1999). The infinitesimal strain tensor is $\underset{\sim}{\varepsilon}$. The generalized strain rates $\nabla \dot{\underline{u}} - \dot{\underset{\sim}{\chi}}$ and $\nabla \dot{\underset{\sim}{\chi}}$ are invariant with respect to (infinitesimal) changes of observers[1]. The balance equations of momentum and of generalized moment of momentum take the form

$$div\,(\underset{\sim}{\sigma} + \underset{\sim}{s}) + \rho \underline{f} = 0, \quad div\, \underset{\approx}{S} + s = 0 \tag{161}$$

which shows an explicit coupling between both balance equations via the relative stress tensor $\underset{\sim}{s}$. Such a coupling was not explicit in the general formulation of section 2.1 (see equations (134) and (135), but it finally becomes evident through the constitutive coupling in equation (155).

The microstrain theory proposed in Forest and Sievert (2006) is an application of the micromorphic approach when taking

$$\phi \equiv \underset{\sim}{C} = \underset{\sim}{F}^T . \underset{\sim}{F}, \quad {}^{\chi}\phi \equiv {}^{\chi}\underset{\sim}{C} \tag{162}$$

or

$$\phi \equiv \underset{\sim}{\varepsilon}, \quad {}^{\chi}\phi \equiv {}^{\chi}\underset{\sim}{\varepsilon} \tag{163}$$

within the small strain approximation. Because of the symmetry of the microstrain tensor, it is not necessary to introduce a relative stress tensor in the enriched power of internal forces. Instead, the standard form (129) is adopted:

$$p^{(i)} = \underset{\sim}{\sigma} : \dot{\underset{\sim}{\varepsilon}} + \underset{\sim}{a} : {}^{\chi}\dot{\underset{\sim}{\varepsilon}} \tag{164}$$

The coupling between macro and microstrain arises at the constitutive level since the free energy density was proposed in Forest and Sievert (2006) to be a function

$$\psi(\underset{\sim}{\varepsilon}^e, \quad T, \quad \alpha, \quad \underset{\sim}{e} := \underset{\sim}{\varepsilon} - {}^{\chi}\underset{\sim}{\varepsilon}, \quad \underset{\approx}{K} := \nabla^{\chi}\underset{\sim}{\varepsilon}) \tag{165}$$

[1] The reasoning holds true at finite deformation as done in Mindlin (1964); Eringen and Suhubi (1964).

The additional arguments can be limited to their non–dissipative parts $\underset{\sim}{e}^e, \underset{\approx}{\boldsymbol{K}}^e$ according to Forest and Sievert (2006). The existence of dissipative micromechanisms is considered in Forest and Sievert (2003). When the following linearized constitutive equations are adopted:

$$\underset{\sim}{a} = H_\chi \underset{\sim}{e}, \quad \underset{\approx}{b} = A \boldsymbol{\nabla}^\chi \underset{\sim}{\varepsilon} \tag{166}$$

the extra balance equation takes the following simple form:

$${}^\chi\underset{\sim}{\varepsilon} - l_c^2 \Delta^\chi \underset{\sim}{\varepsilon} = \underset{\sim}{\varepsilon}, \quad with \quad l_c^2 = \frac{A}{H_\chi} \tag{167}$$

as shown in Dillard et al. (2006). This represents an extension of the scalar partial differential (155) to a tensor valued micromorphic variable. The Laplace operator applies here to each individual tensor component, within a Cartesian orthonormal coordinate system.

The full micromorphic continuum can be regarded as the combination of the microstrain and Cosserat continua Forest and Sievert (2006). The Cosserat continuum itself can be interpreted in terms of the proposed methodology. It corresponds to the choice:

$$\phi = \underset{\sim}{\boldsymbol{R}}, \quad {}^\chi\phi \equiv {}^\chi\underset{\sim}{\boldsymbol{R}} \tag{168}$$

where $\underset{\sim}{\boldsymbol{R}}$ is the material rotation in the polar decomposition of $\underset{\sim}{\boldsymbol{F}}$. The associated quantity ${}^\chi\underset{\sim}{\boldsymbol{R}}$ is nothing but the micropolar rotation representing the rotation of a triad of directors attributed to each material point. The generalized stress $\underset{\sim}{a}$ in the enriched power of internal forces is then related to the skew–symmetric part of the Cosserat stress tensor. Cosserat models are known to be able to account for some size effects in the softening plastic behavior of granular materials Mühlhaus and Vardoulakis (1987) and in the hardening behaviour of metals Forest et al. (2000).

Microstrain gradient plasticity

The proposed methodology is now applied to the simplest models available for the isothermal elastic–plastic behaviour of materials. Different generalized models are worked out and compared to the existing models in literature. It turns out that several existing strain gradient plasticity models are recovered using the proposed systematic procedure. Some differences are evidenced due to the precise thermodynamical background of the approach which is not always present in the earlier approaches. In particular the approach can be used to tackle coupled problems, thus provided coupled strain gradient equations not present in the literature. A simple example of this coupling in given to account for a possible dependence of material parameters with temperature. The framework is applicable to plasticity and viscoplasticity.

The reference state space corresponding to a classical elastoplasticity model retained in this subsection is

$$DOF0 = \{\underline{\boldsymbol{u}}\}, \quad STATE0 = \{\underset{\sim}{\boldsymbol{\varepsilon}}^e, \quad p, \quad \alpha\} \tag{169}$$

where $\underset{\sim}{\boldsymbol{\varepsilon}}^e$ is the (infinitesimal) elastic strain tensor, p the cumulated plastic strain variable and α denotes another possible internal variable of any tensorial rank. The selected variable for the micromorphic approach is

$$\phi \equiv p \tag{170}$$

which the micromorphic variable ${}^{\chi}p$ is associated to. The classical power of internal and contact forces are extended in the following way:

$$p^{(i)} = \underset{\sim}{\boldsymbol{\sigma}} : \dot{\underset{\sim}{\boldsymbol{\varepsilon}}} + a\, {}^{\chi}\dot{p} + \underline{\boldsymbol{b}}.\boldsymbol{\nabla}^{\chi}p, \quad p^{(c)} = \underline{\boldsymbol{t}}.\dot{\underline{\boldsymbol{u}}} + a^c\, {}^{\chi}\dot{p} \tag{171}$$

in which generalized stresses a and $\underline{\boldsymbol{b}}$ have been introduced. The application of the method of virtual power leads to the following additional local balance equation and boundary conditions in addition to the classical local balance of momentum and traction condition at the outer boundary:

$$div\, \underline{\boldsymbol{b}} - a = 0, \quad \forall \underline{\boldsymbol{x}} \in \Omega, \quad a^c = \underline{\boldsymbol{b}}.\underline{\boldsymbol{n}}, \quad \forall \underline{\boldsymbol{x}} \in \partial\Omega \tag{172}$$

Generalized body forces a^e and $\underline{\boldsymbol{b}}^e$ could be introduced in case of necessity in the balance equations. The extended state space on which constitutive functions may depend is

$$STATE = \{\underset{\sim}{\boldsymbol{\varepsilon}}^e, \quad p, \quad \alpha, \quad {}^{\chi}p, \quad \boldsymbol{\nabla}^{\chi}p\} \tag{173}$$

The total strain is still split into its elastic and plastic parts:

$$\underset{\sim}{\boldsymbol{\varepsilon}} = \underset{\sim}{\boldsymbol{\varepsilon}}^e + \underset{\sim}{\boldsymbol{\varepsilon}}^p \tag{174}$$

The free energy density function ψ is assumed to be a function of the previous set $STATE$. The Clausius–Duhem inequality takes then the form

$$(\underset{\sim}{\boldsymbol{\sigma}} - \rho\frac{\partial\psi}{\partial\underset{\sim}{\boldsymbol{\varepsilon}}^e}) : \dot{\underset{\sim}{\boldsymbol{\varepsilon}}}^e + (a - \rho\frac{\partial\psi}{\partial{}^{\chi}p}){}^{\chi}\dot{p} + (\underline{\boldsymbol{b}} - \rho\frac{\partial\psi}{\partial\boldsymbol{\nabla}^{\chi}p}).\boldsymbol{\nabla}^{\chi}\dot{p} + \underset{\sim}{\boldsymbol{\sigma}} : \dot{\underset{\sim}{\boldsymbol{\varepsilon}}}^p - \rho\frac{\partial\psi}{\partial p}\dot{p} - \rho\frac{\partial\psi}{\partial\alpha}\dot{\alpha} \geq 0 \tag{175}$$

from which the following state laws and residual dissipation are derived:

$$\underset{\sim}{\sigma} = \rho \frac{\partial \psi}{\partial \underset{\sim}{\varepsilon}^e}, \quad a = \rho \frac{\partial \psi}{\partial {}^{\chi}p}, \quad \underline{b} = \rho \frac{\partial \psi}{\partial \nabla^{\chi} p}, \quad R = \rho \frac{\partial \psi}{\partial p}, \quad X = \rho \frac{\partial \psi}{\partial \alpha} \tag{176}$$

$$D^{res} = \underset{\sim}{\sigma} : \dot{\underset{\sim}{\varepsilon}}^p - R\dot{p} - X\dot{\alpha} \geq 0 \tag{177}$$

The plastic behaviour is characterized by the yield function $f(\underset{\sim}{\sigma}, R, X)$. In the micromorphic model, the yield function can still be treated as the dissipation potential providing the flow and evolution rules for internal variables. This corresponds to the hypothesis of maximal dissipation or normality rule:

$$\dot{\underset{\sim}{\varepsilon}}^p = \dot{\lambda} \frac{\partial f}{\partial \underset{\sim}{\sigma}}, \quad \dot{p} = -\dot{\lambda} \frac{\partial f}{\partial R}, \quad \dot{\alpha} = -\dot{\lambda} \frac{\partial f}{\partial X} \tag{178}$$

At this stage, a coupling between the macroscopic and microscopic variables must be introduced, for instance via the relative cumulative plastic strain $p - {}^{\chi}p$. An example of such a possible coupling is given in the next paragraph.

Example

A quadratic form is proposed to model the free energy density function, with respect to elastic strain, cumulative plastic strain, relative plastic strain and micromorphic plastic strain gradient:

$$\rho\psi(\underset{\sim}{\varepsilon}^e, p, {}^{\chi}p, \nabla^{\chi}p) = \frac{1}{2}\underset{\sim}{\varepsilon}^e : \underset{\approx}{\Lambda} : \underset{\sim}{\varepsilon}^e + \frac{1}{2}Hp^2 + \frac{1}{2}H_{\chi}(p - {}^{\chi}p)^2 + \frac{1}{2}\nabla^{\chi}p.\underset{\sim}{A}.\nabla^{\chi}p \tag{179}$$

The corresponding classical model describes an elastoplastic material behaviour with linear elasticity characterized by the tensor of elastic moduli $\underset{\approx}{\Lambda}$ and the linear hardening modulus H. Two additional material parameters are introduced in the micromorphic extension of this classical model, namely the coupling modulus H_{χ} (unit MPa) and the micromorphic "stiffness" $\underset{\sim}{A}$ (unit MPa.m^2). The thermodynamic forces associated with the state variables are given by the relations (176):

$$\underset{\sim}{\sigma} = \underset{\approx}{\Lambda} : \underset{\sim}{\varepsilon}^e, \quad a = -H_{\chi}(p - {}^{\chi}p), \quad \underline{b} = \underset{\sim}{A}.\nabla^{\chi}p, \quad R = (H + H_{\chi})p - H_{\chi}{}^{\chi}p \tag{180}$$

Note that when the relative plastic strain $e = p - {}^{\chi}p$ is close to zero, the linear hardening rule retrieves its classical form and the generalized stress a vanishes. Only the strain gradient effect ∇p remains in the enriched work of internal forces (171). This is the situation encountered in the strain gradient plasticity models developed in Fleck and Hutchinson (2001). When inserted in the additional balance equation (172), the previous states laws lead to the following partial differential equation:

$$ {}^{\chi}p - \frac{1}{H_{\chi}} div\,(\underset{\sim}{\boldsymbol{A}}.\boldsymbol{\nabla}^{\chi}p) = p \tag{181} $$

Let us specialize this equation to the case of isotropic materials, for which the second order tensor of micromorphic stiffness reduces to

$$ \underset{\sim}{\boldsymbol{A}} = A\underset{\sim}{\boldsymbol{1}} \tag{182} $$

which involves then a single additional material parameter. The equation (183) then becomes

$$ {}^{\chi}p - \frac{A}{H_{\chi}}\Delta^{\chi}p = p \tag{183} $$

which is identical to the second partial differential equation used in the so–called implicit gradient gradient–enhanced elastoplasticity in Engelen et al. (2003). The microstrain ${}^{\chi}p$ is called there the "non local strain measure" $\bar{p}$. Note however that the latter model involves only one additional material parameter, namely $l_c^2 = A/H_{\chi}$ instead of two in the micromorphic approach. It will turn out to be a special case of the micromorphic model for a specific value of the coupling modulus H_{χ}. No thermodynamical framework was proposed for the elastoplasticity model in the original contribution Engelen et al. (2003). Such a framework has been sketched in the reference Peerlings et al. (2004) where a quadratic potential similar to (179) is introduced which involves in particular the same coupling term. In contrast to the micromorphic approach, however, no additional contribution is introduced in the power of internal forces so that the additional partial differential equation is derived as a sufficient condition to identically fulfill the global form of the entropy inequality. In the micromorphic approach, the coupling modulus H_{χ} plays a central role and makes it possible to have a fully consistent thermomechanical basis for the model. When its value is high enough, it acts as a penalty term forcing the micromorphic plastic strain to follow the macroscopic one as close as possible.

The necessity of an additional boundary condition associated with the nonlocal strain measure is recognized in Engelen et al. (2003). The associated Neumann condition is used in the form:

$$ \boldsymbol{\nabla}^{\chi}p.\underline{\boldsymbol{n}} = 0 \quad on \quad \partial\Omega \tag{184} $$

It coincides with the more general boundary condition derived in the micromorphic approach:

$$ \underline{\boldsymbol{b}}.\underline{\boldsymbol{n}} = a^c \quad on \quad \partial\Omega \tag{185} $$

when $a^c = 0$ and when $\underline{b}$ is linear with respect to $\nabla^\chi p$, as it is the case for the quadratic potential (179).

The yield function is now chosen as

$$f(\underset{\sim}{\sigma}, R) = \sigma_{eq} - \sigma_Y - R \tag{186}$$

where σ_{eq} is an equivalent stress measure and σ_Y the initial yield stress. The hardening rule then takes the following form:

$$R = \frac{\partial \psi}{\partial p} = (H + H_\chi)p - H_\chi {}^\chi p \tag{187}$$

After substituting the balance (172) into the hardening law, yielding takes place when

$$\sigma_{eq} = \sigma_Y + H^\chi p - A(1 + \frac{H}{H_\chi})\Delta^\chi p \tag{188}$$

This expression coincides with the enhanced yield criterion originally proposed in Aifantis (1987) and used for strain localization simulations in Aifantis (1987); Borst et al. (1993) when the micromorphic variable remains as close as possible to the plastic strain: ${}^\chi p \simeq p$. In the latter references, the Laplace operator is directly introduced in the yield function as a postulate, whereas its presence is derived here from the combination of the additional balance equation and the linear generalized constitutive equations.

In the reference Engelen et al. (2003), after introducing the partial differential equation (183) in addition to the classical balance and constitutive equations, the authors propose to substitute the classical hardening law $R(p)$ by the same function $R({}^\chi p)$ where the argument is replaced by the nonlocal equivalent plastic strain. If such a hardening law is adopted, this model turns out to be a special case of the present microstrain theory for the following specific value of the hardening modulus:

$$H_\chi = -H \tag{189}$$

which follows from the identification $(H + H_\chi)p - H_\chi {}^\chi p = H^\chi p$ according to equation $(180)_4$. This assumption indeed reduces the number of free additional parameter to one, namely the choice of A related to the intrinsic length of the material. Such a choice, however, is acceptable only for softening materials for which $H < 0$. Otherwise, the additional contribution to the free energy associated with ${}^\chi p$ in (179) will act as a destabilizing term in the material behaviour. Furthermore the type of the partial differential equation (183) would be changed. The authors also point out the limitations of the simplistic method consisting in substituting the microstrain ${}^{\chi}p$ in the classical hardening law instead of p, especially regarding the subsequent evolution inside plastic strain localization bands.

Keeping both H_χ and A as free parameters of the theory makes it possible, in principle, to envisage applications to strain localization phenomena in softening materials, as done in Engelen et al. (2003); Dillard et al. (2006) but also to size effects in hardening plasticity, as done in Hutchinson (2000); Forest et al. (2000); Dillard et al. (2006). For a detailed discussion of the pros and the cons of various available strain gradient plasticity models for both types of applications, the reader is referred to Engelen et al. (2006).

The consistency condition and extension to viscoplasticity are described in Forest (2009).

Full microstrain gradient plasticity

The approach is not restricted to scalar micromorphic variables. As in Eringen's micromorphic model where the selected variable is the full deformation gradient itself, it can be applied to the full plastic strain tensor:

$$\phi \equiv \underset{\sim}{\varepsilon}^p, \quad {}^{\chi}\phi \equiv {}^{\chi}\underset{\sim}{\varepsilon}^p \tag{190}$$

This corresponds to 5 additional degrees of freedom if the micromorphic plastic strain tensor ${}^{\chi}\underset{\sim}{\varepsilon}^p$ is treated as a deviatoric tensor like the macroscopic plastic strain tensor. The generalized stresses are symmetric second and third order tensors, respectively:

$$p^{(i)} = \underset{\sim}{\sigma} : \dot{\underset{\sim}{\varepsilon}} + \underset{\sim}{a} : {}^{\chi}\dot{\underset{\sim}{\varepsilon}}^p + \underset{\approx}{b} \,\vdots\, \nabla^{\chi}\varepsilon^p \tag{191}$$

The symmetry condition applies only to the two first indices of b_{ijk}. The power of internal forces is indeed invariant with respect to (infinitesimal) changes of observers, due to the invariance of $\underset{\sim}{\varepsilon}$ and ${}^{\chi}\underset{\sim}{\varepsilon}^p$ themselves. The initial and extended sets of state variables are:

$$STATE0 = \{\underset{\sim}{\varepsilon}^e, \quad \underset{\sim}{\varepsilon}^p\}, \quad STATE = \{\underset{\sim}{\varepsilon}^e, \quad \underset{\sim}{\varepsilon}^p, \quad {}^{\chi}\underset{\sim}{\varepsilon}^p, \quad \nabla^{\chi}\underset{\sim}{\varepsilon}^p\} \tag{192}$$

When the micromorphic variable is constrained to remain as close as possible to the macroscopic one, the theories of gradient of plastic strain presented in Forest and Sievert (2003); Gurtin (2003); Abu Al-Rub et al. (2007) are recovered. In these works, generalized stresses are associated with the plastic strain rate tensor in the extended power of internal forces.

As an illustration, we adopt the following quadratic form for the free energy potential:

$$\begin{aligned}\rho\psi(\underset{\sim}{\varepsilon}, \underset{\sim}{\varepsilon}^p, {}^\chi\underset{\sim}{\varepsilon}^p, \nabla^\chi\underset{\sim}{\varepsilon}^p) &= \frac{1}{2}\underset{\sim}{\varepsilon}^e : \underset{\approx}{\Lambda} : \underset{\sim}{\varepsilon}^e + \frac{1}{3}C\underset{\sim}{\varepsilon}^p : \underset{\sim}{\varepsilon}^p \\ &+ \frac{1}{2}(\underset{\sim}{\varepsilon}^p - {}^\chi\underset{\sim}{\varepsilon}^p) : \underset{\approx}{C}_\chi : (\underset{\sim}{\varepsilon}^p - {}^\chi\underset{\sim}{\varepsilon}^p) \\ &+ \frac{1}{2}\nabla^\chi\underset{\sim}{\varepsilon}^p \,\vdots\, \underset{\approx}{\underset{\approx}{A}} \,\vdots\, \nabla^\chi\underset{\sim}{\varepsilon}^p \end{aligned} \tag{193}$$

from which the state laws are derived:

$$\underset{\sim}{\sigma} = \underset{\approx}{\Lambda} : \underset{\sim}{\varepsilon}^e, \quad \underset{\sim}{X} = \frac{2}{3}C\underset{\sim}{\varepsilon}^p + \frac{2}{3}C_\chi(\underset{\sim}{\varepsilon}^p - {}^\chi\underset{\sim}{\varepsilon}^p) \tag{194}$$

$$\underset{\sim}{a} = -\frac{2}{3}\underset{\approx}{C}_\chi : (\underset{\sim}{\varepsilon}^p - {}^\chi\underset{\sim}{\varepsilon}^p), \quad \underset{\approx}{b} = \underset{\approx}{\underset{\approx}{A}} \,\vdots\, \nabla^\chi\underset{\sim}{\varepsilon}^p \tag{195}$$

In the simplified situation for which

$$\underset{\approx}{C}_\chi = C_\chi\underset{\approx}{1}, \quad \underset{\approx}{\underset{\approx}{A}} = A\underset{\approx}{\underset{\approx}{1}} \tag{196}$$

where $\underset{\approx}{1}$ and $\underset{\approx}{\underset{\approx}{1}}$ are the fourth rank and sixth rank identity tensors operating respectively on symmetric second order tensors and symmetric (w.r.t. the two first indices) third rank tensors, the combination of the additional balance equation and state laws leads to the following partial differential equation:

$$\underset{\sim}{a} = div\,\underset{\approx}{b} = A\Delta^\chi\underset{\sim}{\varepsilon}^p = -\frac{2}{3}C_\chi(\underset{\sim}{\varepsilon}^p - {}^\chi\underset{\sim}{\varepsilon}^p) \quad \Longrightarrow \quad {}^\chi\underset{\sim}{\varepsilon}^p - \frac{3A}{2C_\chi}\Delta^\chi\underset{\sim}{\varepsilon}^p = \underset{\sim}{\varepsilon}^p \tag{197}$$

The differential operators act in the following way w.r.t. to a Cartesian frame $(\underline{e}_i)_{i=1,3}$:

$$div\,\underset{\approx}{b} = b_{ijk,k}\underline{e}_i \otimes \underline{e}_j, \quad \Delta^\chi\underset{\sim}{\varepsilon}^p = (\Delta^\chi\varepsilon^p_{ij})\underline{e}_i \otimes \underline{e}_j \tag{198}$$

The associated boundary conditions on the boundary of the body are given by a set of 6 equations:

$$\underset{\approx}{b}.\underline{n} = \underset{\sim}{a}^c \tag{199}$$

The internal variable $\underset{\sim}{\alpha} = \underset{\sim}{\varepsilon}^p$ is the proper state variable for a plasticity theory incorporating linear kinematic hardening, $\underset{\sim}{X}$ being the back–stress tensor. The retained isotropic yield function for extended J_2–plasticity is

$$f(\underset{\sim}{\sigma}, \underset{\sim}{X}) = J_2(\underset{\sim}{\sigma} - \underset{\sim}{X}) - \sigma_Y \quad (200)$$

$$= J_2(\underset{\sim}{\sigma} - \frac{2}{3}(C + C_\chi)\underset{\sim}{\varepsilon}^p - \frac{2}{3}C_\chi {}^\chi\underset{\sim}{\varepsilon}^p) - \sigma_Y \quad (201)$$

$$= J_2(\underset{\sim}{\sigma} - \frac{2}{3}C_\chi {}^\chi\underset{\sim}{\varepsilon}^p + A(1 + \frac{C}{C_\chi})\Delta^\chi\underset{\sim}{\varepsilon}^p) - \sigma_Y \quad (202)$$

where $J_2(\underset{\sim}{\sigma}) = \sqrt{3(\underset{\sim}{\sigma}^{dev} : \underset{\sim}{\sigma}^{dev})/2}$ is the von Mises second invariant for symmetric second rank tensors. The normality rule is adopted:

$$\dot{\underset{\sim}{\varepsilon}}^p = \dot{\lambda}\frac{\partial f}{\partial \underset{\sim}{\sigma}} = -\dot{\lambda}\frac{\partial f}{\partial \underset{\sim}{X}} = \dot{p}\underset{\sim}{N} \quad (203)$$

The intrinsic dissipation then takes its classical form:

$$D^{res} = \underset{\sim}{\sigma} : \dot{\underset{\sim}{\varepsilon}}^p - \underset{\sim}{X} : \dot{\underset{\sim}{\varepsilon}}^p = f\dot{p} + \sigma_Y\dot{p} \geq 0 \quad (204)$$

with energy storage associated with kinematic hardening.

The fact that the gradient of the plastic strain tensor (or part of it, in models retaining only the rotational part) mainly impacts on the kinematic hardening of the material has been recognized in Steinmann (1996); Forest et al. (2002); Forest and Sievert (2003); Gurtin (2003). In these references, the divergence of the higher order generalized stress tensor acts as a backs–stress.

Plasticity in single crystals

The proposed micromorphic approach can be applied to crystal plasticity in several ways. For small strain and small rotations, the plastic strain increments and the lattice rotation rate are

$$\dot{\underset{\sim}{\varepsilon}}^p = \sum_{i=1}^{N} \dot{\gamma}^s \{\underline{\boldsymbol{m}}^s \otimes \underline{\boldsymbol{n}}^s\}, \quad \underset{\sim}{\omega}^e = \underset{\sim}{\omega} - \underset{\sim}{\omega}^p = \underset{\sim}{\omega} - \sum_{i=1}^{N} \dot{\gamma}^s \}\underline{\boldsymbol{m}}^s \otimes \underline{\boldsymbol{n}}^s\{ \quad (205)$$

Definitions of total, plastic and elastic rotation rates are given by (84). Which variable ϕ should be chosen to introduce a micromorphic variable? There are several possibilities depending on the physical nature of the deformation mechanism to be accounted for. If ϕ is the lattice rotation vector given by (84)

$$\phi \equiv \underline{\boldsymbol{\Phi}}^e, \quad {}^\chi\phi \equiv \underline{\boldsymbol{\Phi}} \quad (206)$$

the Cosserat theory proposed in chapter 1 is recovered. The gradient of $\underline{\boldsymbol{\Phi}}^e$ is nothing but the lattice curvature tensor.

One may think also of the following choice:

$$\phi \equiv \gamma^s \tag{207}$$

as proposed in Gurtin (2000); Gurtin and Needleman (2005); Bayley et al. (2006). Such a model is indeed applicable for a single crystal but it seems to have severe limitations when applied to bicrystals or multicrystals for the following reason. The variables γ^s are not uniquely defined in each domain. They are defined only up to a symmetry belonging to the crystal symmetry group of the material. As a result, jump conditions at an interface between two such materials (like two grains) cannot be uniquely defined. Similar difficulties arise in anisotropic damage of composites Germain et al. (2007). An acceptable alternative choice is to consider the full plastic strain $\phi \equiv \underset{\sim}{\varepsilon}^p$ as already illustrated in 2.1.

Internal constraints

The coupling parameter H_χ can be seen as a penalty factor to enforce the internal constraint:

$$\phi \simeq {}^\chi\phi \tag{208}$$

Indeed, the generalized stress a was taken proportional to the relative strain $\phi - {}^\chi\phi$ so that parameter H_χ acts as a penalty factor in the power of internal forces (129). The constraint (208) can be strictly enforced if H_χ is treated as a Lagrange multiplier in the variational formulation Tian-Hu (2004). Such a model then incorporates the effect of the first gradient $\nabla\phi$ of the selected variable in the constitutive theory.

Second gradient crystal plasticity When $\phi \equiv \underset{\sim}{F}$ is taken as the full deformation gradient, the considered theory is that of the full micromorphic continuum as recognized in 2.1. The internal constraint (208) imposes that the microdeformation coincides with the macrodeformation. As a result, the effect of the gradient of $grad\,\underline{\mathbf{u}}$ is introduced in the theory, which corresponds to the second gradient model developed by Mindlin (1965); Mindlin and Eshel (1968); Germain (1973a).

Fleck and Hutchinson have proposed a crystal plasticity theory based on the second grade continuum in Fleck and Hutchinson (1997); Shu and Fleck (1999). This theory is sketched in the appendix 2.2. It incorporates the effect of the dislocation density tensor but also of gradient of slip normal to the slip plane. The latter effect is introduced therefore in addition to the additional hardening associated with the densities of so–called geometrically necessary dislocations. The second gradient and the Cosserat theories of crystal plasticity are compared in a simple situation of single slip in section 2.2.

Aifantis and Gurtin models When ϕ is related to plastic deformation, this situation corresponds to the strain gradient plasticity models presented in the literature. For instance, if $\phi \equiv p$, the cumulative plastic strain, the following evolution of the stress under plastic loading is deduced from (188), when the internal constraint is enforced ($H \ll H_\chi$):

$$\sigma_{eq} = \sigma_Y + Hp - A\Delta p \tag{209}$$

which is identical to the strain gradient model proposed by Aifantis (1987). It is also closely related to the corresponding deformation theory exposed in Fleck and Hutchinson (2001). There is also a direct connection to the so–called "explicit gradient" theory presented in Engelen et al. (2003). These authors introduce the micromorphic equivalent plastic strain:

$$^{\chi}p = p - c^2 \Delta p \tag{210}$$

If this expression is inserted in a hardening law of the form $\sigma_{eq} = \sigma_Y + H\,^{\chi}p$, the partial differential equation (209) is obtained.

The strain gradient plasticity models proposed in Gurtin (2003); Forest and Sievert (2003) result from the choice $\phi \equiv \underset{\sim}{\varepsilon}^p$, i.e. the full plastic strain tensor. In these works, the corresponding generalized balance equation and boundary conditions are clearly stated, based on the method of virtual power.

2.2 Application to simple shear in a two–phase single crystal

In the current endeavour to connect the different scales involved in plasticity of metals from atomistics to continuum plasticity Forest et al. (2001c), there is a strong research stream trying to replace detailed descriptions of the collective behaviour of dislocations by continuum mechanical models. When the size of the investigated microstructure is of the order of magnitude of typically 1 to 10 μm, or below, classical continuum crystal plasticity theory, as settled in Asaro (1983), for instance, ceases to be relevant since it is not able to reproduce the observed size effects. In contrast, nonlocal or generalized–continuum plasticity models incorporating intrinsic length scales account for the size effects, at least in a qualitative manner Fleck and Hutchinson (1997).

The aim of the present section is to directly compare the governing equations and analytical solutions resulting from dislocation-based, Cosserat and strain–gradient models of crystal plasticity. For this purpose, a two-phase single crystal material oriented for single slip and subjected to shear loading is investigated. A one-dimensional (1D) laminate microstructure consisting of periodically alternating layers of plastically soft and hard phases is considered. The hard phase is supposed to behave only elastically, and the soft channel as an elastic-ideally plastic material. Classical continuum crystal plasticity predicts homogeneous deformation and stress in each phase, possibly with a jump of some quantities at the interface. In contrast, the interest here is focused on the non-homogeneous distribution of plastic slip and lattice rotations in the soft phase arising from the analysis based on dislocation mechanics, as well as on Cosserat and strain gradient continua. Intrinsic length scales which are responsible for size effects in plasticity of such layered microstructures arise naturally in the dislocation analysis. The present simple examples enlighten the importance and physical relevance of the additional boundary or interface conditions usually introduced in generalized plasticity models Shu et al. (2001).

Comparisons between dislocation-based and strain gradient models exist in the literature, at least in two-dimensional cases. Generally, results of 2D discrete dislocation dynamics are compared with nonlocal continuum models. For instance in Bassani et al. (2001), the hardening behavior induced by a periodic distribution of hard precipitates in a crystalline matrix oriented for single slip is analysed for the case of simple shear. For that purpose, dislocation dynamics is resorted to on the discrete level, whereas a nonlocal model is used at the continuum level. Localized or diffuse plastic deformation patterns resulting from the dislocation simulations are compared with the predictions of the nonlocal model. In Shu et al. (2001), shearing of a single crystal layer is studied in detail for both single slip and double slip orientations. Dislocations are not allowed to cross the boundaries of the layer, which leads to non-homogeneous plastic deformation patterns. For both dislocation and continuum models, the analysis is based on numerical simulations, which makes it difficult to interpret clearly the continuum intrinsic length scales present in the model in terms of the corresponding elementary dislocation processes.

The dislocation simulations in Cleveringa et al. (1998); Bassani et al. (2001); Shu et al. (2001) are two-dimensional in the sense that only straight dislocations in a plane perpendicular to them are considered. Formation of dislocation pileups and dipoles are then the most active elementary deformation and hardening mechanisms. However the dislocations which are constrained to glide in the small volumes or narrow channels are generally required to bow-out. Obviously, the bowing of dislocations cannot be taken into account in the 2D dislocation dynamics framework. By contrast, the continuum line-tension dislocation-based model presented in section 2.2 is able to account for the Orowan bowing which is frequently encountered in the plastic deformation of various dislocation substructures (cells, subgrains, ladders, ...), single-crystal nickel-based superalloys, passivated thin films or microlaminates. The model to be presented here shows that the bowing of dislocations in the narrow channels leads to size effects. For completeness, an analysis of double-ended pileups of straight dislocations that can form in the soft channels under the considered loading conditions are investigated in section 2.2. Pileups are usually regarded as an appropriate illustration of the dislocation accumulation at interfaces or grain boundaries. They classically illustrate the effect of the so–called geometrically necessary dislocations Ashby (1970). As such, they have been the source of inspiration of several strain gradient models. Both line–tension and pileup dislocation-based models will be compared with the response of generalized continuum models. This attempt to directly identify the line-tension dislocation-based model with a Cosserat model has been presented in Sedláček and Forest (2000); Forest and Sedláček (2003); Forest (2008).

The Cosserat elastoplasticity model proposed in section 2.2 mimics the local response of the considered material in various situations, without explicitly introducing dislocation distributions. Analytical solutions under periodicity conditions are found and compared with the above dislocation approach. Especially, it is shown under which circumstances the Cosserat rotation coincides with the lattice rotation. Finally, in section 2.2, the plastic strain gradient models of Shu, Fleck and Hutchinson Fleck and Hutchinson (1997); Shu and Fleck (1999) and of Aifantis Aifantis (1987, 1999) are recalled and applied to the present simple case to compare them with the dislocation and Cosserat models. It will be shown that the interface conditions to be fulfilled in each model play a major role in the modelling of non–homogeneous deformation patterns.

Reference dislocation models

First, the line-tension model presented in Sedláček and Forest (2000); Forest and Sedláček (2003) is briefly reviewed and extended in the following two sections. Then, in section 2.2, an alternative configuration for which an explicit expression of dislocation distribution in the channels under shear is available, viz. that of the double-ended pileup, is briefly presented.

The considered periodic laminate microstructure with direction of the applied stress indicated and the coordinate system used, is sketched in figure 3. Instead of individual dislocations, a continuous field of curved glide dislocations in the soft channel is considered. For simplicity, all quantities are assumed to be independent of y and z. The equilibrium position of a

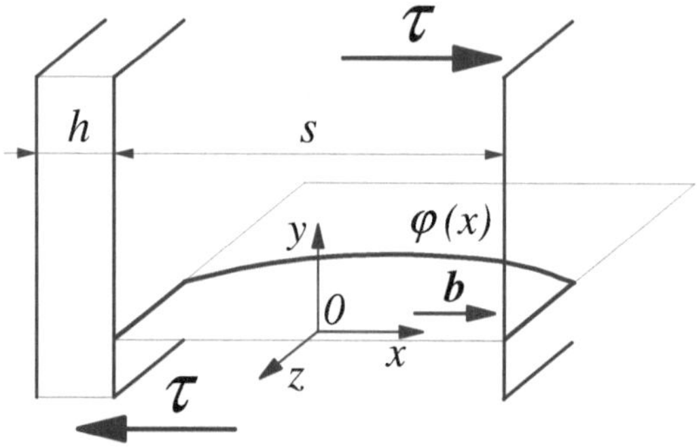

Figure 3. Dislocation bowing in the soft phase. A part of the loop gliding in the xOz plane is shown, with the curved (originally screw) section $\varphi(x)$ and edge segments at the soft / hard phase interface. The resolved shear stress τ and Burgers vector $\boldsymbol{b}$ are indicated. Labels s and h are used to designate the soft and hard phase, respectively.

representative bowing screw dislocation $\varphi(x)$ with Burgers vector magnitude b and constant line tension T in a shear stress field $\tau(x)$ is considered,

$$\tau(x)\, b + T\, \partial_{xx}\, \varphi(x) = 0\,. \tag{211}$$

The second derivative of the dislocation displacement $\varphi(x)$ follows from the linearized dislocation curvature. Strictly speaking, the linearized dislocation model is restricted to anelasticity (bowing of dislocations in the soft phase) since it is unable to describe fully plastic flow i.e. the glide of critically bowed dislocations depositing edge segments at the interfaces. To be able to deal with the plasticity, at least in an approximate way, one can introduce the Orowan stress in the soft channel of width s, $\tau_{Or} \approx 1.5\mu b/s$, as a threshold stress : if the mean shear stress in the channel reaches the value of the Orowan stress, the dislocation shape does not change any more and the bowed-out dislocations glide in the channels, depositing edges at the interfaces. This transition from anelasticity to plasticity, which is introduced here in an ad hoc manner, arises naturally in the framework of the full-curvature model Sedláček and Forest (2000); Sedláček et al. (2003).

Increment of plastic shear caused by the displacement $\varphi(x)$ of the mobile dislocations with density ϱ_m follows from the Orowan relation,

$$\gamma^p(x) = \varrho_m b\, \varphi(x)\,. \tag{212}$$

The non-homogeneous plastic deformation is accommodated by the 'geometrically necessary' content of the bowing dislocations $\varphi(x)$ which is described by the Nye-Kröner dislocation density tensor $\underset{\sim}{\alpha}$. Its only non-vanishing component $\alpha = \alpha_{xz}$ can be directly derived from the plastic shear,

$$\alpha(x) = \partial_x \gamma^p(x) \,. \tag{213}$$

It corresponds formally to continuously distributed 'pileups' of edge dislocations aligned with the z axis of figure 3, which is exactly the edge content of the bowing (originally screw) mobile dislocations with the scalar density ρ_m. By utilizing Hooke's law for elastic shear strain $\varepsilon^e = \varepsilon^e_{xy}$,

$$\tau(x) = 2\mu\varepsilon^e(x) \,, \tag{214}$$

a differential equation for the elastic and plastic shear strains results,

$$2\varepsilon^e(x) + \lambda^2 \partial_{xx} \gamma^p(x) = 0 \,, \tag{215}$$

with the intrinsic length scale λ given approximately by the average distance between the mobile dislocations,

$$\lambda = \sqrt{\frac{T}{\mu \varrho_m b^2}} \approx \frac{1}{\sqrt{\varrho_m}} \,. \tag{216}$$

We note in passing that the ratio between the channel width and intrinsic length s/λ is crucial for the size effect appearing during plastic deformation of narrow channels, thin films, microlaminates, etc., cf. Sedláček and Forest (2000).

To be able to derive equations for the lattice rotation based on eq. (215), relations for the elastic and plastic strains and rotations which will be extracted from the stress equilibrium and strain compatibility conditions are needed.

A compatible material displacement field $\boldsymbol{u} = (\bar{\gamma} y, u_y(x), 0)^T$ is considered, leading to material displacement gradient, $\boldsymbol{\beta} = grad\,\boldsymbol{u}$, of the form

$$\boldsymbol{\beta} = \begin{bmatrix} 0 & \bar{\gamma} & 0 \\ \partial_x u_y & 0 & 0 \\ 0 & 0 & 0 \end{bmatrix} . \tag{217}$$

The non-homogeneous plastic shear (212) causes a generally incompatible plastic distortion

$$\boldsymbol{\beta}^{\mathrm{p}} = \gamma^{\mathrm{p}} \underline{\boldsymbol{m}} \otimes \underline{\boldsymbol{n}} = \begin{bmatrix} 0 & \gamma^{\mathrm{p}} & 0 \\ 0 & 0 & 0 \\ 0 & 0 & 0 \end{bmatrix} , \tag{218}$$

The slip direction $\underline{m}$ and the normal to the slip plane $\underline{n}$ coincide with axes x and y, respectively, see figure 3. Finally, the elastic distortion $\boldsymbol{\beta}^{\mathrm{e}} = \boldsymbol{\beta} - \boldsymbol{\beta}^{\mathrm{p}}$ results in the form

$$\boldsymbol{\beta}^{\mathrm{e}} = \begin{bmatrix} 0 & \gamma^{\mathrm{e}} & 0 \\ \partial_x u_y & 0 & 0 \\ 0 & 0 & 0 \end{bmatrix}, \tag{219}$$

where $\gamma^{\mathrm{e}}(x) = \bar{\gamma} - \gamma^{\mathrm{p}}(x)$. The symmetric part of $\boldsymbol{\beta}^{\mathrm{e}}$ determines the elastic strain which enters Hooke's law (214),

$$\varepsilon^{\mathrm{e}}(x) = \frac{\gamma^{\mathrm{e}} + \partial_x u_y}{2}. \tag{220}$$

The skew-symmetric part of $\boldsymbol{\beta}^{\mathrm{e}}$ can be represented by the axial vector of lattice rotation $\boldsymbol{\phi}^{\mathrm{e}} = (0, 0, \phi^{\mathrm{e}} = \phi^{\mathrm{e}}_z)^T$,

$$\phi^{\mathrm{e}}(x) = -\frac{\gamma^{\mathrm{e}} - \partial_x u_y}{2}. \tag{221}$$

To solve the boundary value problem outlined here, a homogeneous strain and a homogeneous stress approximation are considered successively.

Periodic solution Stress equilibrium ($\operatorname{div} \boldsymbol{\sigma} = 0$) requires that the shear stress is constant and equal to the applied shear stress,

$$\tau(x) = \bar{\tau}. \tag{222}$$

With Hooke's law (214) in the form

$$\varepsilon^{\mathrm{e}} = \frac{\bar{\tau}}{2\mu}, \tag{223}$$

eq. (215) yields the following equation for the plastic shear strain,

$$\lambda^2 \, \partial_{xx} \, \gamma^P(x) = -\frac{\bar{\tau}}{\mu}. \tag{224}$$

Due to the fact that the soft phase is elastic - ideally plastic and from the discussion following equation (211), it is clear that $\bar{\tau} \leq \tau_{\mathrm{Or}}$. To obtain a unique solution to the problem at $\bar{\tau} = \tau_{\mathrm{Or}}$ (plastic regime), one has to fix the mean material strain by requiring that e.g. for the mean applied material shear strain

$$\frac{\bar{\gamma}}{2} = \frac{1}{s+h} \left(\bar{\varepsilon}_{\mathrm{h}} h + \int_{(s)} \varepsilon_{\mathrm{s}}(x) \, dx \right). \tag{225}$$

From equations (217), (220), (223) and (225), the amount of plastic deformation can be found,

$$\int_{\langle s \rangle} \gamma^{\mathrm{p}}(x)\, dx = \left(\bar{\gamma} - \frac{\bar{\tau}}{\mu}\right)(s+h)\,. \tag{226}$$

The lattice rotation can be obtained from eqs. (220), (221) and (223) as

$$\phi^{\mathrm{e}}(x) = \frac{\bar{\tau}}{2\mu} - \bar{\gamma} + \gamma^{\mathrm{p}}(x)\,. \tag{227}$$

As a consequence of (227), $\partial_x \gamma^{\mathrm{p}} = \partial_x \phi^{\mathrm{e}}$, and the governing equation for the lattice rotation in the soft phase $\phi^{\mathrm{e}}_{\mathrm{s}}$ becomes,

$$\lambda^2\, \partial_{xx}\, \phi^{\mathrm{e}}_{\mathrm{s}}(x) = -\frac{\bar{\tau}}{\mu}\,. \tag{228}$$

Accordingly, the lattice rotation and plastic slip (equation (224)) displays a parabolic profile,

$$\phi^{\mathrm{e}}_{\mathrm{s}}(x) = ax^2 + bx + c\,, \tag{229}$$

with $a = -\bar{\tau}/(2\lambda^2\mu)$. Note that, in the iso-stress framework, the exact solution of the full-curvature problem is well known: it is a circular arc. Nevertheless, we have linearized the curvature in equation (211), to be consistent with the rest of the paper. From symmetry reasons (periodicity),

$$\partial_x \phi^{\mathrm{e}}(0) = 0\,, \tag{230}$$

which implies $b = 0$. In the anelastic regime ($\bar{\tau} < \tau_{Or}$), the requirement for continuity of lattice rotation,

$$\phi^{\mathrm{e}}_{\mathrm{s}}(\pm\frac{s}{2}) = \phi^{\mathrm{e}}_{\mathrm{h}}\,, \quad \bar{\tau} < \tau_{Or}\,, \tag{231}$$

determines the constant c. Lattice rotation in the hard phase results from eq. (227),

$$\phi^{\mathrm{e}}_{h} = \frac{\bar{\tau}}{2\mu} - \bar{\gamma}\,. \tag{232}$$

Accordingly, the solution takes the form

$$\phi^{\mathrm{e}}_{\mathrm{s}}(x) = \left(\frac{\bar{\tau}}{2\mu} - \bar{\gamma}\right) - \frac{\bar{\tau}}{2\mu}\left(\frac{x^2}{\lambda^2} - \frac{s^2}{4\lambda^2}\right), \quad \bar{\tau} < \tau_{Or}\,. \tag{233}$$

In the plastic regime ($\bar{\tau} = \tau_{Or}$), there is a discontinuity of the lattice rotation $\Delta\phi^{\mathrm{e}}$ caused by the discontinuity of the plastic strain $\Delta\gamma^{p}$ which is

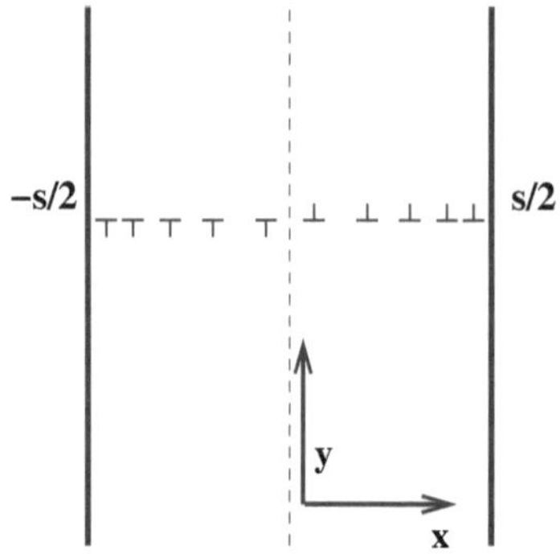

Figure 4. Double–ended dislocation pile–up in the soft phase under simple shear.

accommodated by the edge dislocations deposited at the interfaces. The solution is

$$\phi^{\rm e}_{\rm s}(x) = \left(\frac{\bar{\tau}}{2\mu} - \bar{\gamma}\right) - \frac{\bar{\tau}}{2\mu}\left(\frac{x^2}{\lambda^2} - \frac{s^2}{4\lambda^2}\right) + \Delta\phi^{\rm e}\,, \quad \bar{\tau} = \tau_{Or}\,. \tag{234}$$

The magnitude of the discontinuity $\Delta\phi^{\rm e}$ is determined from the equations (226) and (227),

$$\Delta\phi^{\rm e} = \frac{s+h}{s}\left(\bar{\gamma} - \frac{\bar{\tau}}{\mu}\right) - \frac{\bar{\tau}}{12\mu}\frac{s^2}{\lambda^2}\,. \tag{235}$$

We will return to the meaning of the step in lattice rotation later in the text.

Even though the stress equilibrium and strain compatibility are fulfilled exactly in the 1D iso-stress framework, the model is not realistic enough, for it cannot account for the internal stresses (i.e. kinematic hardening) which would arise in 2D or 3D structures. In the 1D model, the internal stresses are fully relaxed by the lattice rotations.

Pileup model This subsection presents a different model of dislocation structure that can form under the same applied loading conditions and that leads to a one-dimensional distribution of plastic slip in the macroscopic limit, that is dislocation pileups. This model is often advocated for the motivation of nonlocal theories and is recalled here, although its ingredients are different from the models in section 2.2 based on line tension effects.

A configuration for which an explicit expression of dislocation distribution in the channels under shear is available, is that of the double–ended pileup Hirth and Lothe (1982); Tanaka and Mura (1981), caused by the simple shearing of a single Frank–Read source at the centre of the channel of width s (figure 4). For simplicity, we take $h \simeq 0$ so that any image force due to the presence of the hard phase h acting on dislocations in the channel can be neglected. The approach based on the continuum theory of dislocations is briefly recalled here, in the case of pileups of edge dislocations with Burgers vector of magnitude b. The equilibrium of dislocations under the applied stress $\bar{\tau}$ can be written :

$$\bar{\tau} + \tau_d + \tau_c = 0\,, \tag{236}$$

where τ_d is the stress at x due to all present dislocations and τ_c the (assumed constant) threshold for the onset of dislocation motion. If $n(x)$ denotes the number of dislocations per unit length, the dislocation stress takes the form :

$$\tau_d(x) = A \int_{-s/2}^{s/2} \frac{n(x')}{x - x'} dx' \quad with \quad A = \frac{\mu b}{2\pi(1-\nu)}\,. \tag{237}$$

A solution $n(x)$ of the integral equation (236) exists under the condition of unbounded density at two tips of the pileup, viz. $x = \pm s/2$ in figure 4 :

$$n(x) = \frac{\bar{\tau} - \tau_c}{\pi A} \frac{x}{\sqrt{\left(\frac{s}{2}\right)^2 - x^2}}\,. \tag{238}$$

The total number of dislocations in each pileup is

$$N = \int_0^{s/2} n(x) dx = \frac{\bar{\tau} - \tau_c}{\pi A} \frac{s}{2}\,. \tag{239}$$

The displacement of material above the slip plane with respect to that below is given by

$$u_x(x, y) = b H(y) \int_x^{s/2} n(x') dx'\,, \tag{240}$$

where H is the Heaviside function and the pileups are assumed to lie at $y = 0$ in the volume considered here. Differentiating the previous equation with respect to y yields :

$$\partial_y u_x(x, y) = b\delta(y) \int_x^{s/2} n(x') dx'\,. \tag{241}$$

with δ the Dirac distribution. The corresponding amount of plastic slip is defined by

$$\gamma^{\mathrm{p}}(x) = \frac{1}{l}\int_{-l/2}^{l/2} \partial_y u_x(x,y)dy = \frac{b}{l}\int_{x}^{s/2} n(x)dx \tag{242}$$

$$= \frac{b}{l}\frac{\bar{\tau}-\tau_c}{\pi A}\sqrt{\left(\frac{s}{2}\right)^2 - x^2}, \quad for \quad -s/2 \le x \le s/2 \tag{243}$$

It is assumed that the pileups are periodically distributed along direction y perpendicular to slip plane with period l. The length l is assumed to be large enough for the interaction between parallel pileups to be neglected. As a result, the distribution of plastic slip in the channel is the arc of an ellipse. It vanishes at the tips of the pileups and takes its maximal value Nb/l at the centre. This distribution is therefore different from the parabolic profile found in section 2.2. The mean value of the plastic slip is

$$\bar{\gamma}^{\mathrm{p}} = \frac{2}{s}\int_0^{s/2} \gamma(x)dx = \frac{b}{l}\frac{\bar{\tau}-\tau_c}{2\pi A}\frac{s}{2}\,. \tag{244}$$

It can be shown that the dislocation stress τ_d introduced in equation (236) does not depend on x, as required by the equilibrium condition

$$\tau_d = -(\bar{\tau}-\tau_c) = -\frac{2\mu l}{\pi(1-\nu)s}\bar{\gamma}^{\mathrm{p}}\,. \tag{245}$$

This proves that the double–ended dislocation pileup produces a hardening component of linear kinematic type :

$$X = -\tau_d = C\bar{\gamma}^{\mathrm{p}} \quad with \quad C = \frac{2\mu l}{\pi(1-\nu)s}\,. \tag{246}$$

We note that that the pileup model is formulated at a different level of approximation than the line-tension models of subsection 2.2. Especially, a periodic distribution of pileups is assumed along the y direction. This enables the presence of the dislocation stress τ_d in the 1D model. Furthermore, each pileup is completely embedded in an elastic matrix, so that the stress cannot relax. If the pileups were distributed continuously along the y direction, as the bowed dislocations in the previous sections are, the stress would then relax exactly as in section 2.2 and the local internal stresses and thus the macroscopic hardening would disappear.

Application of the Cosserat model

The two-phase material now is a heterogeneous Cosserat continuum. In the first paragraph, both phases will be assumed to have a linearized behaviour, with different moduli. In the two following ones, the soft phase exhibits elastoplastic behaviour. We look here for solutions fulfilling all compatibility and equilibrium requirements with the only constraint that all fields must be periodic along x with period $s+h$ (see figure 3).

Linear approximation Both phases are Cosserat linear materials with constants μ_h, μ_{ch}, β_h and μ_s, μ_{cs}, β_s (see section 1.2 for the definition of isotropic Cosserat elasticity). The moduli of phase s can be also treated as secant elastoplastic moduli and the elastic and plastic parts will not be distinguished. Phase h must be thought of as almost classical, which means that β_h is small, but the solution is given here in the general case.

A mean shear deformation $\overline{\gamma}$ is applied along the direction x. We look again for a displacement field of the form :

$$u_x = \overline{\gamma} y, \quad u_y(x), \quad u_z = 0\,. \tag{247}$$

The deformation of a Cosserat material is described also by the micro–rotation axial vector field :

$$\phi_x = \phi_y = 0, \quad \phi_z = \phi(x)\,. \tag{248}$$

Thus, the Cosserat deformation and curvature tensors take the form :

$$\boldsymbol{e} = \begin{bmatrix} 0 & \overline{\gamma}+\phi & 0 \\ \partial_x u_y - \phi & 0 & 0 \\ 0 & 0 & 0 \end{bmatrix}, \quad \boldsymbol{\kappa} = \begin{bmatrix} 0 & 0 & 0 \\ 0 & 0 & 0 \\ \partial_x \phi & 0 & 0 \end{bmatrix}. \tag{249}$$

The associated non–vanishing components of the force and couple–stress tensors are :

$$\sigma_{xy} = \mu(\overline{\gamma} + \phi + e_{yx}) + \mu_c(\overline{\gamma} + \phi - e_{yx})\,, \tag{250}$$

$$\sigma_{yx} = \mu(\overline{\gamma} + \phi + e_{yx}) - \mu_c(\overline{\gamma} + \phi - e_{yx})\,, \tag{251}$$

$$\mu_{zx} = 2\beta \partial_x \phi\,. \tag{252}$$

The balance equations give

$$\partial_x \sigma_{yx} = 0, \quad \partial_x \mu_{zx} - (\sigma_{xy} - \sigma_{yx}) = 0\,, \tag{253}$$

Taking the elasticity relations into account, these equations become :

$$\mu(\partial_x \phi + \partial_x e_{yx}) - \mu_c(\partial_x \phi - \partial_x e_{yx}) = 0\,, \tag{254}$$

$$\beta \partial_{xx} \phi - \mu_c(\overline{\gamma} + \phi - e_{yx}) = 0\,. \tag{255}$$

Equation (254) can be rearranged to give

$$\partial_x e_{yx} = -\frac{\mu - \mu_c}{\mu + \mu_c}\, \partial_x \phi \,. \tag{256}$$

The equation for ϕ follows then from (255) :

$$\beta \partial_{xxx}\phi - \frac{2\mu\mu_c}{\mu + \mu_c}\, \partial_x \phi = 0 \,. \tag{257}$$

We define for each phase :

$$\omega_s^2 = \frac{2\mu_s\mu_{cs}}{\beta_s(\mu_s + \mu_{cs})}, \quad \omega_h^2 = \frac{2\mu_h\mu_{ch}}{\beta_h(\mu_h + \mu_{ch})}, \tag{258}$$

Each ω is the inverse of a length. The solution of (257) takes the form :

$$\phi_s = a_s \cosh(\omega_s x) + d_s, \quad \mathit{for} \quad -s/2 < x < s/2 \,, \tag{259}$$

$$\phi_h^+ = a_h \cosh(\omega_h (x - \frac{s+h}{2})) + d_h, \quad \mathit{for} \quad s/2 < x < (s+h)/2 \,, \tag{260}$$

$$\phi_h^- = a_h \cosh(\omega_h (x + \frac{s+h}{2})) + d_h, \quad \mathit{for} \quad -(s+h)/2 < x < -s/2 \,. \tag{261}$$

To reduce the number of integration constants in equations (259) to (261), the periodicity of ϕ has been used, together with the following symmetry conditions :

$$\partial_x \phi(0) = \partial_x \phi(-\frac{s+h}{2}) = \partial_x \phi(\frac{s+h}{2}) = 0 \,. \tag{262}$$

Then, e_{yx} can now be determined from (256) as follows :

$$e_{yx}^s = -\frac{\mu_s - \mu_{cs}}{\mu_s + \mu_{cs}} \phi_s + e_s \,,$$

$$e_{yx}^{h+} = -\frac{\mu_h - \mu_{ch}}{\mu_h + \mu_{ch}} \phi_h^+ + e_h \,, \tag{263}$$

$$e_{yx}^{h-} = -\frac{\mu_h - \mu_{ch}}{\mu_h + \mu_{ch}} \phi_h^- + e_h \,. \tag{264}$$

where e_h and e_s are integration constants. Furthermore, equation (255) implies that :

$$e_{yx} = -\frac{\beta}{\mu_c} \partial_{xx}\phi + \overline{\gamma} + \phi \,, \tag{265}$$

$$e_{yx}(0) = -\frac{\beta_s}{\mu_{cs}} a_s \omega_s^2 + \overline{\gamma} + a_s + d_s = -\frac{\mu_s - \mu_{cs}}{\mu_s + \mu_{cs}}(a_s + d_s) + e_s \tag{266}$$

$$e_{yx}(\frac{s+h}{2}) = -\frac{\beta_h}{\mu_{ch}} a_h \omega_h^2 + \overline{\gamma} + a_h + d_h = -\frac{\mu_h - \mu_{ch}}{\mu_h + \mu_{ch}}(a_h + d_h) + e_h \tag{267}$$

from which the values of e_s and e_h are deduced :

$$e_s = \overline{\gamma} + d_s \frac{2\mu_s}{\mu_s + \mu_{cs}}, \quad e_h = \overline{\gamma} + d_h \frac{2\mu_h}{\mu_h + \mu_{ch}}. \tag{268}$$

For the determination of the four integration constants a_s, d_s, a_h, d_h, certain conditions at the interface must be enforced. In the Cosserat theory, the degrees of freedom u_i and ϕ_i are continuous if one excludes cracks and kinks. Therefore, the corresponding traction and couple–stress vectors must also be transmitted at the interface. Alternative conditions would be to prescribe specific values or jumps at the interface for displacement and micro–rotation at the interface, as done for instance in Shu and Fleck (1999). We note that in the dislocation-based model, the value of the step in lattice rotation given by eq. (235), follows from the condition of mean prescribed glide $\bar{\gamma}$. Here, this condition is automatically satisfied by the periodicity requirement to be enforced by equation (273). Thus, continuity requirements are imposed in this work in the absence of a more specific interface model. The interface conditions require :

- continuity of ϕ at $s/2$:

$$a_s \cosh(\omega_s \frac{s}{2}) + d_s = a_h \cosh(\omega_h \frac{h}{2}) + d_h . \tag{269}$$

- continuity of μ_{zx} at $s/2$:

$$\beta_s a_s \omega_s \sinh(\omega_s \frac{s}{2}) = -\beta_h a_h \omega_h \sinh(\omega_h \frac{h}{2}). \tag{270}$$

- continuity of σ_{yx} at $s/2$. Rearranging equation (251) as follows :

$$\sigma_{yx} = (\mu + \mu_c) e_{yx} + (\mu - \mu_c)(\overline{\gamma} + \phi) = 2\mu(\overline{\gamma} + d) \tag{271}$$

  one obtains the third equation :

$$\mu_s(\overline{\gamma} + d_s) = \mu_h(\overline{\gamma} + d_h). \tag{272}$$

- periodicity of u_y implies

$$< e_{yx} > = < \partial_x u_y - \phi > = < -\phi >,$$

  where the brackets denote averaging over x from $-(s+h)/2$ to $(s+h)/2$. One finds :

$$< \phi + e_{yx} > = < \frac{2\mu_c}{\mu + \mu_c}\phi + e > = < \frac{2\mu_c}{\mu + \mu_c}\phi + \overline{\gamma} + \frac{2\mu}{\mu + \mu_c} d >,$$

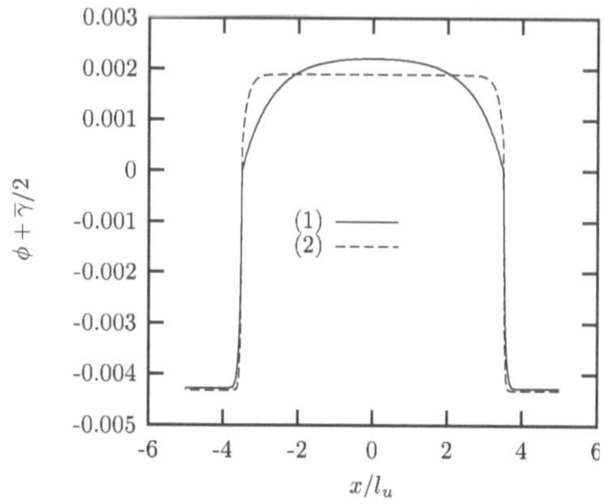

Figure 5. Distribution of microrotation of a two-phase linear elastic Cosserat material undergoing simple glide, using two different sets of parameters : (1) $\mu_h = 10\mu_s = 26923$MPa, $\mu_{ch} = 20\mu_{cs} = 100000$MPa, $\beta_h = \beta_s/30. = 100$MPa.$l_u^2$; and (2) $\beta_h = \beta_s/100 = 1$MPa.$l_u^2$; l_u is the chosen length unit (mm, μm...) and $s + h = 10l_u$, $\bar{\gamma} = 0.01$.

which gives the fourth equation

$$\frac{4\mu_{cs}}{\mu_s + \mu_{cs}} \frac{a_s}{\omega_s} \sinh(\omega_s \frac{s}{2}) + \frac{4\mu_{ch}}{\mu_h + \mu_{ch}} \frac{a_h}{\omega_h} \sinh(\omega_h \frac{h}{2}) + \bar{\gamma}(s+h) + 2d_s s + 2d_2 h = 0\,. \quad (273)$$

The four equations (269), (270), (272) and (273) represent a linear system of equations for the unknowns a_s, a_h, d_s and d_h. For conciseness, the final expressions are not given explicitly. Instead, the profiles of ϕ are plotted in figure 5 for two different sets of material parameters. It can be seen that, for an appropriate choice of the material parameters ($\mu_h > \mu_s$, $\mu_{ch} > \mu_{cs}$, $\beta_s > \beta_h$), ϕ_h is almost constant and ϕ_s displays a cosh profile with a characteristic length $1/\omega_s$. This profile mimics the distribution found in subsection 2.2 and suggests that ϕ can be interpreted as a lattice rotation, providing that ω_s is taken to be of the order of the magnitude of λ.

In the limiting case when $\beta \longrightarrow 0$, it can be shown that stresses and strains are constant in each phase, as expected for the solution of this simple glide problem for the classical Cauchy continuum. This can also be inferred from figure 5.

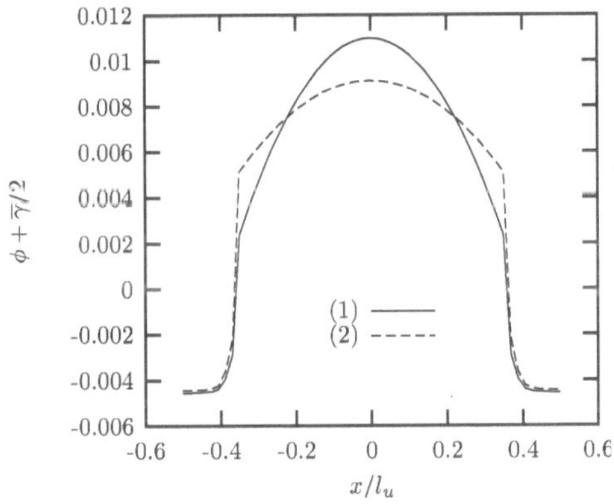

Figure 6. Distribution of microrotation of a two-phase elastoplastic Cosserat material undergoing simple glide, using two different sets of parameters : (1) $\mu_h = \mu_s = 26923$MPa, $\mu_{ch} = 5\mu_{cs} = 500000$MPa, $\beta_h = \beta_s/10. = 10$MPa.$l_u^2$, $\tau_c = 10$ MPa; and (2) $\beta_h = \beta_s/30 = 1$MPa.l_u^2; $s + h = l_u$, $\overline{\gamma} = 0.01$.

It can be checked also that $< \sigma_{yx} - \sigma_{xy} >= 0$, so that the macroscopic stress is of course symmetric. This point indicates that we are implicitly considering the problem of the homogenization of heterogeneous Cosserat media. This general problem is tackled in Forest et al. (1999, 2001b). The definition of the effective stress is

$$\overline{\sigma}_{ij} =< \sigma_{ij} > .$$

If neither mean curvature nor relative rotation is prescribed to the unit cell, the effective stress is symmetric.

Elastoplastic case Deformation in the soft phase is now decomposed into its elastic and plastic parts and the Schmid law is used as the yield criterion. The threshold shear stress τ_c is taken as constant (thus no hardening is considered). In contrast, we still do not distinguish between elastic and plastic curvature and keep a linearized relation between couple–stresses and total curvature. In this case, he total deformation in the cell is split into elastic and plastic parts :

$$\underset{\sim}{e} = \underset{\sim}{e}^e + \underset{\sim}{e}^p . \tag{274}$$

Only single slip is considered. The normal $\boldsymbol{\nu}$ to the slip plane is supposed to be parallel to the y-direction (figure 3) and the slip direction $\boldsymbol{s}$ is parallel to x :

$$\boldsymbol{e}^p = \gamma^p \underline{\boldsymbol{m}} \otimes \underline{\boldsymbol{n}} = \begin{bmatrix} 0 & \gamma^p & 0 \\ 0 & 0 & 0 \\ 0 & 0 & 0 \end{bmatrix}, \quad \boldsymbol{e}^e = \begin{bmatrix} 0 & e^e_{xy} & 0 \\ e^e_{yx} & 0 & 0 \\ 0 & 0 & 0 \end{bmatrix},$$

so that $e^e_{yx} = e_{yx}$. The non-vanishing force–stress components are :

$$\sigma_{xy} = \mu_s(e^e_{xy} + e_{yx}) + \mu_{cs}(e^e_{xy} - e_{yx}), \tag{275}$$

$$\sigma_{yx} = \mu_s(e_{yx} + e^e_{xy}) + \mu_{cs}(e_{yx} - e^e_{xy}). \tag{276}$$

The driving force to activate plastic slip is taken as the projection τ of the symmetric part of the force–stress tensor on the normal to the slip plane, and in the slip direction. Kröner (1956) suggests to take the full non–symmetric force stress in the computation of the resolved shear stress. This is not done here since the main Cosserat effects shown in this work do not come from the asymmetry of stress but rather from the presence of lattice curvature and the associated couple–stresses. Additional effects associated with the asymmetric character of the stress tensor have not yet been studied. Comments on the role of the skew–symmetric part of the force–stress tensor are given in subsection 2.2. The yield criterion therefore gives :

$$\tau = (\sigma_{xy} + \sigma_{yx})/2 = \tau_c = \mu_s(e^e_{xy} + e_{yx}). \tag{277}$$

From the first balance equation, viz. $\partial_x \sigma_{yx} = 0$, and (277), it can be seen that e_{yx} and e^e_{xy} are constant. The second balance equation reads :

$$\beta_s \partial_{xx}\phi - \mu_{cs}(e^e_{xy} - e_{yx}) = 0, \tag{278}$$

which gives

$$\partial_{xxx}\phi = 0. \tag{279}$$

Thus, ϕ displays a parabolic profile in the cell. The wall of width h in figure 3 is taken as an elastic Cosserat solid. The profile is the same as in section 2.2 :

$$\phi_s = a_s x_1^2 + d_s, \quad for \quad -s/2 < x < s/2, \tag{280}$$

$$\phi_h = a_h \cosh(\omega_h(x - \frac{s+h}{2})) + d_h \quad for \quad s/2 < x < (s+h)/2 \tag{281}$$

where ω_h is still given by (258). To determine the integration constants, the following conditions must be accounted for :

- continuity of ϕ at $x = s/2$:

$$a_s \frac{s^2}{4} + d_s = a_h \cosh(\omega_h \frac{h}{2}) + d_h. \tag{282}$$

- continuity of μ_{zx}

$$\beta_s a_s s = -\beta_h a_h \omega_h \sinh(\omega_h \frac{h}{2}) \,. \tag{283}$$

- continuity of σ_{yx}; e_{yx}, e^e_{xy} and σ_{yx} are determined in each phase :

$$e_{yx} + e^e_{xy} = \frac{\tau_c}{\mu_s}, \quad e^e_{xy} - e_{yx} = \frac{2\beta_s}{\mu_{cs}} a_s \,,$$

$$\sigma^s_{yx} = \tau_c - 2\beta_s a_s, \quad \sigma^h_{yx} = 2\mu_h(\overline{\gamma} + d_h) \,,$$

so that

$$2\mu_h(\overline{\gamma} + d_h) = \tau_c - 2\beta_s a_s \,. \tag{284}$$

- periodicity of u_y; we use again the property

$$< e_{yx} > = < \partial_x u_y - \phi > = < -\phi > \,,$$

$$e^s_{yx} = \frac{\tau_c}{2\mu_s} - \frac{\beta_s}{\mu_{cs}} a_s, \quad e^h_{yx} = -\frac{\mu_h - \mu_{ch}}{\mu_h + \mu_{ch}} \phi_h + \overline{\gamma} + d_h \frac{2\mu_h}{\mu_h + \mu_{ch}} \,, \tag{285}$$

$$e^s_{yx} + \phi_s = \frac{\tau_c}{2\mu_s} - \frac{\beta_s}{\mu_{cs}} a_s + a_1 x_1^2 + d_1 \,,$$

$$e^h_{yx} + \phi_h = \frac{2\mu_{ch}}{\mu_h + \mu_{ch}} \phi_h + \overline{\gamma} + \frac{2\mu_h}{\mu_h + \mu_{ch}} d_h \tag{286}$$

$$= \frac{2\mu_{ch}}{\mu_h + \mu_{ch}} a_h \cosh(\omega_h (x - \frac{s+h}{2})) + \overline{\gamma} + 2d_h \,, \tag{287}$$

from which the last equation for the determination of the integration constants is deduced :

$$(\frac{\tau_c}{2\mu_s} - \frac{\beta_s}{\mu_{cs}} a_s + d_s) s + \frac{a_s s^3}{12} + 2d_h h + \overline{\gamma} h + \frac{4\mu_{ch}}{\mu + \mu_{ch}} \frac{a_h}{\omega_h} \sinh(\omega_h \frac{h}{2}) = 0 \,. \tag{288}$$

The linear system (282), (283), (284), and (288) can be solved for the four unknowns (a_s, d_s, a_h, b_h), as in the previous section.

Once the micro–rotation $\phi(x)$ is known, the amount of plastic slip can be found from :

$$\gamma^p = \overline{\gamma} + \phi - \frac{\tau_c}{2\mu_s} - \frac{\beta_s a_s}{\mu_{cs}} \,. \tag{289}$$

The resulting parabolic distribution of Cosserat micro–rotation and plastic strain are given in figure 6 for two different sets of parameters. Note that for the classical limiting case when β_s tends to zero (and μ_c to infinity, see next section), the classical relation (227) is retrieved.

Limiting case for constrained Cosserat single crystal plasticity At this point, a precise discussion of the links between the Cosserat micro–rotation ϕ and the lattice rotation ϕ^e must be given. The previous analysis of two different situations has shown the similarity between the profiles of ϕ and ϕ^e, as deduced from the combination of classical continuum mechanics and equilibrium of a dislocation line. In the continuum framework of classical crystal plasticity, in which dislocations are not considered individually, a clear definition of lattice rotation exists. For instance, in the specific case of the shear test, it is given by equation (221). Such continuum description of lattice rotation will now be compared with the Cosserat micro–rotation computed in section (2.2). By definition, the lattice rotation is related to the skew–symmetric part of the elastic distortion :

$$\boldsymbol{e}^e = \begin{bmatrix} 0 & \overline{\gamma} - \gamma^p + \phi & 0 \\ \partial_x u_y - \phi & 0 & 0 \\ 0 & 0 & 0 \end{bmatrix}, \qquad (290)$$

the corresponding axial vector being $(0, \quad 0, \quad \dfrac{\overline{\gamma} - \gamma^p - \partial_x u_y}{2} + \phi)^T$.
If the skew–symmetric part of the Cosserat elastic deformation tensor vanishes, the following relation is found :

$$\phi = -\frac{\overline{\gamma} - \gamma^p - \partial_x u_y}{2} \qquad (291)$$

which is exactly that given by equation(221). Thus, the Cosserat micro–rotation is found to coincide exactly with the standard definition of lattice rotation when the Cosserat elastic deformation is symmetric. It is recalled that the skew–symmetric part of the stress and elastic deformation are linked by the elastic modulus μ_c (see equations (275) and (276)). Thus the value of μ_c controls the difference between the Cosserat micro–rotation and lattice rotation. The physical meaning of ϕ in the case of crystal plasticity is therefore clear only when it is close or equal to the lattice rotation. This is the case when μ_c is large compared to the other moduli. The condition $\mu_c \to \infty$ enforces indeed the symmetry condition for the elastic deformation. The resulting finite skew–symmetric part of the stress can be regarded as reaction stresses (similarly to pressure in classical incompressible materials). This is a kind of constrained Cosserat continuum for which the Cosserat micro–rotation follows strictly lattice rotation, which is different from the well–known couple–stress medium, for which the Cosserat rotation is forced to follow the material rotation, namely pure rotation component of the overall deformation gradient Koiter (1963).

Consider the analysis of section 2.2 in the case when μ_c goes to infinity. The equations are indeed simplified and a clearer connection between material and geometric constants can be derived. The characteristic length in the hard phase becomes :

$$\frac{1}{\omega_h} = \sqrt{\frac{\beta_h}{2\mu_h}}\,. \tag{292}$$

In the previous system of four equations (282),(283), (284) and (288), the fourth one now becomes :

$$(\frac{\tau_c}{2\mu_s} + d_s)s + \frac{a_s s^3}{12} + (2d_h + \gamma)h - \frac{2\beta_s}{\mu_h} a_s s = 0\,. \tag{293}$$

The solution can then be given in a rather concise form, at least for a_s which characterizes the parabolic profile of the lattice rotation distribution :

$$a_s = -6\frac{\overline{\gamma}(s+h) - \frac{\tau_c}{2\mu_h}(s + 2h + s\frac{\mu_h}{\mu_s})}{s^3 + \frac{\beta_s}{\mu_h}(18s + 12h + 3s^2\omega_h cotanh(\omega_h \frac{h}{2}))}\,. \tag{294}$$

It can be seen that, contrary to the result of the analysis in the purely linear case of subsection 2.2, both material and geometric parameters contribute to the shape of the distribution ϕ in the elastoplastic channel. It can also be noted that when $\beta_h \to 0$, i.e. $\omega_h \to \infty$, the coefficient a_s vanishes and the classical homogeneous distribution of rotation and plastic slip is retrieved. Interestingly, when a non–zero and constant value of β_h is assumed and when $\beta_s \to 0$, the coefficient a_s does not vanish but rather reaches the limit :

$$a_s = -6(\overline{\gamma} - \frac{\tau_c}{\mu})\frac{s+h}{s^3} \tag{295}$$

where $\mu_h = \mu_s = \mu$ has been assumed for simplicity. The fact that the classical homogeneous solution is not found in this case can be interpreted as follows. Letting β_s vanish in the moment of momentum balance equation (278) makes this second partial differential equation disappear, so that in principle we are left with the usual force–stress balance equation and a classical solution could be expected. However, if equation (278) is multiplied by $1/\beta_s$ and differentiated again, we are left with (279) that can be assumed to hold in the limit for $\beta_s \to 0$. It amounts then to finding a parabolic distribution of ϕ fulfilling the classical equation (227) and continuity conditions at the interface. Indeed for very low values of β_s finite element simulations provide this limit solution. Identifying the present result (295) and the corresponding solution of the dislocation model (216), the link between the

Cosserat model and the intrinsic length scale is

$$\frac{1}{\lambda^2} = 24(\frac{\overline{\gamma}\mu}{\tau_c} - 1)\frac{s+h}{s} \simeq \varrho_m \tag{296}$$

which provides a reasonable but very approximate estimate of the density of mobile dislocation. The complete solution with all available material constants is of course more elaborate and incorporates the length scale associated with a non–vanishing β_s. In the latter case, the complete identification of the Cosserat model with the corresponding dislocation model can be done using (294) in order to calibrate β_s, as a function of ϱ_m.

Comparison with strain gradient plasticity models

Alternative nonlocal continuum theories are available to model size effects in crystal plasticity, usually called strain gradient plasticity models. In this section, the response predicted by two of them is investigated in the simple case of shearing of laminate microstructures. In particular, the shape of the non–homogeneous plastic slip profiles are compared to the previous Cosserat results.

Second gradient formulation Based on the continuum framework introduced by Mindlin and Germain Mindlin and Eshel (1968); Germain (1973a), a strain gradient plasticity constitutive framework has been developed by Shu, Fleck and Hutchinson Fleck and Hutchinson (1997); Shu and Fleck (1999). Two main models have been proposed by the authors. The first one deals with isotropic plasticity, the second one with crystal plasticity. The main constitutive and balance equations are recalled in section 2.2 in a simplified case of single slip. Moreover the originally viscoplastic framework is translated into a purely elastoplastic one, for which analytical solutions can be found in the shear test considered in this work.

As in section 2.2, the displacement field in a periodic two–phase laminate microstructure subjected to mean shear deformation $\bar{\gamma}$ is considered. The solution must fulfill balance and constitutive equations in each phase and interface conditions. First, a solution is obtained assuming a linear approximation for the constitutive behaviour without distinction between elastic and plastic parts. The second solution is valid for the nonlinear elastoplastic regime.

Using the same coordinate frame as in section 2.2, the solution still takes the form $u_x = (\bar{\gamma}y, u_y(x), 0)^T$. The non–vanishing components of the strain gradient tensor $\boldsymbol{\eta}$, as defined in Appendix 2.2, are

$$\eta_{xyx} = \eta_{yxx} = \frac{1}{2}\partial_{xx}u_y\,. \tag{297}$$

Elastoplastic case

Adopting the additive decomposition (1) of total strain and strain gradient into elastic and plastic parts, the elastic laws become

$$\sigma_{xy} = 2\mu\varepsilon^{e}_{xy} = \mu(\bar{\gamma} + u_{y,x} - \gamma^{p})\,, \tag{298}$$

$$m_{yxx} = 2l_e^2\mu(\eta_{yxx} - \eta^{p}_{yxx}) = l_e^2\mu(\partial_{xx}u_y - \gamma^{S})\,, \tag{299}$$

where γ^S is the slip gradient variable (see Appendix 2.2). The yield condition is a generalized Schmid law involving resolved shear stress and hyperstress :

$$\sigma_{eq} = |\tau| + |m|/l_p = \sigma_{xy} + m_{yxx}/l_p = \tau_c\,. \tag{300}$$

It must be noted that a second characteristic length l_p enters the yield criterion. The yield condition combined with the balance equation

$$\partial_x\sigma_{xy} - \partial_{xx}m_{yxx} = 0\,. \tag{301}$$

can then be shown to lead to the following partial differential equations

$$\partial_{xx}u_y - l_p^2\partial_{xxxx}u_y = 0, \quad \partial_x\gamma^p - l_p^2\partial_{xxx}\gamma^p = 0\,. \tag{302}$$

Since the definition of lattice rotation (227) still holds, one is led to the following partial differential equation for ϕ :

$$\partial_x\phi - l_p^2\partial_{xxx}\phi = 0\,, \tag{303}$$

which again gives a *cosh*−profile but associated with the characteristic length l_p. This result is different from the parabolic profile found in the same analysis with the Cosserat model. The reason is however clear : it stems from the modified yield condition (300). A consequence of this choice is that shear stress σ_{xy} does not remain constant in space, contrary to all investigated cases in the previous sections. It introduces hardening associated with strain gradient.

The strain gradient model can however be slightly modified to be closer to the simple Cosserat constitutive equations used in section 2.2. We let l_p go to infinity and do not distinguish elastic and plastic parts in $\boldsymbol{\eta}$:

$$\sigma_{eq} = \tau = \sigma_{xy} = \tau_c, \quad m_{yxx} = 2\mu l_e^2\eta_{yxx}\,. \tag{304}$$

l_e is in fact here the characteristic length of the elastoplastic secant moduli in the second elasticity law, since $\boldsymbol{\eta}$ has not been partitioned. The balance equation (301) now yields :

$$\partial_{xxxx}u_y = 0, \quad \partial_{xxx}\gamma^p = 0, \quad \partial_{xxx}\phi = 0\,, \tag{305}$$

and a parabolic profile is retrieved.

Crystal plasticity with gradient of internal variable One of the earliest proposals to introduce higher–order gradients in the relevant constitutive variables is Aifantis' model that belongs to a class of generalized continua different from the theories used in the previous sections. The models presented and illustrated in Aifantis (1987, 1999) do not introduce additional rotational degrees of freedom like the Cosserat theory, nor higher order derivatives of the displacement field. Instead, the constitutive behaviour of the material is assumed to depend on an internal variable γ^p and its gradient $grad\,\gamma^p$. The free energy is a function of temperature, elastic strain, plastic slip and its gradient. The classical expression of work of internal forces is in fact complemented by terms related to the internal variable and its gradient Forest et al. (2002) :

$$p^{int} = \sigma_{ij}\dot{\varepsilon}_{ij} + \alpha\dot{\gamma}^p + B_i\partial_i\dot{\gamma}^p \,, \qquad (306)$$

where α and B_i are the thermodynamical forces associated with the internal variable and its gradient. The principle of virtual work can be used to derive the balance equations :

$$div\,\boldsymbol{\sigma} = 0, \qquad \alpha = div\,\boldsymbol{B} \,. \qquad (307)$$

The classical balance equation is conserved, whereas the second one can be regarded as a definition of the introduced generalized force α. The degrees of freedom and associated reaction forces are therefore the pairs $(u_i, t_i = \sigma_{ij}n_j)$ and $(\gamma^p, B = B_in_i)$. The dissipation rate takes the form

$$D = \sigma_{ij}\dot{\varepsilon}^p_{ij} + (div\,B)\dot{\gamma}^p = (\tau + div\,B)\dot{\gamma}^p \,. \qquad (308)$$

The latter expression leads one to propose a yield condition of the form :

$$\sigma_{eq} = \tau + div\,B = \tau_c \,. \qquad (309)$$

In Aifantis model, $\boldsymbol{B}$ is assumed to be simply proportional to $grad\,\gamma^p$ (quadratic potential) :

$$B_i = c\partial_i\gamma^p \,, \qquad (310)$$

where c is a constitutive parameter. The yield condition (309) then becomes :

$$\tau = \tau_c - c\Delta\gamma^p \,, \qquad (311)$$

which is the well–known gradient–enhanced yield criterion, with Δ being the Laplace operator. This theemodynamics of Aifantis model has also been investigated by Bardella (2007); Gurtin and Anand (2009); Forest and Aifantis (2010).

This theory is now applied to the simple situation investigated in the present article. The yield condition reduces to the simple form :

$$\tau = \sigma_{xy} = \tau_c - c\,\partial_{xx}\gamma^p \,. \tag{312}$$

The equilibrium equation requires constant shear stress so that we must have :

$$\partial_{xxx}\gamma^p = 0 \,. \tag{313}$$

The elasticity relation (298) implies again $\partial_{xx}u_y = \partial_x\gamma^p$ so that lattice rotation ϕ must fulfill the same partial differential equation :

$$\phi = \frac{\partial_x u_y + \gamma^p - \bar{\gamma}}{2}, \quad \partial_{xxx}\phi = 0 \,. \tag{314}$$

This model therefore gives the same answer for the lattice rotation and plastic slip distribution as the Cosserat model of section 2.2, namely a parabolic one. The interface conditions are dictated by the chosen degrees of freedom and associated forces :

- displacement u_y and traction vector $\sigma_{xy}n_x$ are continuous across the interface;
- plastic slip γ^p and force $B = (c\,grad\,\gamma^p).n = c\partial_x\gamma^p$ are continuous across the interface.

In contrast to the Cosserat theory, plastic slip is assumed to be continuous, whereas lattice rotation need not necessarily to be continuous (this comes from the fact that $\partial_x u_y$ is not necessarily continuous in contrast to the strain gradient theory).

If linear hardening is introduced in the yield condition (309), it can be shown that a *cosh* distribution is obtained. However, this situation cannot be compared to that of section 2.2 and 2.2 since there no hardening was taken into account.

Direct comparison between the dislocation and generalized continuum frameworks

A direct comparison between the dislocation models and the continuum frameworks is difficult insofar as the dislocation models considered in this work are very specific whereas the continuum models can be used in very general situations. The line tension model considers a representative bowing screw dislocation which is a convenient and idealized situation. As for the pile–up model, it deals with periodic arrays of edge dislocations. However, a parallel between the line tension dislocation model and the Cosserat, second gradient and Aifantis models can be drawn. All models share the main variables plastic slip γ^p and lattice rotation ϕ^e, even though the two gradient models more explicitly introduce the gradient of plastic slip. The classical divergence equation for the stress tensor is valid for all theories. The generalized continuum involve an additional (or a higher order) balance equation which reflects in a continuum way the dislocation line tension equilibrium equation. A one–to–one identification is however not possible since the continuum theories do not explicitly introduce densities of line defects. The consequence is that the governing partial differential equations in the case of shearing of the elastoplastic laminate microstructure have almost the same structure. The gradient theories are somewhat different, in the sense that they predict that the shear stress τ does not remain equal to the critical resolved shear stress τ_c. In the strain gradient theory, the shear stress τ is equal to $\tau_c - m_{yxx}/l_p$ and to $\tau_c - c\partial_{xx}\gamma^p$ in Aifantis model. This strain–gradient induced hardening behaviour is not introduced nor predicted in the idealized line tension dislocation and Cosserat models. The strain gradient theory can be simplified to get rid of this hardening component, as shown at the end of section 2.2. The similarity of the governing p.d.e. makes it possible to identify some parameters of the generalized continuum models with the dislocation based quantities, especially the involved characteristic lengths.

As a consequence of the choice of the primary variables, the interface conditions are different for all generalized continuum models. In the line tension model, at least in the plastic regime, plastic slip and lattice rotation are not necessarily continuous, which corresponds to a density of edge dislocations deposited at the interface. In contrast, the Cosserat theory assumes the continuity of lattice rotation whereas plastic slip is continuous in the gradient of internal variable theory. The second gradient model introduces the continuity of the normal gradient of displacement, the physical meaning of which is more difficult to assess. These differences in interface conditions make the previous identification of characteristic lengths at the level of the governing equations insufficient. That is why, we must also compare the final solution, namely the predicted distribution of slip and lattice rotation in the considered simple glide test. As shown in section 2.2, the lattice distributions predicted by the Cosserat and line tension models can be identified. The phenomenological moduli are then related to dislocation material parameters but also to the geometry of the microstructure, especially the layer thicknesses s and h.

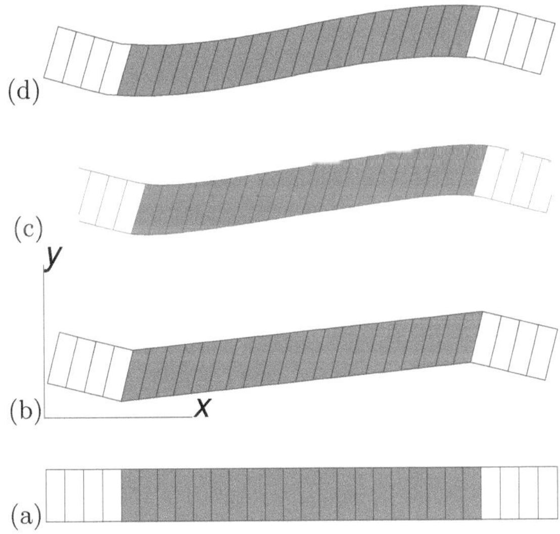

Figure 7. Deformed states of a two–phase material element under simple shear : (a) initial state; (b) solution according to classical crystal plasticity; (c) Cosserat elasticity (see subsection 2.2); (d) Cosserat elastoplasticity (see subsection 2.2). The solutions have been computed using the finite element method and each phase is divided into elements : hard phase in white, soft phase in grey; the mean deformation is the same in (b,c,d).

Non–homogeneous plastic deformation in channels

In the simple situation investigated in this work, several available nonlocal models of crystal plasticity - ranging from Cosserat, second grade to gradient of internal variable theories - predict development of a non–homogeneous distribution of plastic slip and lattice rotation in the soft channels of the laminate microstructure under shear. Each of the nonlocal models has its own advantages and drawbacks. The Cosserat model naturally comes out of almost thirty years old theoretical reflexions on crystal plasticity Mandel (1973) but the physical meaning of the skew–symmetric part of the stress still remains unclear. In the strain gradient model recalled in Appendix 2.2, the constitutive links between slip gradient variables γ^S and the gradient of slip $grad\,\gamma^p$ remains phenomenological. This model gives also the opportunity of incorporating effects that are not associated with geometrically necessary dislocations (in particular the gradient of slip in the direction normal to the slip plane, the physical interpretation of which remains to be explained). Regarding Aifantis' model, the Laplace term in yield condition (311) can in some cases be derived from the physics of dislocations (see for instance Estrin et al. (1998) in the case of double slip).

Anyway, the use of Cosserat, strain gradient or Aifantis models invariably leads to parabolic or *cosh*- distributions of plastic slip and lattice rotation, depending on specific constitutive assumptions. Each profile is characterized by a length that is directly related to the constitutive length(s) introduced in each model. This intrinsic length enters the elastic or the plastic part of the constitutive equations, or both. Figure 7 summarizes and illustrates the different deformed states of a sample of such two–phase material, according to the classical and Cosserat models.

The parabolic or *cosh*- distributions of plastic slip and lattice rotation result also from the line–tension dislocation–based model that incorporates the bowing of screw dislocations in narrow channels into a simple one–dimensional continuum–mechanics description. Loosely speaking, the anelastic regime for which dislocations move over short distances can be associated with a *cosh*- distribution, whereas the fully plastic regime corresponds to a parabolic profile.

Thus it appears that the plastic slip and lattice rotation distribution obtained within the proposed Cosserat framework accurately mimics the results of the dislocation–based line–tension models. On the generalized–continuum level, simple linear and/or perfectly plastic constitutive equations that are usual in phenomenological modelling can be used. The balance of moment of momentum equation (253)$_2$, that does not exist in the classical continuum framework, turns out to be the continuum counterpart of the equilibrium condition (211) for a representative dislocation bowing in the channel.

Even a direct identification of the dislocation and Cosserat models is possible. In particular, the wavelength λ of the dislocation model (216) enables one to compute the corresponding value of the Cosserat parameter β_s from (292) :

$$\beta_s \simeq \frac{2\mu}{\varrho_m} \tag{315}$$

Since it depends on the density of mobile dislocations which may vary during deformation, β_s should not be seen as a constant material parameter. In the case of the parabolic profile, not only the constant β_s, but also explicitly the channel width s are the determinant parameters (see equation (295)). The identification for the iso–strain model has been proposed in Sedláček and Forest (2000). The present work focuses on the detailed solution of the iso–stress periodic problem. Other models like strain gradient and Aifantis' model have been shown to reproduce the line–tension effects as well.

Interestingly, the mentioned nonlocal theories were not originally designed for the modelling of line tension effects but rather hardening effects due to so–called geometrically necessary dislocations as in the pileup model. Admittedly, several nonlocal models have shown their ability to account for particle or grain size effects that can be related to the presence of dislocation pileups (see Forest et al. (2000); Acharya and Beaudoin (2000) for the simulation of Hall–Petch effects in polycrystals). However one should not hastily associate strain gradient plasticity and dislocation pileup effects. In fact, the distribution of plastic slip in dislocation pileups in the soft phase is not correctly described by any of the mentioned nonlocal models. Double–ended pileups are dislocation structures that can also form in the laminate microstructure under the prescribed loading conditions, for instance because of periodically distributed Frank–Read sources, or as the result of passage of many bowed screw dislocations. It could be argued that the pileup model includes internal stresses and associated hardening, which has not been taken into account in the Cosserat model. However, the strain gradient model used in section 2.2 incorporates hardening associated with hyperstresses and still provides a *cosh*- distribution. A Cosserat model including classical linear hardening would lead in fact also to a *cosh*- distribution. It shows that the main ingredients of the current nonlocal crystal plasticity models are not really best–suited for the description of dislocation pileup effects in crystals, but rather of dislocation line tension effects. This is surprising since the size effects arising from dislocation bowing did not belong explicitly to the initial main motivations that have led to the development of the nonlocal theories. Conversely, this fact can be regarded as an important property of the nonlocal models since the line tension effects have proved to dominate the mechanical response of many engineering materials. The case of single-crystal nickel-based superalloys is of special interest, since they display a periodic microstructure of hard precipitates and soft channels. A description of precipitate size effects in single crystal superalloys based on the Cosserat theory can be found in Forest et al. (2000). An alternative model including gradients of internal variables has also been applied to this single crystal material Busso et al. (2000). Size effects associated with the channel width are predicted by the models. For example, it can be checked that the solution of the elastoplastic Cosserat model tends towards the classical one when the relative size s/h goes to infinity (coefficient a_s of the parabolic profile then vanishes according to equation (294)).

The presented results are applicable to cyclic loading conditions. Further effort must now be concentrated on dislocation interaction and hardening, which remains challenging since the line tension model becomes difficult to handle explicitly in the three–dimensional case Sedláček et al. (2003). Dislocation dynamics and finite element simulations are then useful tools to go towards more realistic multi–slip situations, as initiated in Shu and Fleck (1999); Shu et al. (2001). The situation of double slip has been shown to be dramatically different from the single slip case in Shu et al. (2001), from the dislocation dynamics point of view.

Role of interface conditions

In the present study, nonlocal mechanical models incorporating additional higher order boundary or interface conditions have been considered. It appears clearly in Shu et al. (2001) that models that keep the classical structure of the boundary value problem unchanged, like Acharya and Bassani (2000); Busso et al. (2000), predict homogeneous glide as the classical continuum mechanics does for the shearing of a crystalline layer. Non–homogeneous distributions can be obtained with a nonlocal model by applying higher order boundary conditions at the boundary of the sheared layer. The higher order boundary conditions concern the additional degrees of freedom of higher order derivatives, introduced in the model, and their associated forces. Similarly, in the two–phase microstructure considered in this work, the enriched interface conditions are responsible for the development of a non–homogeneous plastic slip pattern in the soft phase. Since neither special constitutive properties nor direct boundary conditions have been applied to the interface, both phases must be treated as generalized continua, and not only the soft one. If one considers in the Cosserat model of section 2.2, the limit case for which $\mu_{ch} = \infty$ and $\beta_h = 0$, constant values for ϕ are obtained in each phase with a jump. It means that, in order to get a non–constant distribution, the wall cannot be regarded as purely classical : it must be able to carry the surface couples produced in phase s at the boundary. This can be achieved by setting a relatively low value of β_h and a 100 time bigger β_s. In that case, the distribution in the hard phase is then almost flat with a steep rise close to the boundary, that mimics a jump of the considered variable. The same holds for the two other nonlocal models handled in section 2.2. An alternative method could be to consider the hard phase as a classical material and to set directly boundary conditions at the interface (see for instance Shu and Fleck (1999) for the interface between two crystallites). This has not been done here because the conditions to be prescribed are not necessarily known a priori. Instead, in the Cosserat theory, lattice rotation, the additional degree of freedom, is continuous at the interface and so does the traction vector and the couple–stress vector. Similarly in the second gradient theory, the normal gradient of the displacement and the associated force must be continuous at the interface. So does plastic slip and associated force in Aifantis' model. These conditions are therefore slightly different for each theory, even if the governing partial differential equation may be the same in the bulk. It is however difficult to assess which continuity requirement is the most realistic from the physical point of view. The main point is that these continuity constraints are the origin of the non–uniform lattice rotation field in the soft phase.

More general grain boundary behavior could include discontinuities of lattice rotation, plastic strain, or even grain boundary sliding and opening for damage Acharya (2007); Gurtin and Anand (2008).

Description of internal stresses and hardening

The attention has been focused on the continuum description of plastic slip and lattice rotation distribution, but the question of hardening is also an important point from the macroscopic point of view. The pileup model of subsection 2.2 leads to the existence of linear hardening due to internal stresses, described by equation (246). This kinematic hardening component can be seen as a sort of nonlocal hardening law since it depends on the mean value of plastic slip and not on the local value of γ. It has the classical form used in phenomenological constitutive equations for single crystal under cyclic loading Méric et al. (1991), although nonlinear kinematic hardening is usually observed experimentally. It appears also that this hardening term does not depend on lattice curvature or dislocation density tensor and therefore has no direct relation to the density of geometrically necessary dislocations (see also Mughrabi (2001)). A size effect is expected from the dependence of the hardening modulus C on the channel width s, see equation (246).

Such a hardening term should be introduced in the continuum model of section 2.2 to account for arising internal stresses. It can be easily done by replacing τ_c in (277) by a term of the form $\tau_c + C\bar{\gamma}$. The form of the solution is not affected by this term which does not depend on position x. It is clear however that the continuum Cosserat model does not account for the exact distribution of plastic strain in a double–ended pileup (parabolic profile instead of an elliptic one). The introduction of the linear hardening component then keeps a phenomenological character.

In Shu et al. (2001), the local hardening modulus entering the constitutive equations of the second grade model has been identified numerically using the mean response of the discrete dislocation dynamics model. The hardening modulus links the equivalent stress and plastic strain rates $\dot{\sigma}_{eq}$ and $\dot{q}$ (see section 2.2). However the plastic multiplier q is a measure of cumulated plastic strain and strain gradient without distinction, because of the use of a single coupled yield criterion (300) in the spirit of Borst (1991) :

$$\dot{q} = |\dot{\gamma}^p| + l_p|\dot{\gamma}^S|\,. \tag{316}$$

The pileup model indicates that hardening originates from mean plastic slip and not from local slip gradient. Accordingly, a distinction between both contributions should be preferable in the continuum model. Such a distinction exists in the full Cosserat model used in Forest et al. (2000) by considering two different plastic multipliers, one for plastic slip, the other one for lattice curvature. This means also that the single yield criterion (300) is replaced by two yield conditions : one involves resolved shear stresses, the second one involves resolved couple–stresses. Coupling between both comes then from the hardening laws. It must be noticed that this hardening law for the two–phase material must be of kinematic type for application to cyclic plasticity.

In many generalized crystal plasticity models, the higher order stresses generate kinematic hardening, generallly linear. Different classes of such models are compared in Forest (2008); Cordero et al. (2010).

Appendix: Second grade continuum

The strain and its gradient are decomposed into their elastic and plastic parts :

$$\varepsilon_{ij} = \varepsilon^e_{ij} + \varepsilon^p_{ij}, \quad \eta_{ijk} = \partial_{x_k}\varepsilon_{ij} = \eta^e_{ijk} + \eta^p_{ijk}\,. \tag{1}$$

The associated stresses are the classical Cauchy stress σ_{ij} and the hyperstress tensor m_{ijk}. The following simplified form of the elastic relations has been chosen in Shu and Fleck (1999) :

$$\sigma_{ij} = C_{ijkl}\varepsilon^e_{kl}, \quad m_{ijk} = l_e^2 C_{ijpq}\eta^e_{pqk}\,, \tag{2}$$

where the usual four–rank elasticity tensor is denoted by C_{ijkl} and l^e is a characteristic length associated with the higher order elasticity law. For the structure to be in equilibrium, the stress tensors must fulfill the following balance equation :

$$\partial_{x_j}\sigma_{ij} - \partial_{x_j x_k} m_{ijk} = 0\,, \tag{3}$$

where volume simple and double forces have been excluded. Note that equilibrium is governed by a single higher order partial differential equation whereas two balance equations must be fulfilled for the Cosserat continuum. For this continuum, a boundary value problem is well–posed provided that boundary conditions are prescribed to one element in each pair (u_i, t_i) and (Du_i, r_i). The unknown displacement is denoted by u_i and Du_i is the normal gradient of u_i defined by

$$Du_i = (\partial_{x_j} u_i)n_j \tag{4}$$

for a unit vector $\boldsymbol{n}$ normal to a surface. The corresponding tangent gradient operator is

$$D_j(.) = \partial_{x_j}(.) - D(.)n_j \tag{5}$$

The traction vector $\boldsymbol{t}$ and double traction vector $\boldsymbol{r}$ on a surface element of normal $\boldsymbol{n}$ are respectively defined by

$$t_i = (\sigma_{ij} - \partial_{x_k} m_{ijk})n_j + (D_l n_l)m_{ijk}n_j n_k - D_j(m_{ijk}n_k) \tag{6}$$

$$r_i = m_{ijk} n_j n_k \,. \tag{7}$$

In the case of single slip in a system $(\underline{\boldsymbol{m}}\,, \underline{\boldsymbol{\nu}})$ (slip direction $\boldsymbol{s}$ and slip plane normal $\boldsymbol{\nu}$), the plastic strain and strain gradient are related to the amount of slip γ^p and the slip gradient variable γ^S by :

$$\dot{\varepsilon}^p_{ij} = \dot{\gamma}^p \, s_{(i}\nu_{j)}, \quad \dot{\eta}^p_{ijk} = \dot{\gamma}^S \, s_{(i}\nu_{j)} s_k \,, \tag{8}$$

where the parentheses around indices stand for symmetrization. Note that in the general theory proposed in Shu and Fleck (1999), additional slip gradient variables γ^T, γ^M are introduced, that we do not include in the simple case investigated here. It must be noted also that in this theory, the slip gradient variable γ^S does not necessarily coincide with the gradient of slip γ^p.

The plastic yield criterion is a generalized Schmid law involving resolved shear stresses and hyperstresses :

$$\sigma_{eq} = |\tau| + |m|/l_p, \quad with \quad \tau = \sigma_{ij} s_i \nu_j, \quad and \quad m = m_{ijk} s_i \nu_j s_k \,. \tag{9}$$

Plastic deformation can then occur when σ_{eq} reaches the threshold τ_c. The authors in Shu and Fleck (1999) propose a viscoplastic formulation of the constitutive framework. In contrast, an elastoplastic formulation is used here, for the simple case of single slip. Associative plastic flow is assumed. Thus, the normality rule is given as :

$$\dot{\gamma}^p = \dot{q}\partial_\tau \sigma_{eq} = \dot{q}\, sign\, \tau \,, \tag{10}$$

$$\dot{\gamma}^S = \dot{q}\partial_m \sigma_{eq} = \frac{\dot{q}}{l^p} sign\, m \,, \tag{11}$$

where $\dot{q}$ denotes the plastic multiplier in both equations.

Two characteristic lengths appear in the theory, namely l^e in equations (2) and l^p in equation (9).

3 Continuum modelling of size effects in the mechanics of single and polycrystals

3.1 Plasticity at the crack tip in single crystals

The stress–strain field at the crack tip in a single crystal can be simulated using dislocation dynamics and continuum crystal plasticity Van der Giessen et al. (2001). The problem of the crack in an infinite medium has been addressed very often during the development of the mechanics of generalized continua in the 60s and 70s. Solutions are available for the elastic couple stress medium and fully non local elasticity. In the case of crystal plasticity, numerical solutions with the Cosserat continuum have been compared to Rice's solution for the classical continuum in Forest et al. (2001a). The classical solution, derived for an elastoplastic single crystal without hardening, predicts the existence of sectors with constant stress separated by localization bands. The physical nature of the bands depends on crystal structure and orientation. Intense slip bands, kink bands and multislip bands can coexist. The use of the Cosserat continuum leads to the weakening and even elimination of kink bands at the crack tip when additional hardening associated to lattice curvature is introduced. The figure 8 shows the profiles of equivalent plastic slip along half a circle close to the crack tip for three different values of extra–hardening modulus H assuming a simple extra–hardening model linear in the variable θ (see equation (106)). For the considered crystal orientation, kink bands lie at $55°$ and $125°$ from the horizontal axis, whereas the vertical band is a pure slip band. The intensity of the kink bands decreases for increasing H values. Note that asymptotic analyses of the crack tip field are possible within classical crystal plasticity that exclude kink banding Drugan (2001) but they are difficult to implement numerically. Recent experimental investigations on nickel–base superalloys at room temperature confirm in fact the existence of kink bands at the crack tip under monotonous loading (cf. figure 9 and reference Flouriot et al. (2003)). Extra–hardening parameter H can be identified from the found intensity of the kink bands.

The use of physically more realistic models at the crack tip than classical crystal plasticity is an important issue for the prediction of subsequent crack growth, especially in fatigue Deshpande et al. (2001).

3.2 Strain heterogeneities in polycrystals

The continuum crystal plasticity framework can be used to evaluate the development of plastic strain heterogeneities in thin or bulk polycrystals. We report first computations based on classical continuum crystal plasticity,

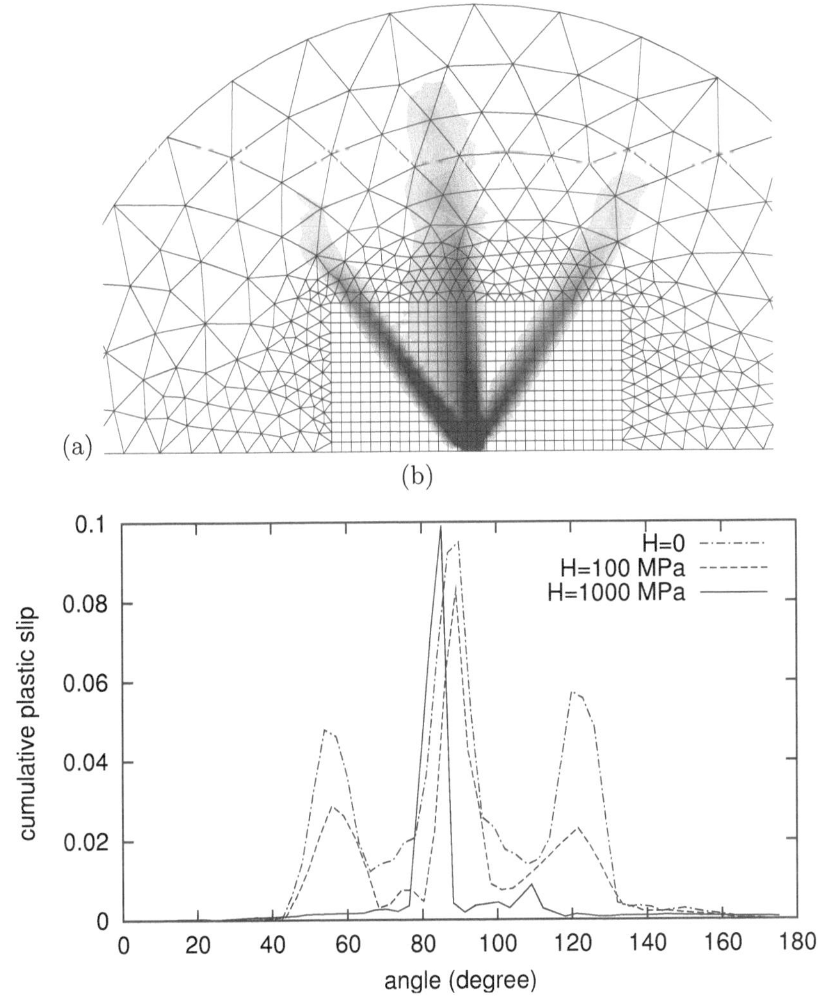

Figure 8. Equivalent plastic strain at a crack tip in an elastic ideally–plastic Cosserat single crystal under mode I loading conditions (F.C.C. crystal, the vertical and horizontal directions are respectively [011] and [100]). The results are obtained from finite element computations of cracked Cosserat single crystals. The equivalent plastic strain is localized in 3 bands (a). The amount of plastic slip is plotted along half a circle close to the crack tip for three different values of the extra–hardening parameter H (b).

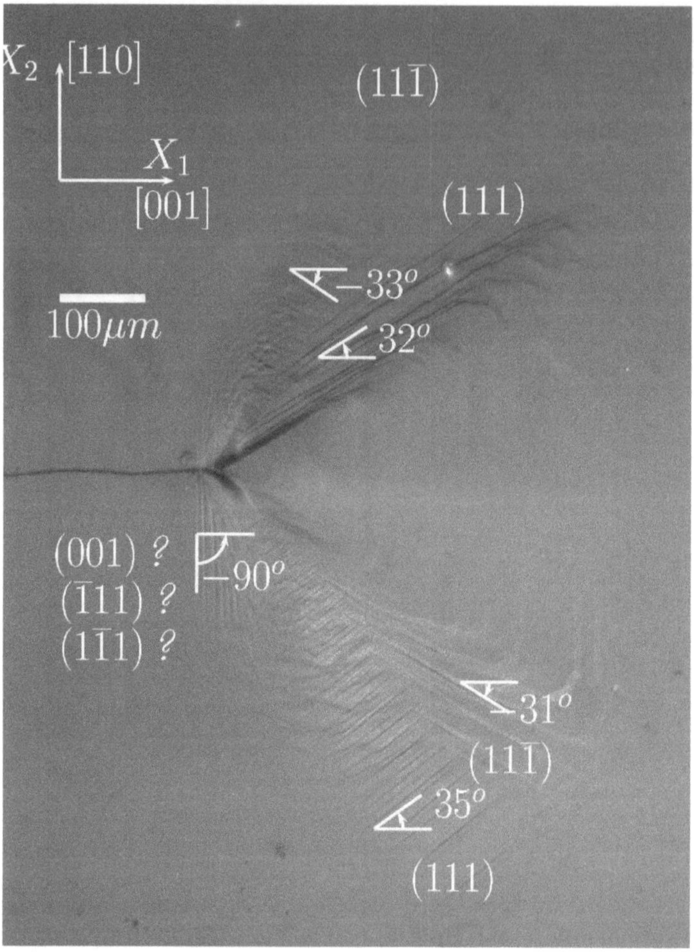

Figure 9. Slip lines at the crack tip of a nickel–base superalloy single crystal in a CT specimen. The crack plane is (110) and the crack propagation direction is [001]. Two kink bands are clearly visible on each side of the crack tip (after Flouriot et al. (2003)).

theory recalled in section 1.1. The presented simulations and the results obtained in Barbe et al. (2001a,b); Cailletaud et al. (2003); Parisot et al. (2004); Šiška et al. (2007) can integrate only relative grain size effects and size–independent strain incompatibilities. The application of the Cosserat model to polycrystals is postponed to section 3.3.

Multicrystalline specimens and coatings Metallic multicrystals have been investigated both experimentally and numerically in several situations in the past ten years: copper bicrystals Méric et al. (1994), copper, nickel, aluminum or iron muticrystals Teodosiu et al. (1993); Eberl et al. (1998); Ziegenbein et al. (1998); Delaire et al. (2000). The specimens may be strained in situ in a SEM. An EBSD analysis provides the lattice rotation field and the use of grids on the surface enables one to derive some components of the strain fields. In all mentioned contributions, the experimental results have been compared with success with realistic 3D computations using crystal plasticity.

The interest of computing samples containing a small number of large grains lies in the fact that the whole framework can be checked experimentally by comparing local and global predictions to strain or stress field measurements. In the case of coatings for instance, the computations are also of industrial interest since they can be used to optimise the microstructure, like grain size for instance.

The case of zinc coatings on steel sheets has been presented in Parisot et al. (2004) and the case of copper thin films in Šiška et al. (2007). The finite element computation of the tension or expansion of a coating on its substrate reveals the following features of the coating behaviour:

- the multiaxial stress state of each grain depending on its orientation;
- activated slip systems in the core of the grains, at the grain boundaries and at the interface coating/substrate;
- the gradient of strain that can develop from the interface to the free surface; this gradient can be shown to increase when the ratio between in–plane grain size and coating thickness decreases;
- forces acting at grain boundaries and at the interface, that can lead to intergranular fracture or interface decohesion;
- the coating roughness induced by local plasticity.

Computation of polycrystalline aggregates The type of geometrical aggregates shown in figure 10 can be used to investigate the intragranular fields, and to contribute to a better knowledge of the state of stress and strain in a current point of a polycrystal, and in more critical areas like the vicinity of the surface or at the grain boundaries. To be significant, the calculation must involve a reasonable number of grains *and* a reasonable number of elements in each grain. According to the literature related to texture effect, one thousand of grains seems to be a good number for representing a given material. The following examples are restricted to 200 grains, in order to have more than 1000 integration points in each grain ($10 \times 10 \times 10$, in 3D). Averaged values can then be considered on the mesh, in order to compare the FE model to the results given by more simple models like self–consistent approaches, on the level of each grain, and for the global mechanical response. The elements used are 20–node bricks with 27 integration points per element.

Boundary conditions and representativity of the volume element Three types of boundary conditions will be considered on a cube containing 200 grains Kanit et al. (2003) :

- Homogeneous strain based boundary conditions: the three components of the displacement at each node of the outer surface are prescribed according to the equations :

$$\underline{\boldsymbol{u}} = \underset{\sim}{\boldsymbol{E}}.\underline{\boldsymbol{x}} \tag{12}$$

  where $\underset{\sim}{\boldsymbol{E}}$ is a given constant strain tensor and $\underline{\boldsymbol{x}}$ the position of the point. We apply in this work a strain $\underset{\sim}{\boldsymbol{E}}$ corresponding to average uniaxial tension. The values of the component are taken from a simulation with a homogenized polycrystal model;
- Mixed boundary conditions for which only the displacement normal to the surface is prescribed according to the previous equation; lateral free surfaces may be considered for the case of simple tension for instance.
- Homogeneous stress based boundary conditions for which the traction vector is prescribed at each point of the boundary.

In this section, the local and global behaviour of the polycrystal is investigated. No special property is attributed to the grain boundaries. In particular, grain boundary sliding or damage are not considered.

The representativity of the considered polycrystalline volume element is a central issue. The number of grains must be large enough for the volume to be sufficiently representative. On the other hand, the number of elements inside each grain must be large enough for a sufficiently accurate description of the local intragranular strain field. That is why the presented computations belong to the largest computations of polycrystals available in literature and require parallel computing. With the number of processors used, the compromise is a volume containing 200 grains. The representativity of the volume element depends on the contrast of phase properties, on the type of boundary conditions and on the wanted accuracy of the estimation of the effective property Kanit et al. (2003). It can be assessed by applying strain based and stress based boundary conditions on the same volume Huet (1990). The difference between the apparent properties found for both conditions is due to the lack of representativity of the sample. For large volumes, the choice of the boundary conditions should not matter any more. The tensile curves obtained for both conditions on the considered sample of 200 grains differ by about 6%. The tensile curves can be found in Barbe et al. (2001b). The interest of periodic boundary conditions compared to homogeneous ones is illustrated in Kanit et al. (2003).

Intragranular fields Figure 10 shows the finite element mesh, and the strain and stress field in the polycrystal in overall simple tension. The material is IN600 Barbe et al. (2001b), the cube is submitted to a 1.5% overall tension, and the orientations of the grains are randomly distributed, in order to simulate an initially isotropic material. The slip activity is mainly due to the gradients of stress at the grain boundaries and not related to a kind of propagation of slip across boundaries. The development of deformation bands is observed independently of the grain morphology. In contrast the stress level is generally higher close to grain boundaries.

Free surface effect Contradictory results can be read in the literature concerning the effect of a free surface on local plastic behaviour Mughrabi (1992). After extensive TEM studies, some authors Pangborn et al. (1981) observe a hardened surface, due to the increase of dislocation density, but other works Fourie (1967); Mughrabi (1970) display an inverse effect, with lower dislocation densities and larger cell sizes. On the other hand, attempts have been made to approach the surface effect from a mechanical point of view, using a crystallographic inclusion in a homogeneous semi–infinite medium, considered as elastoviscoplastic Pilvin (1998) or in full polycrystals Barbe et al. (2001b). In these simulations the presence of a free surface seems to lead to lower stress levels, the perturbed area corresponding to about 3 grains. In the simulation of polycrystalline aggregates in tension with one free lateral surface, the first order surface effect is *scatter*. Relaxing the boundary condition can produce a drop of the local stress, but in some other location, a very low stress level obtained with homogeneous strain conditions can increase when freeing the surface. Averaged quantities can be considered, by plotting the average value of a critical variable in a slice at a given distance of the free surface. In that case, the observed effect is small (less than 10%). In general at a free surface of the cube, the number of active slip systems is significantly smaller (more than 15% less than in

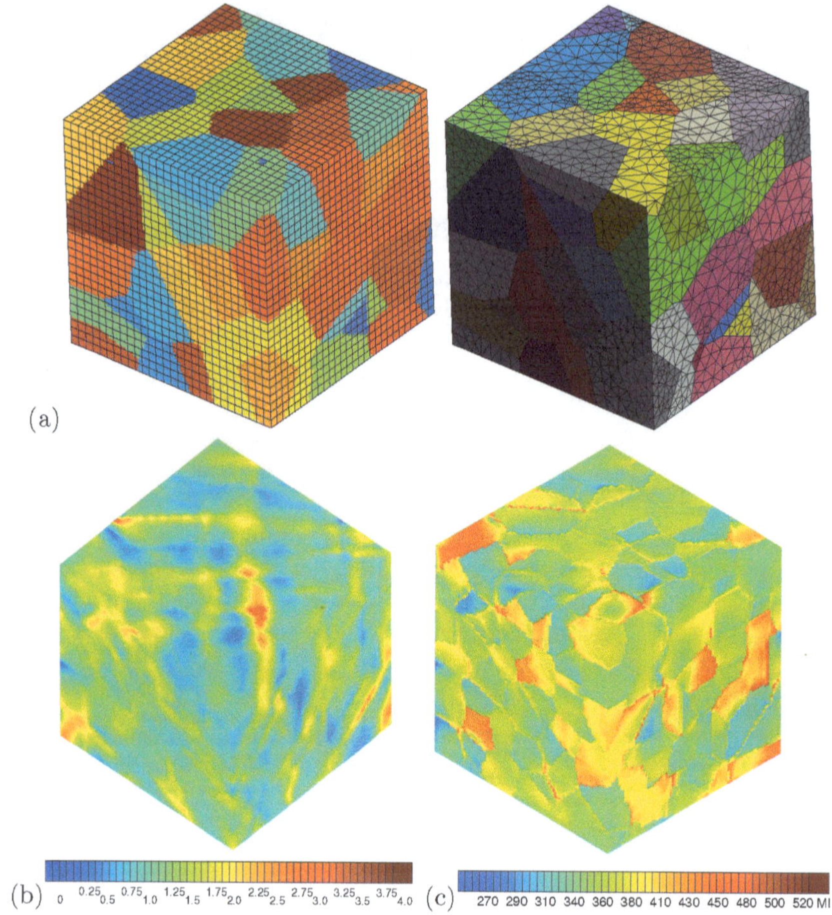

Figure 10. Two meshing strategies for polycrystalline volume elements (a). Contour (b) of the total strain in the tensile direction, (c) of von Mises equivalent stress, for a FCC polycrystal subjected to a mean tensile strain of 1.5%.

the core of the specimen), then the sum of the plastic slip is also smaller.

A systematic study of the 3D morphology of grains on the strain field at a free surface can be found in Zeghadi et al. (2007b,a).

Grain boundary effect Like the surface, the presence of a grain boundary will first induce a large scatter, in terms of stress and strain. However, the averaged stress over one or several grains as a function of the distance to the closest grain boundary is rather constant Barbe et al. (2003). The main results obtained by this statistical analysis are the following. The number of active slip systems and the sum of the plastic slips increase near the grain boundary. The von Mises strain is larger in the center of the grain.

The situation is quite different in the case of an aggregate of h.c.p. crystals like zinc. For this class of symmetry, elasticity and plastic slip are strongly anisotropic. In the zinc alloy studied in Parisot et al. (2004), the activated slip system families are mainly basal slip and pyramidal Π_2, the latter having an initial critical resolved shear stress ten times larger than the former. The consequence is that pyramidal slip takes place principally near the grain boundaries whereas basal slip spreads over the entire grain.

3.3 Grain size effects in polycrystals

Generalized homogenization method for polycrystals The polycrystal can also be regarded as a heterogeneous Cosserat material since it is an aggregate of Cosserat single crystal grains. As a result, some homogenization procedures must be designed to study the resulting properties of the polycrystal. The classical homogenization methods well–established for heterogeneous Cauchy media, as described in details in Sanchez-Palencia and Zaoui (1987); Suquet (1997); Jeulin and Ostoja-Starzewski (2001), must be extended to heterogeneous higher order continua. This task has been undertaken for heterogeneous Cosserat media in Forest et al. (1999, 2001b) and for micromorphic continua in Forest (2002). The theory is presented at small deformation for the sake of brevity.

Hill-Mandel approach The aim is to replace a heterogeneous material by a homogeneous substitute medium (HSM) which can be said to be equivalent in a sense to be made precise. Following Hill-Mandel's approach of the mechanics of heterogeneous materials, a condition of macro-homogeneity is required stipulating that the (isothermal for simplicity) free energy of the HSM at point $\underline{\boldsymbol{x}}$ can be identified to the mean value of the free energy over a representative volume element V under the same overall loading conditions at $\underline{\boldsymbol{x}}$. In the case of heterogeneous Cosserat materials for which the local fields are the relative strain, curvature and force and couple stress tensors $\underset{\sim}{e}, \underset{\sim}{\kappa}, \underset{\sim}{\sigma}$ and $\underset{\sim}{\mu}$ inside V, this condition can be generalized in the following forms. If the overall substitute medium (to be used for structural calculations for instance) is treated as a Cauchy continuum, the condition reads :

$$< \underset{\sim}{\sigma} : \underset{\sim}{e} + \underset{\sim}{\mu} : \underset{\sim}{\kappa} >= \underset{\sim}{\Sigma} : \underset{\sim}{E} \tag{13}$$

where $\underset{\sim}{E}$ and $\underset{\sim}{\Sigma}$ are the effective (symmetric) deformation and stress tensors. If the HSM is regarded as a Cosserat continuum itself, it becomes

$$< \underset{\sim}{\sigma} : \underset{\sim}{e} + \underset{\sim}{\mu} : \underset{\sim}{\kappa} >= \underset{\sim}{\Sigma} : \underset{\sim}{E} + \underset{\sim}{M} : \underset{\sim}{K} \tag{14}$$

where $\underset{\sim}{K}$ and $\underset{\sim}{M}$ are effective curvature and couple stress tensors, $\underset{\sim}{E}$ and $\underset{\sim}{\Sigma}$ being not necessarily symmetric any more. The procedure (14) is of course more general and contains (13) as a special case to which it reduces if the effective characteristic length is very small.

The determination of the effective properties then goes through the resolution of a boundary value problem on V. Boundary conditions on ∂V must then be chosen that automatically fulfill conditions (13) or (14). For instance the following homogeneous conditions at the boundary can be considered:

$$\underline{u} = \underset{\sim}{E}.\underline{x} \quad \textit{and} \quad \underline{\Phi} = \underset{\sim}{K}.\underline{x}, \quad \forall \underline{x} \in \partial V \tag{15}$$

where $\underset{\sim}{E}$ and $\underset{\sim}{K}$ are given and constant. It follows that

$$\underset{\sim}{E} =< \underline{u} \otimes \boldsymbol{\nabla} > \quad \textit{and} \quad \underset{\sim}{K} =< \underset{\sim}{\kappa} > \tag{16}$$

Condition (14) then is automatically satisfied for the following definition of the effective stress tensors

$$\underset{\sim}{\Sigma} =< \underset{\sim}{\sigma} > \quad \textit{and} \quad \underset{\sim}{M} =< \underset{\sim}{\mu} + \left(\underset{\approx}{\epsilon} : \underset{\sim}{\sigma}\right) \otimes \underline{x} >=< \mu_{ij} + \epsilon_{imn}\sigma_{mn}x_j > \underline{e}_i \otimes \underline{e}_j \tag{17}$$

Asymptotic methods In the case of periodic microstructures, a unit cell V can be defined and asymptotic methods are well-adapted for deriving the form of the effective balance and constitutive equations Sanchez-Palencia (1974). The key-point is the choice of the small parameter ε introduced in the multiscale asymptotic developments. Two different schemes have been proposed in Forest et al. (2001b) for periodic heterogeneous Cosserat media. Three characteristic lengths must be distinguished : the size l of the unit cell V, a typical characteristic length l_c of the constituents of the heterogeneous Cosserat material, and a typical wave length L_ω associated with the applied loading conditions. In classical homogenization theory, one usually speaks of slowly varying mean fields when $l \ll L_\omega$ and, in this case, the small parameter is $\varepsilon = l/L_\omega$. In the present situation, one may first consider a limiting process H^I with $\varepsilon \longrightarrow 0$, for which the Cosserat length scale l_c varies in the same way as l, so that the small parameter can also be written $\varepsilon = l_c/L_\omega$. In this case, the effective medium can be shown to be a Cauchy continuum. According to a second limiting process H^{II}, l_c is kept constant, which corresponds to $\varepsilon = l/l_c$. The main fields are now treated as functions of the two variables $\underline{x}$ and $\underline{y} = \underline{x}/\varepsilon$. The local fields can be expanded in power series of ε,

$$\underline{u}^{\varepsilon}(\underline{x}) = \underline{u}^{0}(\underline{x}, \underline{y}) + \varepsilon\, \underline{u}^{1}(\underline{x}, \underline{y}) + \varepsilon^2\, \underline{u}^{2}(\underline{x}, \underline{y}) + ... \tag{18}$$

$$\underline{\Phi}^{\varepsilon}(\underline{x}) = \underline{\Phi}^{0}(\underline{x}, \underline{y}) + \varepsilon\, \underline{\Phi}^{1}(\underline{x}, \underline{y}) + \varepsilon^2\, \underline{\Phi}^{2}(\underline{x}, \underline{y}) + ... \tag{19}$$

where the $\underline{u}^{i}$ and $\underline{\Phi}^{i}$ are assumed to have the same order of magnitude and are periodic in $\underline{y}$. Similar expansions exist for the force and couple stresses:

$$\underset{\sim}{\sigma}^{\varepsilon}(\underline{x}) = \underset{\sim}{\sigma}^{0}(\underline{x}, y) + \varepsilon\, \sigma^{1}(\underline{x}, \underline{y}) + \varepsilon^2\, \underset{\sim}{\sigma}^{2}(\underline{x}, \underline{y}) + ... \tag{20}$$

$$\underset{\sim}{\mu}^{\varepsilon}(\underline{x}) = \underset{\sim}{\mu}^{0}(\underline{x}, y) + \varepsilon\, \mu^{1}(\underline{x}, \underline{y}) + \varepsilon^2\, \underset{\sim}{\mu}^{2}(\underline{x}, \underline{y}) + ... \tag{21}$$

The form of the constitutive equations is different for each homogenization procedure and in the case of linear elasticity, they read:

$$H^I: \quad \underset{\sim}{\sigma}^{\varepsilon} = \underset{\approx}{D}(\underline{y}) : \underset{\sim}{e}^{\varepsilon}(\underline{x}) \quad and \quad \underset{\sim}{\mu}^{\varepsilon} = \varepsilon^2 \underset{\approx}{C}(\underline{y}) : \underset{\sim}{\kappa}^{\varepsilon}(\underline{x}) \tag{22}$$

$$H^{II}: \quad \underset{\sim}{\sigma}^{\varepsilon} = \underset{\approx}{D}(\underline{y}) : \underset{\sim}{e}^{\varepsilon}(\underline{x}) \quad and \quad \underset{\sim}{\mu}^{\varepsilon} = \underset{\approx}{C}(\underline{y}) : \underset{\sim}{\kappa}^{\varepsilon}(\underline{x}) \tag{23}$$

After noting that $\nabla = \nabla_x + 1/\varepsilon \nabla_y$ (with obvious notations), we compute the gradient of the kinematic variables and the divergence of the stresses in order to introduce them in the balance and constitutive equations. Ordering the terms according to ε leads to a set of equations from which a series of auxiliary boundary value problems to be solved on the unit cell can be defined (see Boutin (1996) for the classical case). The first auxiliary problem for the procedure H^{II} consists in determining vector fields $\underline{v}$ and $\underline{\psi}$ such that :

$$\underline{u} = \underset{\sim}{E}.\underline{y} + \underline{v} \quad with \quad \underline{\Phi} = \underset{\sim}{K}.\underline{y} + \underline{\psi}, \quad \forall \underline{y} \in V \tag{24}$$

$$\underset{\sim}{\sigma} = \underset{\approx}{D} : (\underline{u} \otimes \nabla_y) \quad and \quad \underset{\sim}{\mu} = \underset{\approx}{C} : (\underline{\Phi} \otimes \nabla_y) \tag{25}$$

$$\underset{\sim}{\sigma}.\nabla_y = 0 \quad and \quad \underset{\sim}{\mu}.\nabla_y = 0 \tag{26}$$

where $\underline{v}$ and $\underline{\psi}$ take the same values on opposite sides of the cell and the traction and surface couple vectors $\underset{\sim}{\sigma}.\underline{n}$ and $\underset{\sim}{\mu}.\underline{n}$ are anti-periodic. The solution of this problem gives in fact the terms $\underline{u}_1, \underline{\Phi}_1, \underset{\sim}{\sigma}^0$ and $\underset{\sim}{\mu}^0$. This leads to the following expression of the mean work of internal forces :

$$< \underset{\sim}{\sigma} : \underset{\sim}{e} + \underset{\sim}{\mu} : \underset{\sim}{\kappa} > = < \underset{\sim}{\sigma} > : < \underline{u} \otimes \nabla > + < \underset{\sim}{\mu} > : < \underset{\sim}{\kappa} > \tag{27}$$

which defines the effective deformation, curvature, force and couple stress tensors. The effective medium then is a Cosserat continuum.

Retained approach for non linear multiphase materials The polycrystal is a heterogeneous material with a disordered distribution of phases, each phase being a crystal orientation, and Hill-Mandel approach has proved to be efficient for deriving effective properties in such cases Sanchez-Palencia and Zaoui (1987). This requires however computations on a large representative volume element V containing many grains. Such aggregates have already been computed in classical crystal plasticity in section 3.2. But for Cosserat materials, the number of degrees of freedom and internal variables increases dramatically in the three-dimensional case so that we will work here on smaller samples of grains. In the latter case, periodic boundary conditions will induce less pronounced boundary effects than the Dirichlet conditions (15) (see Kanit et al. (2003)). That is why a mixed approach between Hill-Mandel and periodic ones is retained here, combining the periodic scheme H^{II} and unaltered local constitutive equations and thus different from (25). It has been shown numerically in Forest et al. (2001b), at least in the case of linear elasticity, that the approach H^{II} works well even if $l_c \sim l$, which will be the case in section 4 to 6.

Accordingly, the following initial boundary value problem P is considered on a single unit cell V :

$$\begin{gathered} \underline{u} = \underset{\sim}{E}.\underline{x} + \underline{v}, \quad \underline{\Phi} = \underset{\sim}{K}.\underline{x} + \underline{\psi}, \\ \textit{constitutive equations}, \\ \underset{\sim}{\sigma}.\nabla = 0, \quad \underset{\sim}{\mu}.\nabla - \underset{\approx}{\epsilon} : \underset{\sim}{\sigma} = 0 \end{gathered} \tag{28}$$

where $\underline{v}$ (resp. $\underline{\psi}$) takes the same value on opposite sides of the cell. The traction and surface couple vectors of two homologous points on opposite sides of V are opposite. This problem is solved on a single cell V and no attempt is made to extrapolate the solution in a regular solution on the entire body, although it may be possible in particular for symmetric $\underset{\sim}{E}$ or $\underset{\sim}{\Sigma}$ and vanishing $\underset{\sim}{K}$.

Scale dependent behaviour of polycrystals Continuum crystal plasticity models can be enhanced to account for size effects in single or multiphase metal polycrystals. Strain gradient, gradient of internal variable and Cosserat models are available to introduce additional hardening primarily associated with densities of so–called geometrically necessary dislocations Kubin and Mortensen (2003). Direct simulations of polycrystalline aggregates including a detailed description of non–homogeneous deformation inside individual grains can then be used to predict grain size effects classically observed in metal polycrystals Acharya and Beaudoin (2000); Forest et al. (2000). The advantage of such models is that the grain size D is not explicitly introduced in the constitutive equations, so that the size effect scaling is the result of complex in–grain stress–strain fields and interactions between grains. As a result, once identified, the same model can be used for other size effects associated with the interaction with other phases (lamellar structures, two–phase materials).

The aim of the next sections is the identification of the material parameters of a Cosserat crystal plasticity model to account for grain size effects in an IF ferritic steel. Tensile curves for grains sizes ranging from 120 μm to 5 μm can be found in Bouaziz et al. (2001) and show strong effects for smaller grains. This hardening is not only due to grain size effects but also to solid solution hardening since it is impossible to produce IF steels with such small grains with exactly the same composition as for larger grains. The solid solution hardening part can be isolated and estimated by suited experiments so that the present work can concentrate on the remaining part of hardening due solely to grain size changes. Scaling laws in $D^{-0.7}$ were found experimentally. Precise modelling of IF steels based on crystal plasticity at a fixed grain size can be found in the referenced Paquin (2001); Hoc et al. (2001).

The Cosserat crystal plasticity model is specialized in section 3.3 to the case of small strain and rotation. The main capabilities of the model are illustrated by two–dimensional finite element simulations showing the development of lattice rotation and curvature inside grains for different grain sizes. The actual identification of the model for IF steels requires the 3D simulations of section 3.3. Some of the presented results were published in Zeghadi et al. (2005). Simulations based on strain gradient plasticity models also exist for polycrystals in miniaturized samples Bayley et al. (2007); Bargmann et al. (2010). The obtained lattice orientation fields can be compared to experimental results like EBSD field measurements St-Pierre et al. (2008).

Cosserat crystal plasticity model at small strain and rotation Cosserat continuum mechanics introduces, at each material point, not only a displacement vector $\underline{u}$ but also an independent rotation field, represented by the axial vector ϕ_i. The Cosserat deformation and curvature tensors are respectively:

$$e_{ij} = u_{i,j} + \epsilon_{ijk}\Phi_k \qquad\qquad \kappa_{ij} = \Phi_{i,j} \tag{29}$$

In the static case,the force and couple stress tensors associated with the previous deformation measures, must fulfill two balance equations Nowacki (1986):

$$\sigma_{ij,j} + f_i = 0, \qquad\qquad \mu_{ij,j} - \epsilon_{ijk}\sigma_{jk} = 0 \tag{30}$$

The isotropic elasticity law involves the 2 Lamé constants and 4 additional moduli:

$$\sigma_{ij} = \lambda\delta_{ij}e^e_{kk}+2\mu e^e_{(ij)}+2\mu_c e^e_{\{ij\}}, \qquad \mu_{ij} = \alpha\delta_{ij}\kappa^e_{kk}+2\beta\kappa^e_{(ij)}+2\gamma\kappa^e_{\{ij\}} \tag{31}$$

where (ij) and $\{ij\}$ respectively denote the symmetric and skew-symmetric parts of the tensor. The Cosserat directors can be identified with lattice directions when the elastic strain tensor e^e_{ij} is almost symmetric. That is why the constant parameter μ_c is used as a penalty factor in the crystal plasticity Cosserat model. An intrinsic length $l_e = \sqrt{\beta/\mu}$ can be defined. Plastic deformation is due to the activation of plastic slip γ^s on slip system s:

$$\dot{e}_{ij} = \dot{e}^e_{ij} + \dot{e}^p_{ij}, \quad \dot{e}^p = \sum_{s=1}^{N} \dot{\gamma}^s P^s_{ij}, \quad P^s_{ij} = m^s_i n^s_j \tag{32}$$

where m^s_i and n^s_i respectively are the slip direction and normal to the slip plane. For IF steels, the 24 slip systems $\{110\} < 111 >$ and $\{112\} < 111 >$ are retained Paquin (2001); Hoc et al. (2001). Similarly, lattice curvature is accommodated by elastic and plastic parts:

$$\dot{\kappa}_{ij} = \dot{\kappa}^e_{ij} + \dot{\kappa}^p_{ij}, \quad \dot{\kappa}^p_{ij} = \sum_{s=1}^{n} \left(\frac{\dot{\theta}^s_\perp}{l_p} Q^s_{ij\perp}\right), \quad Q^s_{ij\perp} = \epsilon_{ikl} n^s_k m^s_l m^s_k \tag{33}$$

where ϵ_{ijk} is the permutation symbol. Plastic lattice bending only is accounted for in the present work, based on Nye's formula. It is represented for each slip system by the angle θ^s divided by a second characteristic length l_p. Viscoplastic flow rules are adopted for both plastic slip and curvature:

$$\dot{\gamma}^s = \langle\frac{\mid \tau^s \mid - r^s}{k}\rangle^n \; sign\,(\tau^s), \quad \dot{\theta}^s = \langle\frac{\mid \nu^s \mid \, - \, l_p \; r^s_{c0}}{l_p \; k^s_c}\rangle^{n_c} \; sign\;(\nu^s) \tag{34}$$

where $\tau^s = \sigma_{ij}P^s_{ij}$ and $\nu^s = \mu_{ij}Q^s_{ij\perp}$ respectively are the resolved shear stress and the resolved couple stress. The brackets $\langle\rangle$ denote the positive

E (MPa)	ν	k $(MPa.s^{1/n})$	n	r_0 (MPa)	q (MPa)	b_1
208000.	0.3	3.	10.	29.	39.	16.

β $(MPa.mm^2)$	r_{c0} (MPa)	H' (MPa)	l_p (mm)	k_c $(MPa.s^{1/n_c})$	n_c	b_2
0.001	0.000001	10.	1.	0.01	1.	5.

Table 1. (a) Classical parameters used in the computations (b) Specific Cosserat parameters used in the computations

part. A simple nonlinear evolution law is adopted for the critical resolved shear stress r^s:

$$r^s = r_0 + Q \sum_{r=1}^{n} h^{sr}(1 - \exp(-bv^r)) + r^{\perp}, \qquad v^s = \mid \dot{\gamma}^s \mid \tag{35}$$

with an interaction matrix h^{rs} which represents a simplified version of the model used in Hoc et al. (2001).

Influence of Cosserat elasticity and curvature hardening The Cosserat bending modulus β was introduced by Kröner in dislocated crystals Kröner (1963) and related to the density of geometrically necessary dislocations and therefore to θ^s/l_p. In the simple case of single slip in periodic layered microstructures analysed in Forest and Sedláček (2003), it has been related to line tension effects associated with the collective bowing of dislocations in narrow channels. As a result, this parameter (and the corresponding intrinsic length l_e) influences the spreading of lattice curvature inside grains. To show that, finite element simulations of the shearing of a 2D aggregate containing nine ferritic grains were performed for two different values of the bending modulus, under plane strain conditions. Non–homogeneous deformation and lattice rotation are due to plastic strain incompatibilities between grains having different (random) orientations. Lattice curvature primarily develops close to grain boundaries. Higher values of β promote a larger spreading of lattice curvature for the same mean prescribed strain value. This feature will play an important role in the following simulations of grain size effects.

In equation (35), an additional hardening term $r_{\perp}$ associated with lattice curvature is introduced. The following evolution rule is proposed

$$r_{\perp} = Q_{\perp}(\sum_{s=1}^{n}(1 - \exp(-b_{\perp}\theta^{s})))^{\frac{1}{2}} \tag{36}$$

Similar additional hardening terms are summarized and commented in Kubin and Mortensen (2003) and involve the square root of a combination of the density of statistically stored dislocations and density of geometrically dislocations. Such rules are relevant for f.c.c. crystals but they are questionable in b.c.c. crystals. The function (36) of plastic lattice curvature angles is of purely phenomenological nature. It has two advantageous features: a vertical tangent at zero plastic curvature and saturation for higher curvature levels. This allows strong effects at incipient plasticity, as recommended by Ashby Ashby (1970). The effect of extra–hardening is illustrated in the 2D case on figure 12 for two different grain sizes. The geometry of the grains is the same for both computations but the absolute sizes of the aggregates differ, which correspond to grain sizes of 120 and 5 μm respectively. At the very beginning of intragranular plastic flow (0.075% overall shear strain), plastic strain and lattice rotation maps are quite similar for both grain sizes. As a consequence, plastic curvature fields strongly differ (figures 12(a)(b) and (c)). Lattice curvature being much larger in the small grains, the curvature hardening rule (36) induces higher stress levels for small grain sizes, as shown on figures 12(d). This will strongly affect subsequent plastic flow in the grains. This mechanism is the main ingredient in the following simulations of grain size effects in ferritic polycrystals.

3D finite element analyses of ferritic steels 2D simulations were used in the previous section to illustrate the different capabilities of the Cosserat crystal plasticity model. 3D simulations are necessary to actually reproduce the experimental results on grain size effects in ferritic steels. A mean tensile strain up to 10% is prescribed to a small aggregate containing 10 grains. Geometrical and kinematic periodicity conditions are chosen because a periodic homogenization scheme is available for Cosserat continua, and also because smaller representative volumes can be considered Forest et al. (2000). The considered volume element is shown in figure 13.

The identification of the material parameters of the Cosserat models for the considered material proceeds as follows. The initial critical resolved shear stress r_0, the hardening parameters Q, b (equation (35)), are deduced from the tensile curve of IF steel with D =120 μm (large grain polycrystal behaviour). The interaction matrix h^{rs} is taken from the references Paquin (2001); Hoc et al. (2001). The Cosserat parameters $\beta, Q_{\perp}, b_{\perp}$ (equation

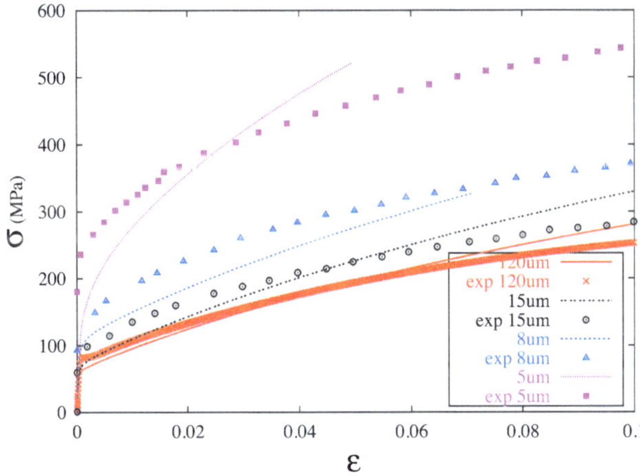

Figure 11. Mean axial stress–mean axial strain curves predicted by the finite element analysis compared with experimental results

(36)) are identified from the dependence of the tensile curve on grain size. For all simulations at different grains sizes, the material parameters, the morphology of the grains and the boundary conditions are the same. The finite element meshes take the actual size of the grains into account. The inverse approach requires large amounts of time since each simulation on one aggregate is time consuming. The best parameters found are given in table 3.3. Reference computations have also been performed with classical polycrystalline model. The Cosserat parameters are such that its response is close to the classical one for grain sizes larger than D =120 μm. The results after identification are shown in figure 11 which gives the average values of axial stress and strain. As expected, the finite element response is significantly harder for the smallest grain size. In both cases, the same grains are activated and the maximal stress is localised near the grains boundaries. The predicted grain size scaling at 0.2% plastic strain can be checked. Grain size effects saturate for grain larger than 120 μm (classical case) and for grains smaller than 1μm with the present parameters.

The effect of grain size on the local fields in 3D is illustrated in figure 13. As expected, plastic strain is higher in larger grains for the same mean total strain. Lattice curvature is higher in small grains and is observed mainly close to grain boundaries where most strain incompatibilities arise.

An enhanced continuum crystal plasticity model was used to simulate grain size effects in a ferritic steel using the finite element method and pe-

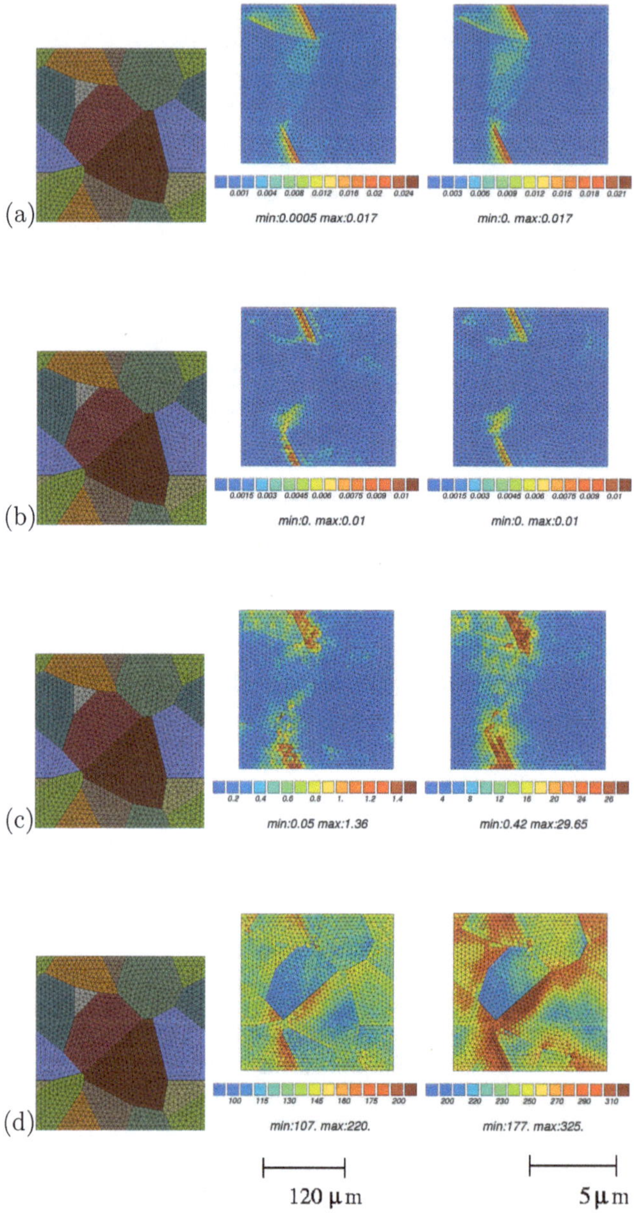

Figure 12. Two–dimensional simulation of microplasticity in a polycrystalline volume element with two grain sizes and periodicity conditions: (a) equivalent plastic strain, (b) norm of the lattice rotation vector (in rad.), (c) norm of lattice curvature, (d) von Mises stress. The mean total strain is 0.00075 in both computations.

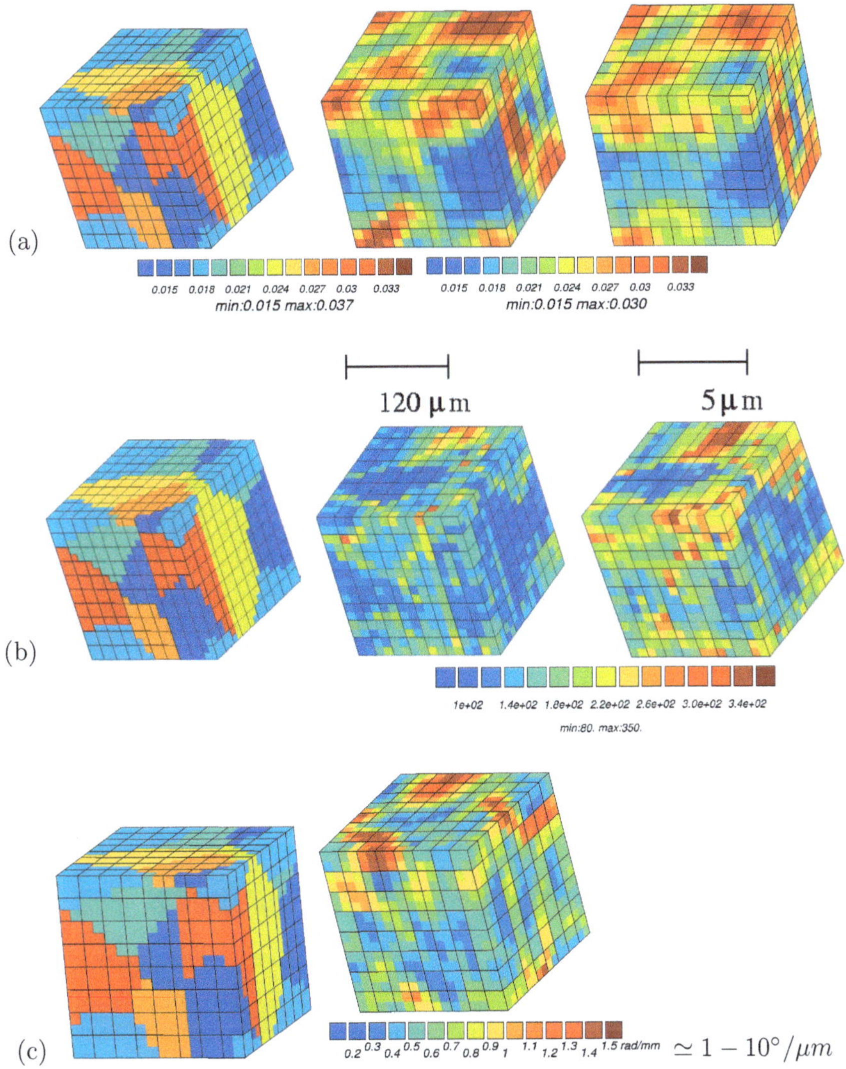

Figure 13. Computation of a periodic volume element of a ferritic steel for two grain sizes at the same mean strain level 0.01: (a) field of equivalent plastic strain, (b) field of equivalent stress, (c) field of equivalent plastic curvature for the smaller grains.

riodic polycrystalline aggregates. In the simulation, the size effect is due to an increase of plastic curvature for small grains associated with additional curvature hardening. A second significant component is the increase of the volume fraction of curved lattice within the grains for decreasing grain size. This also contributes to higher macroscopic stress levels. However, the present simulations still significantly underestimate the grain size effect observed experimentally. This may be due to too coarse meshes, to insufficient lattice curvature spreading at small grains, or to deeper metallurgical physical reasons that remain to be sorted out.

Alternative crystal plasticity models accounting for the overall grain size effects are based on mean field homogenization approaches Pipard et al. (2009).

Bibliography

K. Abu Al-Rub, G.Z. Voyiadjis, and D.J. Bammann. A thermodynamic based higher-order gradient theory for size dependent plasticity. *Int. J. Solids Structures*, 44:2888–2923, 2007.

A. Acharya. Jump condition for GND evolution as a constraint on slip transmission at grain boundaries. *Philosophical Magazine*, 87:1349–1359, 2007.

A. Acharya. New inroads in an old subject: Plasticity, from around the atomic to the macroscopic scale. *Journal of the Mechanics and Physics of Solids*, 58:766–778, 2010.

A. Acharya and J.L. Bassani. Lattice incompatibility and a gradient theory of crystal plasticity. *Journal of the Mechanics and Physics of Solids*, 48: 1565–1595, 2000.

A. Acharya and A.J. Beaudoin. Grain size effects in viscoplastic polycrystals at moderate strains. *Journal of the Mechanics and Physics of Solids*, 48: 2213–2230, 2000.

E.C. Aifantis. The physics of plastic deformation. *International Journal of Plasticity*, 3:211–248, 1987.

E.C. Aifantis. Gradient deformation models at nano, micro and macro scales. *J. of Engineering Materials and Technology*, 121:189–202, 1999.

R.J. Asaro. Crystal plasticity. *J. Appl. Mech.*, 50:921–934, 1983.

R.J. Asaro and J.R. Rice. Strain localization in ductile single crystals. *J. Mech. Phys. Solids*, 25:309–338, 1977.

M.F. Ashby. The deformation of plastically non–homogeneous materials. *Philosophical Magazine*, 21:399–424, 1970.

M.F. Ashby. The deformation of plastically non-homogeneous alloys. In A. Kelly and R.B. Nicholson, editors, *Strengthening Methods in Crystals*, pages 137–192. Applied Science Publishers, London, 1971.

F. Barbe, L. Decker, D. Jeulin, and G. Cailletaud. Intergranular and intragranular behavior of polycrystalline aggregates. Part 1: FE model. *International Journal of Plasticity*, 17:513–536, 2001a.

F. Barbe, S. Forest, and G. Cailletaud. Intergranular and intragranular behavior of polycrystalline aggregates. Part 2: Results. *International Journal of Plasticity*, 17:537–563, 2001b.

F. Barbe, S. Forest, S. Quilici, and G. Cailletaud. Numerical study of crystalline plasticity : measurements of the heterogeneities due to grain boundaries under small strains. *La Revue de Métallurgie*, 101:815–823, 2003.

L. Bardella. Some remarks on the strain gradient crystal plasticity modelling, with particular reference to the material length scales involved. *International Journal of Plasticity*, 23:296–322, 2007.

S. Bargmann, M. Ekh, K. Runesson, and B. Svendsen. Modeling of polycrystals with gradient crystal plasticity: A comparison of strategies. *Philosophical Magazine*, 90(10):1263–1288, 2010.

J.L. Bassani, A. Needleman, and E. Van der Giessen. Plastic flow in a composite : a comparison of nonlocal continuum and discrete dislocation predictions. *Int. J. Solids Structures*, 38:833–853, 2001.

C.J. Bayley, W.A.M. Brekelmans, and M.G.D. Geers. A comparison of dislocation induced back stress formulations in strain gradient crystal plasticity. *International Journal of Solids and Structures*, 43:7268–7286, 2006.

C.J. Bayley, W.A.M. Brekelmans, and M.G.D. Geers. A three–dimensional dislocation field crystal plasticity approach applied to miniaturized structures. *Philosophical Magazine*, 87:1361–1378, 2007.

A. Bertram. *Elasticity and Plasticity of Large Deformations*. Springer, 2005.

D. Besdo. Ein Beitrag zur nichtlinearen Theorie des Cosserat-Kontinuums. *Acta Mechanica*, 20:105–131, 1974.

J. Besson, G. Cailletaud, J.-L. Chaboche, and S. Forest. *Mécanique non linéaire des matériaux*. ISBN 2-7462-0268-9, EAN13 9782746202689, 445 p., Hermès, France, 2001.

B.A. Bilby, R. Bullough, L.R.T. Gardner, and E. Smith. Continuous distributions of dislocations iv: Single glide and plane strain. *Proc. Roy. Soc. London*, A236:538–557, 1957.

R. de Borst. Simulation of strain localization: a reappraisal of the Cosserat continuum. *Engng Computations*, 8:317–332, 1991.

R. de Borst, L.J. Sluys, H.B. Mühlhaus, and J. Pamin. Fundamental issues in finite element analyses of localization of deformation. *Engineering Computations*, 10:99–121, 1993.

O. Bouaziz, T. Iung, M. Kandel, and C. Lecomte. Physical modelling of microstructure and mechanical properties of dual–phase steel. *Journal de Physique IV*, 11:Pr4–223–229, 2001.

C. Boutin. Microstructural effects in elastic composites. *Int. J. Solids Structures*, 33:1023–1051, 1996.

E.P. Busso, F.T. Meissonnier, and N.P. O'Dowd. Gradient–dependent deformation of two–phase single crystals. *Journal of the Mechanics and Physics of Solids*, 48:2333–2361, 2000.

G. Cailletaud. A micromechanical approach to inelastic behaviour of metals. *Int. J. Plasticity*, 8:55–73, 1992.

G. Cailletaud, S. Forest, D. Jeulin, F. Feyel, I. Galliet, V. Mounoury, and S. Quilici. Some elements of microstructural mechanics. *Computational Materials Science*, 27:351–374, 2003.

J.D. Clayton, D.L. McDowell, and D.J. Bammann. Modeling dislocations and disclinations with finite micropolar elastoplasticity. *International Journal of Plasticity*, 22:210–256, 2006.

H. Cleveringa, E. Van der Giessen, and A. Needleman. Discrete dislocation simulations and size dependent hardening in single slip. *Journal de Physique IV*, 8:Pr4–83–92, 1998.

B.D. Coleman and W. Noll. The thermodynamics of elastic materials with heat conduction and viscosity. *Arch. Rational Mech. and Anal.*, 13:167–178, 1963.

N.M. Cordero, A. Gaubert, S. Forest, E. Busso, F. Gallerneau, and S. Kruch. Size effects in generalised continuum crystal plasticity for two–phase laminates. *Journal of the Mechanics and Physics of Solids*, 58:1963–1994, 2010.

E. Cosserat and F. Cosserat. *Théorie des corps déformables.* Hermann, Paris, 1909.

F. Delaire, J.L. Raphanel, and C. Rey. Plastic heterogeneities of a copper multicrystal deformed in uniaxial tension: experimental study and finite element simulations. *Acta mater.*, 48:1075–1087, 2000.

V.S. Deshpande, A. Needleman, and E. Van Der Giessen. A discrete dislocation analysis of near threshold fatigue crack growth. *Acta materialia*, 49:3189–3203, 2001.

T. Dillard, S. Forest, and P. Ienny. Micromorphic continuum modelling of the deformation and fracture behaviour of nickel foams. *European Journal of Mechanics A/Solids*, 25:526–549, 2006.

P.H. Dluzewski. Finite deformation of polar media in angular coordinates. *Arch. Mech.*, 43:783–793, 1991.

P.H. Dluzewski. Finite elastic-plastic deformations of oriented media. In A. Benallal and D. Marquis, editors, *MECAMAT'92, Int. Seminar on Multiaxial Plasticity*. Cachan, 1992.

W.J. Drugan. Asymptotic solutions for tensile crack tip fields without kink–type shear bands in elastic ideally–plastic single crystals. *Journal of the Mechanics and Physics of Solids*, 49:2155–2176, 2001.

M.K. Duszek-Perzyna and P. Perzyna. Adiabatic shear band localization in elastic plastic single crystals. *Int. J. Solids Structures*, 30(1):61–89, 1993.

F. Eberl, F. Feyel, S. Quilici, and G. Cailletaud. Approches numériques de la plasticité cristalline. *J. Phys. IV France*, 8:Pr4–15–25, 1998.

M. Eisenberg. On the relation between continuum plasticity and dislocation theories. *Int. J. Engng. Sci.*, 8:261–271, 1970.

R.A.B. Engelen, M.G.D. Geers, and F.P.T. Baaijens. Nonlocal implicit gradient-enhanced elasto-plasticity for the modelling of softening behaviour. *International Journal of Plasticity*, 19:403–433, 2003.

R.A.B. Engelen, N.A. Fleck, R.H.J. Peerlings, and M.G.D. Geers. An evaluation of higher–order plasticity theories for predicting size effects and localisation. *International Journal of Solids and Structures*, 43:1857–1877, 2006.

A.C. Eringen. *Polar and non local fields theories, in Continuum Physics*, volume IV. Academic Press, 1976.

A.C. Eringen. *Microcontinuum field theories*. Springer, New York, 1999.

A.C. Eringen and W.D. Claus. A micromorphic approach to dislocation theory and its relation to several existing theories. In J.A. Simmons, R. de Wit, and R. Bullough, editors, *Fundamental Aspects of Dislocation Theory*, pages 1023–1062. Nat. Bur. Stand. (US) Spec. Publ. 317, II, 1970.

A.C. Eringen and E.S. Suhubi. Nonlinear theory of simple microelastic solids. *Int. J. Engng Sci.*, 2:189–203, 389–404, 1964.

Y. Estrin, B. Sluys, Y. Bréchet, and A. Molinari. A dislocation based gradient plasticity model. *Journal de Physique IV*, 8:Pr8–135–141, 1998.

M. Fivel and S. Forest. *Plasticité cristalline et transition d'échelle : cas du monocristal.* Techniques de l'Ingénieur, M4016, 23 pages, 2004.

M.C. Fivel, C.F. Robertson, G.R. Canova, and L. Boulanger. Three-dimensional modeling of indent-induced plastic zone at a mesoscale. *Acta materialia*, 46:6183–6194, 1998.

N.A. Fleck and J.W. Hutchinson. A reformulation of strain gradient plasticity. *Journal of the Mechanics and Physics of Solids*, 49:2245–2271, 2001.

N.A. Fleck and J.W. Hutchinson. Strain gradient plasticity. *Adv. Appl. Mech.*, 33:295–361, 1997.

S. Flouriot, S. Forest, G. Cailletaud, A. Köster, L. Rémy, B. Burgardt, V. Gros, S. Mosset, and J. Delautre. Strain localization at the crack tip in single crystal CT specimens under monotonous loading : 3D finite element analyses and application to nickel–base superalloys. *International Journal of Fracture*, 124:43–77, 2003.

S. Forest. Modeling slip, kink and shear banding in classical and generalized single crystal plasticity. *Acta Materialia*, 46(9):3265–3281, 1998.

S. Forest. The micromorphic approach for gradient elasticity, viscoplasticity and damage. *ASCE Journal of Engineering Mechanics*, 135:117–131, 2009.

S. Forest. Some links between cosserat, strain gradient crystal plasticity and the statistical theory of dislocations. *Philosophical Magazine*, 88: 3549–3563, 2008.

S. Forest. Homogenization methods and the mechanics of generalized continua–Part 2. *Theoretical and Applied Mechanics*, 28–29:113–143, 2002.

S. Forest and E. C. Aifantis. Some links between recent gradient thermo-elasto-plasticity theories and the thermomechanics of generalized continua. *International Journal of Solids and Structures*, 47:3367–3376, 2010.

S. Forest and M. Amestoy. Hypertemperature in thermoelastic solids. *Comptes Rendus Mécanique*, 336:347–353, 2008.

S. Forest and G. Cailletaud. Strain localization in single crystals: Effect of boundaries and interfaces. *European Journal of Mechanics A/Solids*, 14 (5):747–771, 1995.

S. Forest and R. Sedláček. Plastic slip distribution in two–phase laminate microstructures: Dislocation–based vs. generalized–continuum approaches. *Philosophical Magazine A*, 83:245–276, 2003.

S. Forest and R. Sievert. Elastoviscoplastic constitutive frameworks for generalized continua. *Acta Mechanica*, 160:71–111, 2003.

S. Forest and R. Sievert. Nonlinear microstrain theories. *International Journal of Solids and Structures*, 43:7224–7245, 2006.

S. Forest, G. Cailletaud, and R. Sievert. A Cosserat theory for elastoviscoplastic single crystals at finite deformation. *Archives of Mechanics*, 49 (4):705–736, 1997.

S. Forest, R. Dendievel, and G.R. Canova. Estimating the overall properties of heterogeneous Cosserat materials. *Modelling Simul. Mater. Sci. Eng.*, 7:829–840, 1999.

S. Forest, F. Barbe, and G. Cailletaud. Cosserat modelling of size effects in the mechanical behaviour of polycrystals and multiphase materials. *International Journal of Solids and Structures*, 37:7105–7126, 2000.

S. Forest, P. Boubidi, and R. Sievert. Strain localization patterns at a crack tip in generalized single crystal plasticity. *Scripta Materialia*, 44: 953–958, 2001a.

S. Forest, F. Pradel, and K. Sab. Asymptotic analysis of heterogeneous Cosserat media. *International Journal of Solids and Structures*, 38:4585–4608, 2001b.

S. Forest, E. van der Giessen, and L. Kubin, Editors. *Scale transitions from atomistics to continuum plasticity*. J. Phys. IV, France, Vol. 11, Pr5, Les Editions de Physique, 2001c.

S. Forest, R. Sievert, and E.C. Aifantis. Strain gradient crystal plasticity : Thermomechanical formulations and applications. *Journal of the Mechanical Behavior of Materials*, 13:219–232, 2002.

J.T. Fourie. The flow stress gradient between the surface and centre of deformed copper single crystals. *Philosophical Magasine*, 15:735–756, 1967.

J. Friedel. *Dislocations*. Pergamon, 1964.

N. Germain, J. Besson, and F. Feyel. Simulation of laminate composites degradation using mesoscopic non–local damage model and non–local layered shell element. *Modelling and Simulation in Materials Science and Engineering*, 15:S425–S434, 2007.

P. Germain. La méthode des puissances virtuelles en mécanique des milieux continus, première partie : théorie du second gradient. *J. de Mécanique*, 12:235–274, 1973a.

P. Germain. The method of virtual power in continuum mechanics. part 2 : Microstructure. *SIAM J. Appl. Math.*, 25:556–575, 1973b.

P. Germain, Q.S. Nguyen, and P. Suquet. Continuum thermodynamics. *J. of Applied Mechanics*, 50:1010–1020, 1983.

P. Gumbsch and R. Pippan. *Multiscale Modelling of Plasticity and Fracture by means of Dislocation Mechanics*. CISM Lecture Notes, Udine, 2005.

M.E. Gurtin. On the plasticity of single crystals: free energy, microforces, plastic-strain gradients. *Journal of the Mechanics and Physics of Solids*, 48:989–1036, 2000.

M.E. Gurtin. On a framework for small–deformation viscoplasticity: free energy, microforces, strain gradients. *International Journal of Plasticity*, 19:47–90, 2003.

M.E. Gurtin and L. Anand. Nanocrystalline grain boundaries that slip and separate: A gradient theory that accounts for grain-boundary stress and conditions at a triple-junction. *Journal of the Mechanics and Physics of Solids*, 56:184–199, 2008.

M.E. Gurtin and L. Anand. Thermodynamics applied to gradient theories involving the accumulated plastic strain: The theories of Aifantis and Fleck & Hutchinson and their generalization. *Journal of the Mechanics and Physics of Solids*, 57:405–421, 2009.

M.E. Gurtin and A. Needleman. Boundary conditions in small–deformation, single–crystal plasticity that account for the burgers vector. *Journal of the Mechanics and Physics of Solids*, 53:1–31, 2005.

H.T. Hahn and W. Jaunzemis. A dislocation theory of plasticity. *Int. J. Engng. Sci.*, 11:1065–1078, 1973.

F. Hehl and E. Kröner. Zum Materialgesetz eines elastischen Mediums mit Momentenspannungen. *Z. Naturforschg.*, 20a:336–350, 1958.

J.P. Hirth and J. Lothe. *Theory of Dislocations.* Wiley Intersciences, 1982.

T. Hoc, C. Rey, and J.L. Raphanel. Experimental and numerical analysis of localization during sequential test for an IF-Ti steel. *Acta Mater.*, 49: 1835–1846, 2001.

C. Huet. Application of variational concepts to size effects in elastic heterogeneous bodies. *J. Mech. Phys. Solids*, 38:813–841, 1990.

J.W. Hutchinson. Plasticity at the micron scale. *Int. J. Solids Structures*, 37:225–238, 2000.

B. Jaoul. *Etude de la plasticité et application aux métaux.* Dunod, Paris, 1965.

D. Jeulin and M. Ostoja-Starzewski. *Mechanics of Random and Multiscale Microstructures.* CISM Courses and Lectures No. 430, Udine, Springer Verlag, 2001.

C.B. Kafadar and A.C. Eringen. Micropolar media: I the classical theory. *Int. J. Engng Sci.*, 9:271–305, 1971.

T. Kanit, S. Forest, I. Galliet, V. Mounoury, and D. Jeulin. Determination of the size of the representative volume element for random composites : statistical and numerical approach. *International Journal of Solids and Structures*, 40:3647–3679, 2003.

S. Kessel. Lineare Elastizitätstheorie des anisotropen Cosserat-Kontinuums. *Abhandlungen der Braunschweig. Wiss. Ges.*, 16, 1964.

S. Kessel. Spannungsfelder einer Schraubenversetzung und einer Stufenversetzung im Cosseratschen Kontinuum. *ZAMM*, 50:547–553, 1970.

W.T. Koiter. Couple-stresses in the theory of elasticity. i and ii. *Proc. K. Ned. Akad. Wet.*, B67:17–44, 1963.

E. Kröner. Die Versetzung als elementare Eigenspannungsquelle. *Z. Naturforschg.*, 11:969–985, 1956.

E. Kröner. *Kontinuumstheorie der Versetzungen und Eigenspannungen.* Ergebnisse der angewandten Mathematik, Vol. 5, Springer-Verlag, 1958.

E. Kröner. On the physical reality of torque stresses in continuum mechanics. *Int. J. Engng. Sci.*, 1:261–278, 1963.

E. Kröner. Initial studies of a plasticity theory based upon statistical mechanics. In M.F. Kanninen, W.F. Adler, A.R. Rosenfield, and R.I. Jaffee, editors, *Inelastic Behaviour of Solids*, pages 137–147. McGraw-Hill, 1969.

E. Kröner. *Statistical Continuum Mechanics.* CISM Courses and Lectures No. 92, Udine, Springer Verlag, Wien, 1971.

Kröner, E. *Mechanics of generalized continua, Proc. of the IUTAM-Symposium on the generalized Cosserat continuum and the continuum theory of dislocations with applications, Freudenstadt, Stuttgart.* Springer Verlag, 1967.

L. Kubin and A. Mortensen. Geometrically necessary dislocations and strain–gradient plasticity. *Scripta materialia*, 48:119–125, 2003.

D. Lachner, H. Lippmann, and L.S. Tóth. On Cosserat plasticity and plastic spin for isotropic materials. *Archives of Mechanics*, 46:531–539, 1994.

R.W. Lardner. Dislocation dynamics and the theory of the plasticity of single crystals. *ZAMP*, 20:514–529, 1969.

H. Lippmann. Eine Cosserat-Theorie des plastischen Fließens. *Acta Mechanica*, 8:255–284, 1969.

J. Mandel. Une généralisation de la théorie de la plasticité de W.T. Koiter. *Int. J. Solids Structures*, 1:273–295, 1965.

J. Mandel. *Plasticité classique et viscoplasticité.* CISM Courses and Lectures No. 97, Udine, Springer Verlag, Berlin, 1971.

J. Mandel. Equations constitutives et directeurs dans les milieux plastiques et viscoplastiques. *Int. J. Solids Structures*, 9:725–740, 1973.

G.A. Maugin. *The Thermomechanics of plasticity and fracture.* Cambridge University Press, 1992.

G.A. Maugin. The method of virtual power in continuum mechanics : Application to coupled fields. *Acta Mechanica*, 35:1–70, 1980.

G.A. Maugin. Internal variables and dissipative structures. *J. Non–Equilib. Thermodyn.*, 15:173–192, 1990.

G.A. Maugin. *Thermomechanics of nonlinear irreversible behaviors.* World Scientific, 1999.

G.A. Maugin and W. Muschik. Thermodynamics with internal variables, Part I. general concepts. *J. Non-Equilib. Thermodyn.*, 19:217–249, 1994.

J.R. Mayeur, D.L. McDowell, and D.J. Bammann. Dislocation–based micropolar single crystal plasticity: Comparison if multi– and single criterion theories. *Journal of the Mechanics and Physics of Solids*, 59:398–422, 2011.

F.A. McClintock, P.A. André, K.R. Schwerdt, and R.E. Stoeckly. Interface couples in crystals. *Nature*, 4636:652–653, 1958.

L. Méric and G. Cailletaud. Single crystal modeling for structural calculations. Part 2: Finite element implementation. *J. of Engng. Mat. Technol.*, 113:171–182, 1991.

L. Méric, P. Poubanne, and G. Cailletaud. Single crystal modeling for structural calculations. Part 1: Model presentation. *J. Engng. Mat. Technol.*, 113:162–170, 1991.

L. Méric, G. Cailletaud, and M. Gaspérini. F.E. calculations of copper bicrystal specimens submitted to tension–compression tests. *Acta metall. mater.*, 42:921–935, 1994.

R.D. Mindlin. Micro–structure in linear elasticity. *Arch. Rat. Mech. Anal.*, 16:51–78, 1964.

R.D. Mindlin. Second gradient of strain and surface–tension in linear elasticity. *Int. J. Solids Structures*, 1:417–438, 1965.

R.D. Mindlin and N.N. Eshel. On first strain gradient theories in linear elasticity. *Int. J. Solids Structures*, 4:109–124, 1968.

H. Mughrabi. On the role of stress gradients and long-range internal stresses in the composite model of crystal plasticity. *Mater. Sci. Eng. A*, 317: 171–180, 2001.

H. Mughrabi. Investigations of plastically deformed copper single crystals in the stress-applied state. *Physica Status Solidi*, 39:317–327, 1970.

H. Mughrabi. Introduction to the viewpoint set on : Surface effect in cyclic deformation and fatigue. *Scripta Metallurgica and Materiala*, 26:1499–1504, 1992.

H.B. Mühlhaus and I. Vardoulakis. The thickness of shear bands in granular materials. *Géotechnique*, 37:271–283, 1987.

Mura, T. On dynamic problems of continuous distribution of dislocations. *Int. J. Engng. Sci.*, 1:371–381, 1963.

Mura, T. Continuous distribution of dislocations and the mathematical theory of plasticity. *Phys. stat. sol.*, 10:447–453, 1965.

W. Nowacki. *Theory of asymmetric elasticity.* Pergamon, 1986.

J.F. Nye. Some geometrical relations in dislocated crystals. *Acta Metall.*, 1:153–162, 1953.

R.N. Pangborn, S. Weissmann, and I.R. Kramer. Dislocations distribution and prediction of fatigue damage. *Metallurgical Transaction*, 12A:109–120, 1981.

C. Papenfuss and S. Forest. Thermodynamical frameworks for higher grade material theories with internal variables or additional degrees of freedom. *Journal of Non-Equilibrium Thermodynamics*, 31:319–353, 2006.

A. Paquin. Micromechanical modeling of the elastic-viscoplastic behavior of polycrystalline steels. *International Journal of Plasticity*, 17:1267–1302, 2001.

R. Parisot, S. Forest, A. Pineau, F. Grillon, X. Démonet, and J.-M. Mataigne. Deformation and Damage Mechanisms of Zinc Coatings on Galvanized Steel Sheets, Part I : Deformation Modes. *Metallurgical and Materials Transactions*, 35A:797–811, 2004.

R.H.J. Peerlings, M.G.D. Geers, R. de Borst, and W.A.M. Brekelmans. A critical comparison of nonlocal and gradient–enhanced softening continua. *Int. J. Solids Structures*, 38:7723–7746, 2001.

R.H.J. Peerlings, T.J. Massart, and M.G.D. Geers. A thermodynamically motivated implicit gradient damage framework and its application to brick masonry cracking. *Comput. Methods Appl. Mech. Engrg.*, 193: 3403–3417, 2004.

H. Petryk. Material instability and strain-rate discontinuities in incrementally nonlinear continua. *J. Mech. Phys. Solids*, 40:1227–1250, 1992.

P. Pilvin. Une approche simplifiée pour schématiser l'effet de surface sur le comportement mécanique d'un polycristal. *J. Phys. IV France*, 4: Pr4–33–38, 1998.

J. M. Pipard, N. Nicaise, S. Berbenni, O. Bouaziz, and M. Berveiller. A new mean field micromechanical approach to capture grain size effects. *Computational Materials Science*, 45(3):604–610, 2009.

J.R. Rice. Inelastic constitutive relations for solids: An internal variable theory and its application to metal plasticity. *J. Mech. Phys. Solids*, 19: 433–455, 1971.

E. Sanchez-Palencia. Comportement local et macroscopique d'un type de milieux physiques hétérogènes. *International Journal of Engineering Science*, 12:331–351, 1974.

E. Sanchez-Palencia and A. Zaoui. *Homogenization techniques for composite media.* Lecture Notes in Physics No. 272, Springer, Berlin, 1987.

C. Sansour. A unified concept of elastic–viscoplastic Cosserat and micromorphic continua. *Journal de Physique IV*, 8:Pr8–341–348, 1998a.

C. Sansour. A theory of the elastic–viscoplastic cosserat continuum. *Archives of Mechanics*, 50:577–597, 1998b.

A. Sawczuk. On yielding of Cosserat continua. *Archives of Mechanics*, 19: 3–19, 1967.

R. Sedláček and S. Forest. Non-local plasticity at microscale : A dislocation-based model and a Cosserat model. *physica status solidi (b)*, 221:583–596, 2000.

R. Sedláček, J. Kratochvil, and E. Werner. The importance of being curved: bowing dislocations in a continuum description. *Philosophical Magazine*, 83:3735–3752, 2003.

J.Y. Shu and N.A. Fleck. Strain gradient crystal plasticity : size–dependent deformation of bicrystals. *Journal of the Mechanics and Physics of Solids*, 47:297–324, 1999.

J.Y. Shu, N.A. Fleck, E. Van der Giessen, and A. Needleman. Boundary layers in constrained plastic flow : comparison of non local and discrete dislocation plasticity. *Journal of the Mechanics and Physics of Solids*, 49:1361–1395, 2001.

R. Sievert. *Eine Systematik für elastisch-plastische Stoffgleichungen.* Schriftenreihe Physikalische Ingenieurwissenschaft, Band 23, Techn. Univ. Berlin, 1992.

F. Šiška, S. Forest, P. Gumbsch, and D. Weygand. Finite element simulations of the cyclic elastoplastic behavior of copper thin films. *Modeling and Simulation in Materials Science and Engineering*, 15:S217–S238, 2007.

F. Šiška, D. Weygand, S. Forest, and P. Gumbsch. Comparison of mechanical behaviour of thin film simulated by discrete dislocation dynamics and continuum crystal plasticity. *Computational Materials Science*, 45: 793–799, 2009.

V.P. Smyshlyaev and N.A. Fleck. The role of strain gradients in the grain size effect for polycrystals. *Journal of the Mechanics and Physics of Solids*, 44:465–495, 1996.

L. St-Pierre, E. Héripré, M. Dexet, J. Crépin, G. Bertolino, and N. Bilger. 3D simulations of microstructure and comparison with experimental microstructure coming from oim analysis. *International Journal of Plasticity*, 24:1516–1532, 2008.

P. Steinmann. A micropolar theory of finite deformation and finite rotation multiplicative elastoplasticity. *Int. J. Solids Structures*, 31:1063–1084, 1994.

P. Steinmann. Views on multiplicative elastoplasticity and the continuum theory of dislocations. *International Journal of Engineering Science*, 34: 1717–1735, 1996.

R. Stojanović. *Recent developments in the theory of polar continua.* CISM Courses and Lectures No. 27, Udine, Springer Verlag, Berlin, 1970.

P. Suquet. *Continuum micromechanics.* CISM Courses and Lectures No. 377, Udine, Springer Verlag, Berlin, 1997.

K. Tanaka and T. Mura. A dislocation model for fatigue crack initiation. *Journal of Applied Mechanics*, 48:97–103, 1981.

V. Taupin, S. Berbenni, C. Fressengeas, and O. Bouaziz. On particle size effects: An internal length mean field approach using field dislocation mechanics. *Acta Materialia*, 58:5532–5544, 2010.

C. Teodosiu. A dynamic theory of dislocations and its applications to the theory of the elastic-plastic continuum. In J.A. Simmons, R. de Wit, and R. Bullough, editors, *Fundamental Aspects of Dislocation Theory*, pages 837–876. Nat. Bur. Stand. (US) Spec. Publ. 317, II, 1970.

C. Teodosiu. *Large plastic deformation of crystalline aggregates.* CISM Courses and Lectures No. 376, Udine, Springer Verlag, Berlin, 1997.

C. Teodosiu and F. Sidoroff. A theory of finite elastoviscoplasticity of single crystals. *Int. J. of Engng Science*, 14:165–176, 1976.

C. Teodosiu, J. Raphanel, and L. Tabourot. Finite element simulation of the large elastoplastic deformation of multi-crystals. In C. Teodosiu and F. Sidoroff, editors, *Large Plastic Deformations MECAMAT'91*, pages 153–158. Balkema, Rotterdam, 1993.

H. Tian-Hu. Lagrange multiplier method of the general variational principles of finite deformation theory of couple stress and its application in the strain gradient plasticity. *International Journal of Engineering Science*, 42:1949–1969, 2004.

E. Van der Giessen, V.S. Deshpande, H.H.M Cleveringa, and A. Needleman. Discrete dislocation plasticity and crack tip fields in single crystals. *J. Mech. Phys. Solids*, 49:2133–2153, 2001.

J. Zarka. Généralisation de la théorie du potentiel plastique multiple en viscoplasticité. *Journal of the Mechanics and Physics of Solids*, 20:179–195, 1972.

A. Zeghadi, S. Forest, A.-F. Gourgues, and O. Bouaziz. Cosserat continuum modelling of grain size effects in metal polycrystals. *Proc. Appl. Math. Mech.*, 5:79–82, 2005.

A. Zeghadi, S. Forest, A.-F. Gourgues, and O. Bouaziz. Ensemble averaging stress–strain fields in polycrystalline aggregates with a constrained surface microstructure–Part 2: Crystal plasticity. *Philosophical Magazine*, 87:1425–1446, 2007a.

A. Zeghadi, F. Nguyen, S. Forest, A.-F. Gourgues, and O. Bouaziz. Ensemble averaging stress–strain fields in polycrystalline aggregates with a constrained surface microstructure–Part 1: Anisotropic elastic behaviour. *Philosophical Magazine*, 87:1401–1424, 2007b.

A. Ziegenbein, H. Neuhäuser, J. Thesing, R. Ritter, H. Wittich, E. Steck, and M. Levermann. Local plasticity of Cu–Al polycrystals - measurements and FEM simulation. *J. Phys. IV France*, 8:Pr8–407–412, 1998.

H. Zorski. Statistical theory of dislocations. *Int. J. Solids Structures*, 4: 959–974, 1968.

Introduction to Discrete Dislocation Dynamics

Hussein. M. Zbib*
School of Mechanical and Materials Engineering
Washington State University, Pullman, WA 99164-2920
(509) 335-7832, zbib@wsu.edu

1 Introduction

This chapter is a review of the dislocation dynamics method and its applications in solving and various problems in crystalline materials. In such materials, a dislocation can be easily understood by considering that a crystal can deform irreversibly by slip, i.e. shifting or sliding along one of its atomic planes. If the slip displacement is equal to a lattice vector, the material across the slip plane will preserve its lattice structure and the change of shape will become permanent. However, rather than simultaneous sliding of two half-crystals, slip displacement proceeds sequentially, starting from one crystal surface and propagating along the slip plane until it reaches the other surface. The boundary between the slipped and still unslipped crystal is a dislocation and its motion is equivalent to slip propagation. In this picture, crystal plasticity by slip is a net result of the motion of a large number of dislocation lines, in response to applied stress. It is interesting to note that this picture of deformation by slip in crystalline materials was first observed in the nineteenth century by Mügge (1883) and Ewing and Rosenhain (1899). They observed that deformation of metals proceeded by the formation of slip bands on the surface of the specimen. Their interpretation of these results was obscure since metals were not viewed as crystalline at that time.

Over the past seven decades, experimental and theoretical developments have firmly established the principal role of dislocation mechanisms in defining material strength. It is now understood that macroscopic properties of crystalline materials are derivable, at least in principle, from the behavior

*The author wishes to acknowledge the financial support of the Office of Basic Energy Science at the US Department of Energy under grant number DE-FG02-07ER46435.

of their constituent defects. However, this fundamental understanding has not been translated into a continuum theory of crystal plasticity based on dislocation mechanisms. The major difficulty in developing such a theory is the multiplicity and complexity of the mechanisms of dislocation motion and interactions that make it impossible to develop a quantitative analytical approach. The problem is further complicated by the need to trace the spatiotemporal evolution of a very large number of interacting dislocations over very long periods of time, as required for the calculation of plastic response in a representative volume element. Such practical intractability of the dislocation-based approaches, on one hand, and the developing needs of material engineering at the nano and micro length scales on the other, have created the current situation when equations of crystal plasticity used for continuum modeling are phenomenological and largely disconnected from the physics of the underlying dislocation behavior.

Bridging the gap between dislocation physics and continuum crystal plasticity has become possible with the advancement in computational technology with bigger and faster computers. To this end, over the past two decades various discrete dislocation dynamics models have been developed. The early discrete dislocation models were two-dimensional (2D) and consisted of periodic cells containing multiple dislocations whose behavior was governed by a set of simplified rules (Lepinoux and Kubin, 1987; Ghoniem and Amodeo, 1988; Groma and Pawley, 1993; der Giessen and Needleman, 1995; Wang and LeSar, 1995; Le and Stumpf, 1996). These simulations, although served as a useful conceptual framework, were limited to 2D and, consequently, could not directly account for such important features in dislocation dynamics as slip geometry, line tension effects, multiplication, certain dislocation intersections and cross-slip, all of which are crucial for the formation of dislocation patterns. In the 90's, development of new computational approaches of dislocation dynamics (DD) in three-dimensional space (3D) generated hope for a principal breakthrough in our current understanding of dislocation mechanisms and their connection to crystal plasticity (Kubin and Canova, 1992; Canova et al., 1993; Hirth et al., 1996; Zbib et al., 1996). In these new models, dislocation motion and interactions with other defects, particles and surfaces are explicitly considered. However, complications with respect to dislocation multiplications, self-interactions and interactions with other defects, and keeping track of complex mechanisms and reactions have provided a new set of challenges for developing efficient computational algorithms.

The DD analysis and its computer simulation modeling devised by many researchers (Ghoniem and Amodeo, 1988; Canova et al., 1992; Kubin, 1993; Schwarz and Tersoff, 1996; Zbib et al., 1996; Zbib et al., 1998) has been advanced significantly over the past two decade. This progress has been further magnified by the idea to couple DD with continuum mechanics analysis in association with computational algorithms such as finite elements. This coupling may pave the way to better understanding of the local response of materials at the nano and micro scales and globally at the macroscale (Zbib and de la Rubia, 2002), increasing the potential for future applications of this method in material, mechanical, structural and process engineering analyses. In the following, the principles of DD analysis will be presented followed by the procedure for the measurement of local quantities such as plastic distortion and internal stresses. The incorporation of DD technique into the three-dimensional plastic continuum mechanics-based finite elements modeling will then be described. Finally, examples are provided to illustrate the applicability of this powerful technique in material engineering analysis.

2 Theoretical Fundamentals of the Method

In order to better describe the mathematical and numerical aspects of the DD methodology, first we will identify the basic geometric conditions and kinetics that control the dynamics of dislocations. This will be followed by discussion of the dislocation equation of motion, elastic interaction equations, and discretization of these equations for numerical implementation.

2.1 Kinematics and Geometric Aspects

A dislocation is a line defect in an otherwise perfect crystal described by its line sense vector $\boldsymbol{\xi}$ and Burgers vector $\boldsymbol{b}$. The Burgers vector has two distinct components: edge, perpendicular to its line sense vector, and screw, parallel to its line sense vector. Under loading, dislocations glide and propagate on slip planes causing deformation and change of shape. When the local line direction becomes parallel to the Burgers vector, i.e. screw character, the dislocation may propagate into other slip planes. This switching of the slip plane, which makes the motion of dislocations three-dimensional, and is better known as cross slip is an important recovery phenomena to be dealt with in dislocation dynamics. In addition to glide and cross slip, dislocations can also climb in a non-conservative three-dimensional motion by absorbing and/or emitting intrinsic point defects, vacancies, and interstitial. Some of these phenomena become important at high load levels or temperatures when point defects become more mobile. In summary, the 3D dislocation dynamics accounts for the following geometric aspects:

- Dislocation topology; 3D geometry, Burgers vector and line sense.

- Identification of all possible slip planes for each dislocation.
- Changes in the dislocation topology when part of it cross-slips and or climbs to another plane
- Multiplication and annihilation of dislocation segments.
- Formation of complex connections and intersections such as junctions, jogs and branching of the dislocation in multiple directions.

2.2 Kinetics and Interaction Forces

The dynamics of the dislocation is governed by a "Newtonian" equation of motion, consisting of an inertia term, damping term, and driving force arising from short-range and long-range interactions. Since the strain field of the dislocation varies as the inverse of distance from the dislocation core, dislocations interact among themselves over long distances. As the dislocation moves, it has to overcome internal drag, and local barriers such as the Peierls stresses (i.e. lattice friction). The dislocation may encounter local obstacles such as stacking fault tetrahedra, defect clusters and vacancies that interact with the dislocation at short ranges and affect its local dynamics. Furthermore, the internal strain field of randomly distributed local obstacles gives rise to stochastic perturbations to the encountered dislocations, as compared with deterministic forces such as the applied load. This stochastic stress field also contributes to the spatial dislocation patterning in the later deformation stages. Therefore, the strain field of local obstacles adds spatially irregular uncorrelated noise to the equation of motion. In addition to the random strain fields of dislocations or local obstacles, thermal fluctuations also provide a stochastic source in dislocation dynamics. Dislocations also interact with free surfaces, cracks, and interfaces, giving rise to what is termed as image stresses or forces. In summary, the dislocation may encounter the following set of forces:

- Drag force, $B\mathbf{v}$, where B is the drag coefficient and v is the dislocation velocity.
- Peierls stress $\mathbf{F}_{Peierls}$.
- Force due to externally applied loads, $\mathbf{F}_{External}$.
- Dislocation-dislocation interaction force $\mathbf{F}_D$.
- Dislocation self-force $\mathbf{F}_{Self}$.
- Dislocation-obstacle interaction force $\mathbf{F}_{Obstacle}$.
- Image force $\mathbf{F}_{Image}$.
- Osmotic force $\mathbf{F}_{Osmotic}$ resulting from non-conservative motion of dislocation (climb) and results in the absorption or emission of intrinsic point defects.
- Thermal force $\mathbf{F}_{Thermal}$ arising from thermal fluctuations.

The DD approach attempts to incorporate all of the aforementioned kinematics and kinetics aspects into a computational traceable framework. In the numerical implementation, three-dimensional curved dislocations are treated as a set of connected segments as illustrated in Figure 1. It is possible to represent smooth dislocations with any desired degree of realism, provided that the discretization resolution is taken high enough for accuracy (limited by the size of the dislocation core radius r0, typically the size of one Burgers vector b). In such a representation, the dynamics of dislocation lines is reduced to the dynamics of discrete degrees of freedom of the dislocation nodes connecting the dislocation segments.

2.3 Dislocation Equation of Motion

The velocity v of a dislocation segment s is governed by a first order differential equation consisting of an inertia term, a drag term and a driving force vector (Hirth, 1992; Hirth et al., 1998; Huang et al., 1999), such that

$$m_s \dot{\mathbf{v}} + \frac{1}{M_s(T,p)} \mathbf{v} = \mathbf{F}_s \quad \text{with} \quad m_s = \frac{1}{v} \left(\frac{dW}{dv} \right) \tag{1}$$

$$\begin{aligned} \mathbf{F}_s &= \mathbf{F}_{Peierls} + \mathbf{F}_{External} + \mathbf{F}_D + \mathbf{F}_{Self} + \\ &\quad + \mathbf{F}_{Obstacle} + \mathbf{F}_{Image} + \mathbf{F}_{Osmotic} + \mathbf{F}_{Thermal} \end{aligned} \tag{2}$$

In the above equation the subscript s stands for the segment, m_s is defined as the effective dislocation segment mass density, M_s is the dislocation mobility which could depend both on the temperature T and the pressure p, and W is the total energy per unit length of a moving dislocation (elastic energy plus kinetic energy). As implied by (1), the glide force vector Fs per unit length arises from a variety of sources described in the previous section. The following relations for the mass per unit dislocation length have been suggested (Hirth et al., 1998) for screw $(m_s)_{screw}$ and edge $(m_s)_{edge}$ dislocations when moving at a high speed.

$$(m_s)_{screw} = \frac{W_0}{\nu^2} \left(-\gamma^{-1} + \gamma^{-3} \right) \tag{3}$$

$$\begin{aligned} (m_s)_{edge} &= \frac{W_0 C^2}{\nu^4} (-16\gamma_l - 40\gamma_l^{-1} + 8\gamma_l^{-3} + 14\gamma + \\ &\quad +50\gamma^{-1} - 22\gamma^{-3} + 6\gamma^{-5}) \end{aligned} \tag{4}$$

where $\gamma_l = \sqrt{1 - \nu^2/C_l^2}$, $\gamma = \sqrt{1 - \nu^2/C^2}$ C_l is the longitudinal sound velocity, C is the transverse sound velocity, ν is Poisson's ratio, $W_0 =$

$\frac{Gb^2}{4\pi}\ln(R/r_0)$ is the rest energy for the screw per unit length, G is the shear modulus. The value of R is typically equal to the size of the dislocation cell (about $1000b$), or in the case of one dislocation is the shortest distance from the dislocation to the free surface (Hirth and Lothe, 1982). In the non-relativistic regime when the dislocation velocity is small compared to the speed of sound, the above equations reduce to the familiar expression $m = \beta\rho b^2 \ln(R/r_0)$, where β is a constant dependent on the type of the dislocation, and ρ is the mass density.

Dislocation Mobility Function. The reliability of the numerical simulation depends critically on the accuracy of the dislocation drag coefficient $B(=1/M)$ which is material dependent. There are a number of phenomenological relations for the dislocation glide velocity v_g (Kocks et al., 1975; Sandstrom, 1977), including relations of power law forms and forms with an activation term in an exponential or as the argument of a sinh form. Often, however (Johnston and Gilman, 1959; Sandstrom, 1977) the simple power law form is adopted for expedience, e.g. $v_g = v_s(\tau_e/\tau_s)^m$, resulting in nonlinear dependence of M on the stress. In a number of cases of pure phonon/electron damping control or of glide over the Peierls barrier a constant mobility (with $m = 1$), predicts the results very well. This linear form has been theoretically predicted for a number of cases as discussed by Hirth and Lothe (1982).

Mechanisms to explain dislocation drag have been studied for long time and the drag coefficients have been estimated in numerous experimental and theoretical works by atomistic simulations or quantum mechanical calculations [see, for example, the review by Al'shitz (1992)]. The determination of each of the two components (phonon and electron drag) that constitute the drag coefficient for a specific material is not trivial, and various simplifications have been made, e.g. the Debye model neglects Van Hove singularities in phonon spectrum (Ashcroft and Mermin, 1976), isotropic approximation of deformation potentials, and so on. Also the values are sensitive to various parameters such as the mean free path or core radius. Nevertheless, in typical metals, the phonon drag B_{ph} range is $30 \sim 80\mu Pa.s$ at room temperature and less than 0.1 Pa.s at very low temperatures around 10K, while for the electron drag Be the range is a few Pa.s and expected to be temperature independent. Under strong magnetic fields at low temperature, macroscopic dislocation behavior can be highly sensitive to orientation relative to the field within accuracy of 1% (McKrell and Galligan, 2000). Except for special cases such as deformation under high strain rate, weak dependencies of drag on dislocation velocity are usually neglected.

Examples of temperature dependence of each component of the drag coefficient can be found for the case of edge dislocation in Copper (Hiratani and Nadgorny, 2001), or in Molybdenum (Jinpeng et al., 1999). Generally, however, the dislocation mobility could be, among other things, a function of the angle between the Burgers vector and the dislocation line sense, i.e. dislocation character, especially at low temperatures. For example, Wasserbäch (1986) observed that at low deformation temperatures (77 to 195 K) the dislocation structure in Ta single crystals consisted of primary and secondary screw dislocations and of tangles of dislocations of mixed characters, while at high temperatures (295 to 470 K) the behavior was similar to that of fcc crystals. In the work of Mason and MacDonald (1971) they measured the mobility of dislocation of an unidentified type in NB as $4.2 \times 10^4 (Pa.s)^{-1}$ near room temperature. A smaller value of $3.3 \times 10^3 (Pa.s)^{-1}$ was obtained by Urabe and Weertman (1975) for the mobility of edge dislocation in Fe. The mobility for screw dislocations in Fe was found to be about a factor of two smaller than that of edge dislocations near room temperature. A theoretical model to explain this large difference in behavior is given in Hirth and Lothe (1982) and is based on the observation that in bcc materials the screw dislocation has a rather complex three-dimensional core structure, resulting in a high Peierls stress, leading to a relatively low mobility for screw dislocations while the mobility of mixed dislocations is higher.

Dislocation Collisions. When two dislocations collide, their response is dominated by their mutual interactions and becomes much less sensitive to the long-range elastic stress associated with external loads, boundary conditions, and all other dislocations present in the system. Depending on the shapes of the colliding dislocations, their approach trajectories and their Burgers vectors, two dislocations may form a dipole, or react to annihilate, or to combine to form a junction, or to intersect and form a jog. In the DD analysis, the dynamics of two colliding dislocations is determined by the mutual interaction force acting between them. In the case that the two dislocation segments are parallel (on the same plane and or intersecting planes) and have the same Burgers vector with opposite sign they would annihilate if the distance between them is equal to the core size. Otherwise, the colliding dislocations would align themselves to form a dipole, a jog or a junction depending on their relative position. A comprehensive review of short-range interaction rules can be found in Zbib et al. (1998).

Discretization of Dislocation Equation of Motion. Equation (1) applies to every infinitesimal length along the dislocation line. In order to solve this equation for any arbitrary shape, the dislocation curve may be discretized into a set of dislocation segments. Then the velocity vector field over each segment may be assumed to be linear and, therefore, the problem is reduced to finding the velocity of the nodes connecting these segments. There are many numerical techniques to solve such a problem. Consider, for example, a straight dislocation segment s bounded by two nodes j and $j+$. Then within the finite element formulation (Bathe, 1982), the velocity vector field is assumed to be linear over the dislocation segment length. This linear vector field $\mathbf{v}$ can be expressed in terms of the velocities of the nodes such that $\mathbf{v} = \left[\mathbf{N}^D\right]^T \mathbf{V}^D$ where $\mathbf{V}^D$ is the nodal velocity vector and $\left[\mathbf{N}^D\right]$ is the linear shape function vector (Bathe, 1982). Upon using the Galerkin method, equation (1) for each segment can be reduced to a set of six equations for the two discrete nodes (each node has three degrees of freedom). The result can be written in the following matrix-vector form.

$$\left[\mathbf{M}^D\right]\dot{\mathbf{V}}^D + \left[\mathbf{C}^D\right]\mathbf{V}^D = \mathbf{F}^D \tag{5}$$

where

$\left[\mathbf{M}^D\right] = m_s \int \left[\mathbf{N}^D\right]\left[\mathbf{N}^D\right]^T \mathrm{d}l$ is the dislocation segment 6×6 mass matrix,

$\left[\mathbf{C}^D\right] = (1/M_s) \int \left[\mathbf{N}^D\right]\left[\mathbf{N}^D\right]^T \mathrm{d}l$ is the dislocation segment 6×6 dumping matrix, and

$\left[\mathbf{F}^D\right] = \int \left[\mathbf{N}^D\right]\mathbf{F}_s \mathrm{d}l$ is the 6×1 nodal force vector

Then, following the standard element assemblage procedure, one obtains a set of discrete system of equations, which can be cast in terms of a global dislocation mass matrix, a global dislocation damping matrix, and a global dislocation force vector. In the case of one dislocation loop and with ordered numbering of the nodes around the loop, it can be easily shown that the global matrices are banded with half-bandwidth equal to one. However, when the system contains many loops that interact among themselves and new nodes are generated and/or annihilated continuously, the numbering of the nodes becomes random and the matrices become un-banded. To simplify the computational effort, one can employ the lumped mass and damping matrix method. In this method, the mass matrix $\left[\mathbf{M}^D\right]$ and damping matrix $\left[\mathbf{C}^D\right]$ become diagonal matrices (half-bandwidth equal to zero), and therefore the only coupling between the equations is through the nodal force vector $\mathbf{F}^D$. The computation of each component of the force vector is described below.

2.4 The Dislocation Stress and Force Fields

The stress induced by any arbitrary dislocation loop at an arbitrary field point P can be computed by the Peach-Koehler integral equation given in Hirth and Lothe (1982). This integral equation, in turn, can be evaluated numerically over many loops of dislocations by discretizing each loop into a series of line segments. If we denote,

$$
\begin{aligned}
N_l &= \text{total number of dislocation loops} \\
N_s^{(l)} &= \text{number of segments of loop } l \\
N_n^{(l)} &= \text{number of nodes associated with the segments of loop } l, \\
&\quad \text{i.e. } N_n^{(l)} = N_s^{(l)} + 1 \\
N_s &= \text{total number of segments} = N_s^{(l)} \times N_l, \text{ where summation} \\
&\quad \text{over } l \text{ is implied} \\
N_n &= \text{total number of nodes} = N_n^{(l)} \times N_l, \text{ where summation over} \\
&\quad l \text{ is implied} \\
l_s &= \text{length of segment } l \\
r &= \text{distance from point } P \text{ to the segment } s
\end{aligned}
$$

Then the discretized form of the Peach-Koehler integral equation for the stress at any arbitrary field point P becomes

$$
\begin{aligned}
\overset{d}{\sigma}_{ij}(P) = & \sum_{l=1}^{N_l}\sum_{s=1}^{N_s^{(l)}}\{-\frac{G}{8\pi}\int_{ls} b_p\epsilon_{mpi}\frac{\partial}{\partial x'_m}\nabla'^2 R\mathrm{d}x'_j - \\
& \frac{G}{8\pi}\int_{ls} b_p\epsilon_{mpi}\frac{\partial}{\partial x'_m}\nabla'^2 R\mathrm{d}x'_j - \\
& \frac{G}{4\pi(1-\nu)}\int_{ls} b_p\epsilon_{mpi}\left(\frac{\partial^3 R}{\partial x'_m\partial x'_i\partial x'_j} - \delta_{ij}\frac{\partial}{\partial x'_m}\nabla'^2 R\right)\mathrm{d}x'_k\}
\end{aligned} \tag{6}
$$

where ϵ_{ijk} is the permutation symbol, and R is the magnitude of the $\mathbf{R} = \mathbf{r}' - \mathbf{r}$ (with $\mathbf{r}$ being the position vector of point P and $\mathbf{r}'$ the position vector of a differential line segment of the dislocation loop or curve). The integral over each segment can be explicitly carried out using the linear element approximation. Exact solution of equation (6) for a straight dislocation segment can be found in DeWit (1960) and Hirth and Lothe (1982). However, evaluation of the above integral requires careful consideration as the integrand becomes singular in cases where point P coincides with one of the nodes of the segment that integration is taken over, i.e., self-segment integration. Thus,

- If P is not part of the segment s, there is no singularity since $R \neq 0$ and the ordinary integration procedure may be performed.
- If P coincides with a node of the segment s where the integration should be carried out, special treatment is required due to the singular nature of the stress field as $R \to 0$. Here, the regularization scheme developed by Zbib and co-workers have been employed.

In general, the dislocation stresses can be decomposed into the following form.

$$\overset{d}{\boldsymbol{\sigma}}(P) \quad = \quad \sum_{s=1}^{N_2-2} \overset{d}{\boldsymbol{\sigma}}{}^{(s)} + \overset{d}{\boldsymbol{\sigma}}{}^{(P+)} + \overset{d}{\boldsymbol{\sigma}}{}^{(P-)} \tag{7}$$

where $\overset{d}{\boldsymbol{\sigma}}{}^{(s)}$ is the contribution to the stress at point P from a segment s, and $\overset{d}{\boldsymbol{\sigma}}{}^{(P+)}$, $\overset{d}{\boldsymbol{\sigma}}{}^{(P-)}$ are the contributions to the stress from the two segments that are shared by a node coinciding with P which will be further discussed below.

Once the dislocation stress field is computed the forces on each dislocation segment can be calculated by summing the stresses along the length of the segment. The stresses are categorized into those coming from the dislocations as formulated above and also from any other externally applied stresses plus the internal friction (if any) and the stresses induced by any other defects or micro-constituents. A model for the osmotic force $\mathbf{F}_{Osmotic}$ is given in Raabe (1998) and its inclusion in the total force is straightforward since it is a deterministic force. However, the treatment of the thermal force $\mathbf{F}_{Thermal}$ is not trivial since this force is stochastic in nature, requiring a special consideration and algorithm leading to what is called Stochastic Dislocation Dynamics (SDD) as developed by Hiratani and Zbib (2002). Therefore, the force acting on each segment can be written as:

$$\mathbf{F}_s = \left(\left(\sum_{m=1}^{N_s} \overset{d}{\boldsymbol{\sigma}}{}^{(m)} \right) + \overset{\alpha}{\boldsymbol{\sigma}}{}^{(s)} + \boldsymbol{\tau}_s \right) \cdot \boldsymbol{b}_s \times \boldsymbol{\xi}_s = \overset{d}{\mathbf{F}}_s + \overset{\alpha}{\mathbf{F}}_s + \mathbf{F}_{Thermal} \tag{8}$$

where $\overset{d}{\boldsymbol{\sigma}}{}^{(m)}$, is the contribution to the stresses along segment s from another segment m (dislocation-dislocation interaction), $\overset{\alpha}{\boldsymbol{\sigma}}{}^{(s)}$ is the sum of all externally applied stresses, internal friction (if any) and the stresses induced by any other defects , and $\boldsymbol{\tau}_s$ is the thermal stress; $\overset{d}{\mathbf{F}}_s$, $\overset{\alpha}{\mathbf{F}}_s$ and $\mathbf{F}_{Thermal}$ are the corresponding total Peach–Koehler (PK) forces.

Using equations (7), the force $\overset{d}{\mathbf{F}}_s$ can also be decomposed into two parts one arising from all dislocation segments and one from the self-segment, which is better known as the self–force, that is,

$$\overset{d}{\mathbf{F}}_s = \sum_{m=1}^{N_s-2} \overset{d}{\mathbf{F}}_s^{(m)} + \overset{d}{\mathbf{F}}_s^{(self)} \tag{9}$$

where $\overset{d}{\mathbf{F}}_s^{(m)}$ and $\overset{d}{\mathbf{F}}_s^{(self)}$ are respectively, the contribution to the force on segment s from segment m and the self-force. In order to evaluate the self-force, a special numerical treatment as given by Zbib et al. (1996) and Zbib and de la Rubia (2002) should be used in which exact expressions for the self-force are given.

The direct computation of the dislocation forces discussed above requires the use of a very fine mesh, especially when dealing with problems involving dislocation-defect interaction. As a rule to capture the effect of the very small defects, the dislocation segment size must be comparable to the size of the defect. Alternatively, one can use large dislocation segments compared to the smallest defect size, provided that the force interaction is computed over many points (Gauss points) over the segment length. In this case, the self-force of segment s would be evaluated first. Then the force contribution from other dislocations and defects is calculated by computing the stresses at several Gauss points along the length of the segment. The summation as in (8) would then follow according to:

$$\mathbf{F}_s = \mathbf{F}_s^{(self)} + \sum_{m=1}^{N_s-2} \frac{1}{n_g} \sum_{g=1}^{n_g} \left(\overset{d}{\boldsymbol{\sigma}}^{(m)} (p_g) + \ldots \right) \cdot \boldsymbol{b}_s \times \boldsymbol{\xi} \tag{10}$$

where p_g is the Gauss point g and n_g is the number of Gauss points along segment s. The number of Gauss points depends on the length of the segment. As a rule the shortest distance between two Gauss points should be larger or equal to $2r_0$, i.e. twice the core size.

2.5 The Stochastic Force and Cross-slip

Thermal fluctuations arise from dissipation mechanism due to collision of dislocations with surrounding particles, such as phonons or electrons. Rapid collisions and momentum transfers result in random forces on dislocations.

These stochastic collisions, in turn, can be regarded as time-independent noises of thermal forces acting on the dislocations. Suppose the exertion of thermal forces follows a Gaussian distribution. Then, thermal fluctuations most likely result in very small net forces due to mutual cancelations. However, they sometimes become large and may cause diffusive dislocation motion or thermal activation events such as overcoming obstacle barriers. Therefore, the DD simulation model should also account not only for deterministic effects but also for stochastic forces; leading to a model called "stochastic discrete dislocation dynamics" (SDD) (Hiratani et al., 2003). The procedure is to include the stochastic force $\mathbf{F}_{thermal}$ in the DD model by computing the magnitude of the stress pulse (τ_s) using a Monte Carlo type analysis.

Based on the assumption of the Gaussian process, the thermal stress pulse has zero mean and no correlation (Ronnpagel et al., 1993; Raabe, 1998) between any two different times. This leads to the average peak height given as (Koppenaal and Kuhlmann-Wilsdorf, 1964; Hiratani and Zbib, 2003)

$$\sigma_s = \sqrt{2kT/M_s(b^2\Delta l\Delta t)} \tag{11}$$

where k denotes Boltzman constant, T absolute temperature of the system, b the magnitude of Burgers vector, Δt time step, and Δl is the dislocation segment length, respectively. Some values of the peak height given by Hiratani and Zbib (2003).

Numerical implementation includes an algorithm where stochastic components are evaluated at each time step of which strengths are correlated and sampled from a bivariate Gaussian distribution (Allen and Tildesley, 1997). With the inclusion of stochastic forces in DD analysis, one can treat cross-slip (a thermally activated process) in a direct manner, since the duration of waiting time and thermal agitations are naturally included in the stochastic process. For example, for the cross-slip in fcc model one can develop a model based on the Escaig-Friedel (EF) mechanism where cross-slip of a screw dislocation segment may be initiated by an immediate dissociation and expansion of Shockley partials. This EF mechanism has been observed to have lower activation energy than Shoeck-Seeger mechanism where the double super kinks are formed on the cross slip plane [this model is used for cross-slip in bcc (Rhee et al., 1998)]. In the EF mechanism, the activation enthalpy ΔG depends on the interval of the Shockley partials (d) and the resolved shear stress on the initial glide plane (σ). The constriction interval L is also dependent on σ. For example, for the case of copper, the activation energy for cross–slip can be computed using an empirical formula fitted to the MD results of Rao et al. (1999).

2.6 Modifications for Long-Range Interactions: The Super Dislocation Principle

Inclusion of the interaction among all the dislocation loops present in a large body is computationally expensive since the number of computations per step would be proportional to N_s^2 where N_s is the number of dislocation segments. A numerical compromise technique termed the super–dislocation method, which is based on the multipolar expansion method (Wang and LeSar, 1995; Hirth et al., 1996; Zbib et al., 1996), reduces the order of computation to $N_s \log N_s$ with a high accuracy. In this approach, the dislocations far away from the point of interest are grouped together into a set of equivalent monopoles and dipoles. In the numerical implementation of the DD model, one would divide the 3D computational domain into sub-domains, and the dislocations in each sub-domain (if there are any) are grouped together in terms of monopoles, dipoles, etc. (depending on the desired accuracy) and the their far stress field is then computed.

2.7 Evaluation of Plastic Strains

The motion of each dislocation segment gives rise to plastic distortion, which is related to the macroscopic plastic strain rate tensor $\dot{\varepsilon}^p$, and the plastic spin tensor $\boldsymbol{W}^P$ via the relations

$$\dot{\varepsilon}^p = \sum_{s=1}^{N_s} \frac{l_s v_{gs}}{2V} (\boldsymbol{n}_s \otimes \boldsymbol{b}_s + \boldsymbol{b}_s \otimes \boldsymbol{n}_s) \tag{12}$$

$$\boldsymbol{W}^p = \sum_{s=1}^{N_s} \frac{l_s v_{gs}}{2V} (\boldsymbol{n}_s \otimes \boldsymbol{b}_s - \boldsymbol{b}_s \otimes \boldsymbol{n}_s) \tag{13}$$

where $\boldsymbol{n}_s$ is a unit normal to the slip plane, v_{gs} is the magnitude of the glide velocity of segment s, V is the volume of the representative volume element (RVE). The above relations provide the most rigorous connection between the dislocation motion (the fundamental mechanism of plastic deformation in crystalline materials) and the macroscopic plastic strain, with its dependence on strength and applied stress being explicitly embedded in the calculation of the velocity of each dislocation. Nonlocal effects are explicitly included into the calculation through long-range interactions. Another microstructure quantity, the dislocation density tensor $\boldsymbol{\alpha}$, can also be calculated according to

$$\boldsymbol{\alpha} = \sum_{s=1}^{N_s} \frac{l_s}{V} \boldsymbol{b}_s \otimes \boldsymbol{\xi}_s \tag{14}$$

This quantity provides a direct measure for the net Burgers vector that gives rise to strain gradient relief (bending of crystal) (Shizawa and Zbib, 1999).

2.8 The DD Numerical Solution: An Implicit–Explicit Integration Scheme

An implicit algorithm to solve the equation of motion (7) with a backward integration scheme may be used, yielding the recurrence equation

$$v^{t+\delta t}\left(1+\frac{\Delta t}{M_s m_s}\right)^{t+\delta t} = v^2+\frac{\Delta t}{m_s}F_s^{t+\delta t} \tag{15}$$

This integration scheme is unconditionally stable for any time step size. However, the DD time step is determined by two factors: i) the shortest flight distance for short-range interactions, and ii) the time step used in the dynamic finite element modeling to be described later. This scheme is adopted since the time step in the DD analysis (for high strain rates) is of the same order of magnitude of the time required for a stable explicit finite element (FE) dynamic analysis. Thus, in order to ensure convergence and stable solution, the critical time tc and the time step for both the DD and the FE ought to be $t_c = l_c/C_l$, and $\Delta t = t_c/20$, respectively, where l_c is the characteristic length scale which is the shortest dimension in the finite element mesh.

In summary, the system of equations given above summarizes the basic ingredients that a dislocation dynamics simulation model should include. There are a number of variations in the manner in which the dislocation curves may be discretized, for example zero order element (pure screw and pure edge), first order element (or piecewise linear segment with mixed character), or higher order nonlinear elements but this is purely a numerical issue. Nonetheless, the DD model should have the minimum number of parameters and, hopefully, all of them should be basic physical and material parameters and not phenomenological ones for the DD result to be predictive. The DD model described above has the following set of physical and material parameters:

- Burgers vectors
- elastic properties,
- core size (equal to one Burgers vector),
- thermal conductivity and specific heat,
- mass density,
- stacking fault energy, and
- dislocation mobility.

Also there are two numerical parameters: the segment length (minimum segment length can't be less that three times the core size) and the time step (as discussed in conjunction with equation (15)), but both are fixed to ensure convergence of the result. In the above list, it is emphasized that in general the dislocation mobility is an intrinsic material property that reflects the local drag mechanisms as discussed above. One can use an "effective" mobility that accounts for additional drag from dislocation-point defect interaction, and thermal activation processes if the defects/obstacles are not explicitly impeded in the DD simulations. However, there is no reason not to include these effects explicitly in the DD simulations (as done in the model described above), i.e. dislocation defect interaction, stochastic processes and inertia effects, which actually permits the prediction of the "effective" mobility from the DD analysis (Hiratani and Zbib, 2002; Hiratani et al., 2003).

3 Integration of DD and Continuum Plasticity

3.1 Continuum Elasto–viscoplasticity

The discrete dislocation model can be coupled with continuum elasto–visoplasticity models, making it possible to correct for *dislocation image stress* and to address a wide range of complex boundary value problems at the microscopic level. In the following, a brief description of this coupling is provided. The coupling is based on a framework in which the material obeys the basic laws of continuum mechanics, i.e. the linear momentum balance

$$\mathrm{div}\boldsymbol{\sigma} = \rho\dot{\boldsymbol{v}} \tag{16}$$

and the energy equation

$$\rho c_v \dot{T} = k\nabla^2 T + \boldsymbol{\sigma}\cdot\dot{\boldsymbol{\varepsilon}}^p \tag{17}$$

where $\boldsymbol{v} = \dot{\boldsymbol{u}}$ is the particle velocity, $\boldsymbol{u}$, rho, c_v and k are the displacement vector field, mass density, specific heat and thermal conductivity respectively. In DD the representative volume cell analyzed can be further discretized into sub-cells or finite elements, each representing a *RVE*. Then the internal stresses field induced by the dislocations (and other defects) $\boldsymbol{\sigma}^D$ and the plastic strain field within each *RVE* can be calculated at any point within each element. However, and to be consistent with the definition of a *RVE*, the heterogeneous internal stress field can be homogenized over each *RVE*, resulting into an equivalent internal stress $\boldsymbol{S}^D$ which is homogenous within the *RVE*, i.e.

$$\boldsymbol{S}^D = <\boldsymbol{\sigma}^D> = \frac{1}{V_{element}} \int_{element} \boldsymbol{\sigma}^D(x)\mathrm{d}v \tag{18}$$

Furthermore, the plastic strain increment results from only the mobile dislocations that exit the *RVE* (or sub-cell) and is computed as in equation (12) but with V being the volume of the element (or sub-cell). The dislocations that are immobile (zero velocity) do not contribute to the plastic strain increment. Therefore, the dislocations within an element induce an internal stress due to their elastic distortion. When some of these dislocations (or all) move and exit the element they leave behind plastic distortion in the element, and the internal stress field should be recomputed by summing the stress from the remaining dislocations in the element. With this homogenization procedure Hooke's law for the *RVE* becomes

$$\boldsymbol{\sigma} + \boldsymbol{S}^D = [\boldsymbol{C}^e][\boldsymbol{\varepsilon} - \boldsymbol{\varepsilon}^p] \tag{19}$$

where $\boldsymbol{C}^e$ is the fourth-order elastic tensor. In the continuum plasticity theory one would need to develop a phenomenological constitutive law for plastic stress-strain. Here, this ambiguity is resolved by using the explicit expressions given by equation (12) for the plastic strain tensor as computed in the dislocation dynamics.

3.2 Modifications for Finite Domains

The solution for the stress field of a dislocation segment (Hirth and Lothe, 1982) is true for a dislocation in an infinite domain and for homogeneous materials. In order to account for finite domain boundary conditions, der Giessen and Needleman (1995) developed a 2D model based on the principle of superposition. The method has been extended by Yasin et al. (2001) and Zbib and de la Rubia (2002) to three–dimensional problems involving free surfaces and interfaces as summarized below.

Interactions with External Free Surfaces In the superposition principle, the two solutions from the infinite domain and finite domain are superimposed. Assuming that the dislocation loops and any other internal defects with self induced stress are situated in the finite domain V bounded by the surface S and subjected to arbitrary external tractions and constraints. Then the stress, displacement, and strain fields are given by the superposition of the solutions for the infinite domain and the actual domain subjected to

$$\boldsymbol{\sigma} = \boldsymbol{\sigma}^\infty + \boldsymbol{\sigma}^* \quad \boldsymbol{u} = \boldsymbol{u}^\infty + \boldsymbol{u}^* \quad \boldsymbol{\varepsilon} = \boldsymbol{\varepsilon}^\infty + \boldsymbol{\varepsilon}^* \tag{20}$$

where $\boldsymbol{\sigma}^{\infty}$, $\boldsymbol{\varepsilon}^{\infty}$ and $\boldsymbol{u}^{\infty}$ are the fields caused by the internal defects as if they were in an infinite domain, whereas $\boldsymbol{\sigma}^{*}$, $\boldsymbol{\varepsilon}^{*}$ and $\boldsymbol{u}^{*}$ are the field solutions corresponding to the auxiliary problem satisfying the following boundary conditions

$$\begin{aligned} \boldsymbol{t} = \boldsymbol{t}^{a} - \boldsymbol{t}^{\infty} \qquad & \textit{on } S \\ \boldsymbol{u} = \boldsymbol{u}^{a} \qquad & \textit{on part of the boundary } S \end{aligned} \tag{21}$$

where $\boldsymbol{t}^{a}$ is the externally applied traction, and $\boldsymbol{t}^{\infty}$ is the traction induced on S by the defects (dislocations) in the infinite domain problem. The traction $-\boldsymbol{t}^{\infty} = \boldsymbol{\sigma} \cdot \boldsymbol{n}$ on the surface boundary S results into an image stress field which is superimposed onto the dislocations segments and, thus, accounting for surface-dislocation interaction.

The treatment discussed above considers interaction between dislocations and external free surfaces, as well as internal free surfaces such as voids. Internal surfaces such as micro-cracks and rigid surfaces around fibers are treated within the dislocation theory framework, whereby each surface is modeled as a pile- up of infinitesimal dislocation loops (Demir et al., 1993; Demir and Zbib, 2001). Hence, defects of these types may be represented as dislocation segments and loops, and their interaction with external free surfaces follows the method discussed above. This subject has been addressed by Khraishi et al. (2001).

Interactions with Interfaces The framework described above for dislocations in homogenous materials can be implemented into a finite element code. The model can also be extended to the case of dislocations in heterogeneous materials using the concept of superposition as outlined by Zbib and de la Rubia (2002). For bi-materials, suppose that domain V is divided into two sub-domains V_1 and V_2 with domain V_1 containing a set of dislocations. The stress field induced by the dislocations and any externally applied stresses in both domains can be constructed in terms of two solutions such that

$$\boldsymbol{\sigma} = \boldsymbol{\sigma}^{\infty 1} + \boldsymbol{\sigma}^{*}, \qquad \boldsymbol{\varepsilon} = \boldsymbol{\varepsilon}^{\infty 1} + \boldsymbol{\varepsilon}^{*} \tag{22}$$

where $\boldsymbol{\sigma}^{\infty 1}$ and $\boldsymbol{\varepsilon}^{\infty 1}$ are the stress and strain fields, respectively, induced by the dislocations (the infinite solution) with the entire domain V having the same material properties of domain V_1 (homogenous solution). Applying Hooke's law for each of the sub-domains, and using (22), one obtains the elastic constitutive equations for each of the materials in each of the sub-domains as:

$$\sigma^* = [C_1^e]\,\varepsilon^*, \qquad in\ V_1$$
$$\sigma^* = [C_2^e]\,\varepsilon^* + \sigma^{*21},\ \sigma^{*21} = [C_2^e - C_1^e]\,\varepsilon^{\infty 1}, \qquad in\ V_2 \tag{23}$$

where C_1^e and C_2^e are the elastic stiffness tensors in V_1 and V_2, respectively. The boundary conditions are:

$$t = t^a - t_1^{\infty}\ on\ S; \qquad u = u^a : on\ part\ of\ S \tag{24}$$

where t^a is the externally applied traction and t_1^{∞} is the traction induced on all of S by the dislocations in V_1 in the infinite-homogenous domain problem. The *"eigenstress"* $\sigma^{\infty 21}$ is due to the difference in material properties. The method described above can be extended to the case of heterogeneous materials with N sub-domains (Zbib and de la Rubia, 2002).

The above system of equations (16-19) can be combined with the boundary corrections given by (20-24). The resulting set of field equations can be solved numerically using the finite element method as described by Zbib and de la Rubia (2002). The end result is a model, which they call a multiscale discrete dislocation dynamics plasticity (*MDDP*), coupling continuum elasto-visoplaticity with discrete dislocation dynamics. The *MDDP* consists of two main modules, the DD module and the continuum finite element module. The DD module computes the dynamics of the dislocations, the plastic strain field they produce, and the corresponding internal stresses field. These field values are passed to the continuum finite element module, in which the stress-displacement-temperature field is computed based on the boundary value problem at hand. The resulting stress field, in turn, is passed to the DD module and the cycle is repeated.

4 Applications of Dislocation Dynamics

So far we have reviewed the main ingredients needed to perform Discrete Dislocation Dynamics (DD) modeling. There all of the essential parameters and quantities to perform the simulations were discussed. In what follows, we present some illustrative examples of the problems that DD can deal with so far to provide both solid mechanists and materials scientists with the data and insight that they might lack when using other simulation methodologies. In particular DD (as will be illustrated) is very good at providing constitutive equations or quantification to relate microstructure to mechanical properties (e.g. flow stress). The power of DD simulations of the underlying dislocation microstructure and evolution can be augmented and enhanced by coupling DD to finite-element modeling in real time. This will be presented first below followed by some typical application examples that DD is capable of.

Over the past decade, the discrete dislocation dynamics has been utilized by a number of researchers to investigation many complicated small-scale crystal plasticity phenomena that occur under a wide range of loading and boundary conditions, and covering a wide spectrum of strain rates. Some of the major phenomena that have been addressed include:

- The role of dislocation mechanisms in strain hardening (Devincre and Kubin, 1997; Zbib et al., 2000; Hiratani and Zbib, 2002).
- Dislocation pattern formation during monotonic and cyclic loading.
- Dislocation-defect interaction problems, including dislocation-void interaction (Ghoniem and Sun, 1999), dislocation-SFT/void-clusters interaction in irradiated materials and the role of dislocation mechanisms in the formation of localized shear bands (de la Rubia et al., 2000; Khraishi et al., 2002, Wirth et al. 2001; Hiratani et al., 2003).
- The effect of particle size on hardening in metal-matrix composites (Khraishi and Zbib, 2002a).
- The effect of thermal stresses associated with inclusions of different coefficient of thermal expansion from the metallic matrix on the composite strengthening (Khraishi et al., 2003).
- The effect of free surfaces on the plasticity of a small material volume (Khraishi and Zbib, 2002b).
- Crack tip plasticity and dislocation-crack interaction (der Giessen and Needleman, 2002; Mastorakos and Zbib, 2006).
- Size effects in polycrystalline metals (Ohashi et al., 2007 and Akasheh et al. (2007c)
- The role of various dislocation patterns such geometrically necessary boundaries (GNB's) in hardening phenomena (Khan et al., 2001, 2003).
- Plastic zone and hardening in Nano-indentation tests (Fivel et al., 1998).
- The role of dislocation mechanisms in increased strength in nano-layered structures (Zbib and de la Rubia, 2002; Schwartz 2003).
- High Strain Rate Phenomena and shock wave interaction with dislocations (Shehadeh et al., 2005a,b, 2006; Zbib et al., 2003, 2011; (Bellou et al., 2011), (Mastorakos et al., 2011))
- Strength in and dislocation mechanisms in nanolayered metallic composites (Akasheh et al., 2007a,b)

In what follows, and in order to illustrate the utility of this approach in investigating a wide range of small-scale plasticity phenomena, representative results for two case–studies are presented. Namely, dislocation behavior during monotonic loading, and the evolution of deformation and dislocation structure during loading of a cracked specimen. The results given below are for both copper single crystals whose materials properties are as follows: mass density $= 8,900kg/m^3$, $G = 54.6GPa$, $\nu = 0.324$, $b = 0.256nm$, $M = 1041/Pa.s$.

4.1 Hardening and Dislocation Structure during Monotonic Loading of small scale specimens

Recent experiments on small scale specimens (micro to sub micrometer, for example see Overman et al., 2009; Groh et al., 2009 and Zbib et al., 2009) shows strong size dependent even when the initial dislocation density is high. This phenomenon goes beyond the nature of strain hardening behavior and the underlying dislocation microstructure evolution during both loading and stress relaxation. Both patterning and the concomitant strain hardening are thought to result from dislocation interaction with surfaces and the mean free path imposed by this interaction. Figure 1 shows a tetragonal cell whose size is 1.25m × 1.25m × 2.5m with FR sources distributed randomly in the crystal. The load is applied in the [001] direction. The cell is meshed into 10 × 10 × 20 finite elements. For the boundary conditions, the bottom of the cell is assumed to be fixed and constant velocity is imposed to the upper surface of the cell; all other surface are assumed to be free surfaces. Results are shown in Figures 1–5.

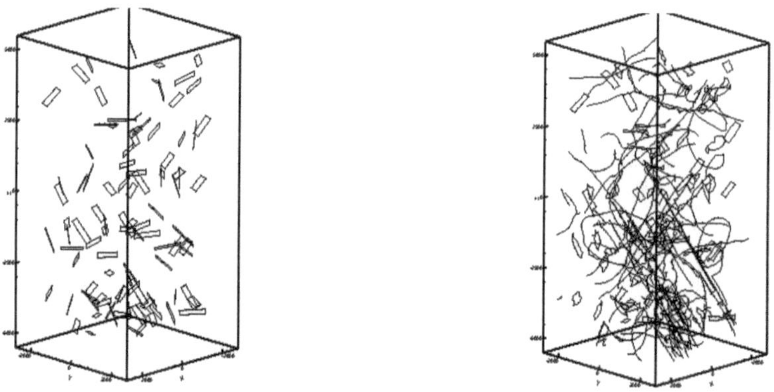

Figure 1. a) Initial dislocation structure, b) dislocation structure at 0.4% strain.

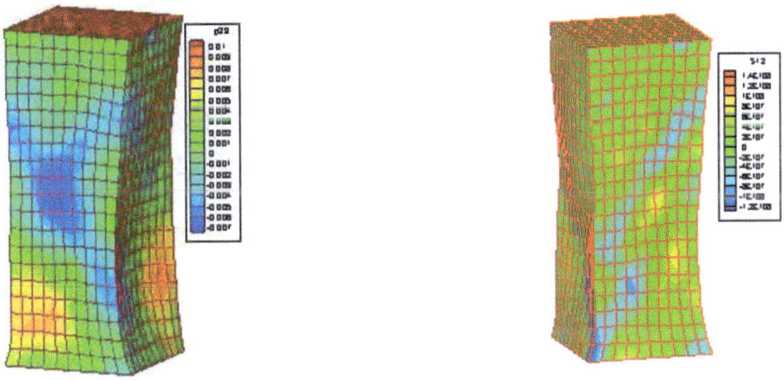

Figure 2. a) Plastic strain distribution, b) stress distribution, both at 0.4% strain and in deformed configuration with displacement magnified by a factor to 20.

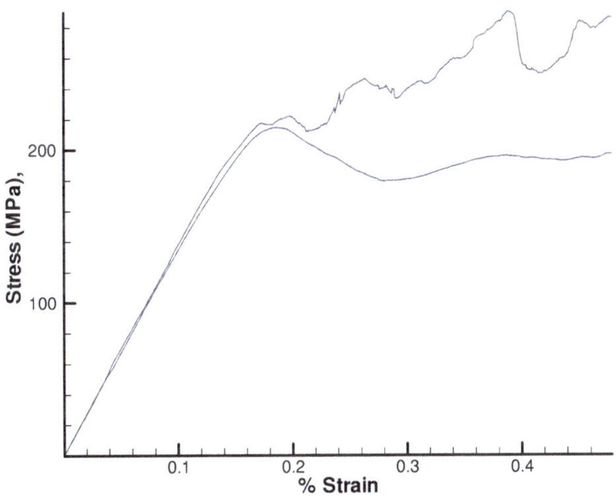

Figure 3. Stress-strain curve predicted by the coupled MDDP with boundary effects (upper curve) as compared to the DD prediction with no boundary effects (lower curve)

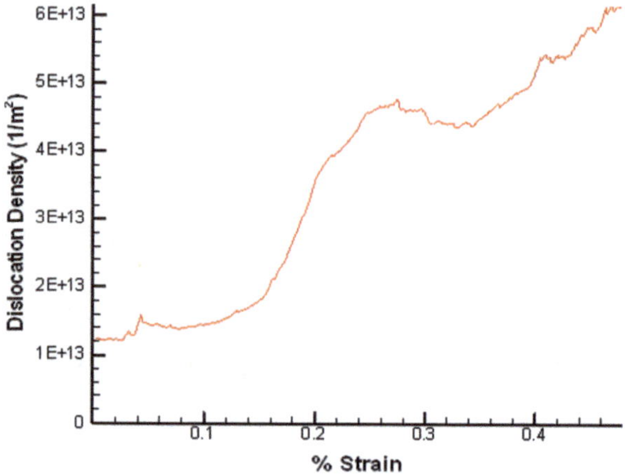

Figure 4. Evolution of dislocation density

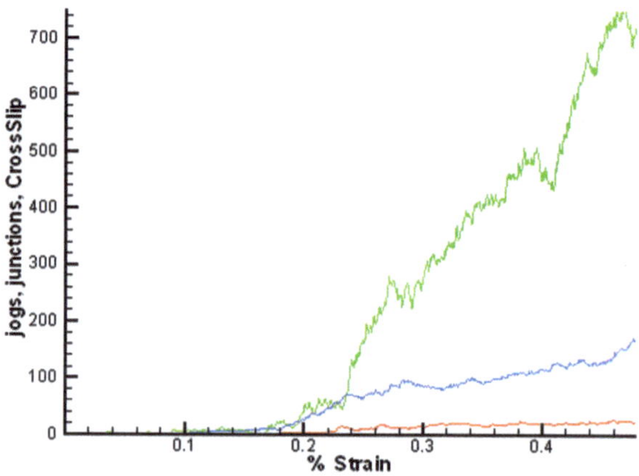

Figure 5. Number of jogs, junctions and cross–slip events during deformation. Upper curve corresponds to number of junctions, middle curve to cross–slip events, and lower curve to number of jogs.

4.2 Dislocation-Crack Interaction

In order to represent a planar crack embedded in an infinite body and subjected to an arbitrary loading, a continuous distribution of infinitesimal dislocation loops, each of area dS and Burgers vector $\boldsymbol{b}(b_x, b_y, b_z)$, is assumed. Because of the arbitrary nature of the loading, the crack will experience both opening and shearing mode displacements. The tractions on the crack plane, induced by the distribution, should be equal and opposite to the applied tractions, thus creating a stress-free surface, thus a crack. The superposition principle leads to the following integral equation:

$$\int_S K_{ij}(\boldsymbol{x}, \boldsymbol{x}')b_j(\boldsymbol{x}')\mathrm{d}S = -t_i^0(\boldsymbol{x}) \tag{25}$$

where $t_i^0(\boldsymbol{x})$ is the traction in a crack-free body induced by external (or internal) loads along the internal surfaces that coincide with the crack in a cracked body. The above integral equation possesses a third degree singularity due to the form of kernel $K_{ij}(\boldsymbol{x}, \boldsymbol{x}')$. This type of singularity is universal, and is encountered in all problems involving crack representation by dislocations, regardless the loading or the geometry of the problem.

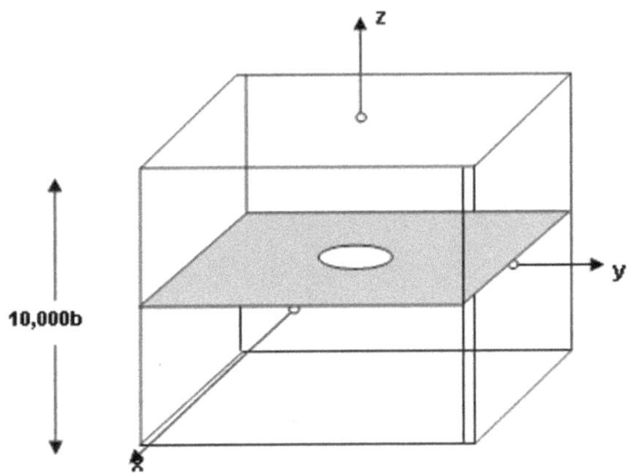

Figure 6. Schematic of the cracked geometry. The plane z=0 contains the crack, and the dislocations produced by a F-R source (not showing) are moving on the (111) plane.

In order to find the numerical solution, the crack plane is divided into a number N of triangular elements. Within each element the unknown distribution b_j is assumed to be the product of a known weight function and an unknown constant. The method leads to a linear algebraic system of $3N \times 3N$ equations. The solution of this system is unknown constant function, which, combined with the weight functions, provide the stress field of the crack. Then, this stress field is implemented within DD framework to study the dislocation behavior under the influence of the crack (Mastorakos and Zbib, 2006, 2007, 2008). The configuration and stress–strain curves for cracked specimens with various cracks are shown in Figures 6 and 7 respectively.

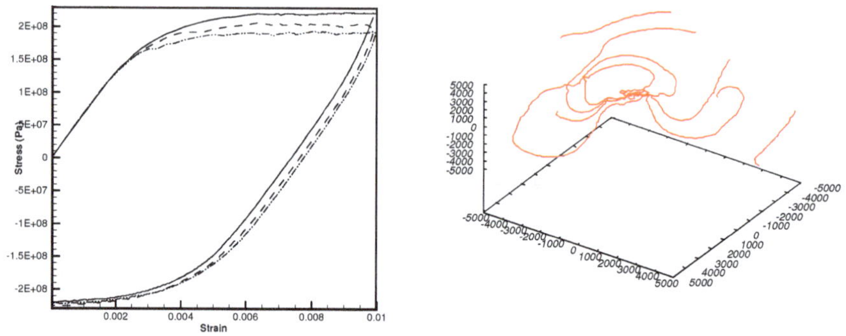

Figure 7. (a) Fatigue plots of three specimens each containing a penny–shape crack of different radius. Solid line: 1,000b; dashed line: 2,000b; dotted line: 2,500b. The behavior of the first specimen (1,000b crack radius) is closer to uncracked specimen than the other two. (b) The dislocation structure in the 2,000b crack case when the total strain was 0.01. The crack is situated in the middle of the specimen. The units of the axes are in Burgers vectors b.

Bibliography

F. Akasheh, H. M. Zbib, J. P. Hirth, R. G. Hoagland, and A. Misra. Dislocation dynamics analysis of dislocation intersections in nanoscale multilayer metallic composites. *J. Appl. Phys.*, 101:084314, 2007a.

F. Akasheh, H. M. Zbib, J. P. Hirth, R. G. Hoagland, and A. Misra. Interactions between gilde dislocations and parallel interfacial dislocations in nanoscale strained layers. *J. Appl. Phys.*, 102, 2007b.

F. Akasheh, H. M. Zbib, and T. Ohashi. Multiscale modeling of size effects in fcc crystals: Discrete dislocation dynamics and density based gradient crystal plasticity. *Phil Mag*, 87:1307–1326, 2007c.

M. P. Allen and D. J. Tildesley. *Computer simulation of liquids.* Oxford Science Publications, 1997.

V. I. Al'shitz. The phonon–dislocation interaction and its role in dislocation dragging and thermal resistivity. In V.L. Indenbom and J. Lothe, editors, *Elastic Strain and Dislocation Mobility*, chapter 11. Elsevier Science Publishers B.V, 1992.

N. W. Ashcroft and N. D. Mermin. *Solid State Physics: Saunders College*, 1976.

K. J. Bathe. *Finite Element Procedures in Engineering Analysis*. Prentice-Hall, 1982.

A. Bellou, C. Overman, H.M. Zbib, D. Bahr, and A. Misra. Strain hardening behavior in nanolayered metallic multilayers. *Scripta Metallurgica*, 64: 641–644, 2011.

G. R. Canova, Y. Brechet, and L. P. Kubin. 3d dislocation simulation of plastic instabilities by work softening in alloys. In S.I. Anderson et al, editor, *Modelling of Plastic Deformation and Its Engineering Applications*. Riso National Laboratory, 1992.

G. R. Canova, Y. Brechet, L. P. Kubin, B. Devincre, V. Pontikis, and M. Condat. 3d simulation of dislocation motion on a lattice: Application to the yield surface of single crystals. In J. Rabiet, editor, *Microstructures and Physical Properties*. CH–Transtech, 1993.

T. Diaz de la Rubia, H. M. Zbib, T. Khraishi, B. D. Wirth, M. Victoria, and M. Caturla. Multiscale modelling of plastic flow localization in irradiated materials. *Nature*, 406:871–874, 2000.

I. Demir and H. M. Zbib. A mesoscopic model for inelastic deformation and damage. *Int. J. Enger. Science*, 39:1597–1615, 2001.

I. Demir, J.P. Hirth, and H. M. Zbib. Interaction between interfacial ring dislocations. *Int. J. Engrg. Sci.*, 31:483–492, 1993.

E. Van der Giessen and A. Needleman. Micromechanics simulations of fracture. *Annu. Rev. Mater. Res.*, 32:141–162, 2002.

E. Van der Giessen and A. Needleman. Discrete dislocation plasticity: A simple planar model. *Mater. Sci. Eng.*, 3:689–735, 1995.

B. Devincre and L. P. Kubin. Mesoscopic simulations of dislocations and plasticity. *Mater. Sci. Eng.*, A234–236:8, 1997.

R. DeWit. The continuum theory of stationary dislocations. *Solid State Phys*, 10:249–292, 1960.

J. A. Ewing and W. Rosenhain. The crystalline structure of metals. *Phil. Trans. Roy. Soc.*, A193:353–375, 1899.

M. C. Fivel, C. F. Roberston, G. Canova, and L. Bonlanger. Three–dimensional modeling of indent-induced plastic zone at a mesocale. *Acta Mater*, 46(17):6183–6194, 1998.

N. M. Ghoniem and R. J. Amodeo. Computer simulation of dislocation pattern formation. *Sol. Stat. Phenom.*, pages 379–406, 1988.

N. M. Ghoniem and L. Sun. A fast sum method for the elastic field of 3–d dislocation ensembles. *Phys. Rev. B*, 60:128–140, 1999.

S. Groh, E. B. Marin, M. F. Horstemeyer, and H. M. Zbib. Multiscale modeling of plasticity in an aluminum single crystal. *Int. Journal of Plasticity*, in press, available online, 2009.

I. Groma and G. S. Pawley. Role of the secondary slip system in a computer simulation model of the plastic behavior of single crystals. *Mater. Sci. Engng*, A164:306–311, 1993.

M. Hiratani and E. M. Nadgorny. Combined model of dislocation motion with thermally activated and drag–dependent stages. *Acta Materialia*, 40:4337–4346, 2001.

M. Hiratani and H. M. Zbib. Stochastic dislocation dynamics for dislocation–defects interaction. *J. Enger. Mater. Tech.*, 124:335–341, 2002.

M. Hiratani and H. M. Zbib. On dislocation–defect interaction and patterning: Stochastic discrete dislocation dynamics. *J. Nuc. Enger.*, 2003, in press, 2003.

M. Hiratani, H. M. Zbib, and M. A. Khaleel. Modeling of thermally activated dislocation glide and plastic flow through local obstacles. *Int. J. Plasticity*, 19:1271–1296, 2003.

J. P. Hirth. Injection of dislocations into strained multilayer structures. In *Semiconductors and Semimetals*, volume 37, pages 267–292. Academic Press, 1992.

J. P. Hirth and J. Lothe. *Theory of dislocations.* Wiley, 1982.

J. P. Hirth, M. Rhee, and H. M. Zbib. Modeling of deformation by a 3d simulation of multipole, curved dislocations. *J. Computer-Aided Materials Design*, 3:164–166, 1996.

J. P. Hirth, H. M. Zbib, and J. Lothe. Forces on high velocity dislocations. *Modeling & Simulations in Maters. Sci. & Enger.*, 6:165–169, 1998.

H. Huang, N. Ghoniem, T. Diaz de la Rubia, M. Rhee, H. M. Zbib, and J. P. Hirth. Development of physical rules for short range interactions in bcc crystals. *ASME–JEMT*, 121:143–150, 1999.

C. Jinpeng, V. V. Bulatov, and S. Yip. Molecular dynamics study of edge dislocation motion in a bcc metal. *Journal of Computer–Aided Materials Design*, 6:165–173, 1999.

W. G. Johnston and J. J. Gilman. Dislocation velocities, dislocation densities, and plastic flow in lithium flouride crystals. *J. Appl. Phys.*, 30: 129–144, 1959.

A. Khan, H. M. Zbib, and D. A. Hughes. Stress patterns of deformation induced planar dislocation boundaries. MRS, 2001.

A. Khan, H. M. Zbib, and D. A. Hughes. Modeling planar dislocation boundaries using a multi–scale approach. *Int. J. Plasticity*, in press, 2003.

T. A. Khraishi and H. M. Zbib. Dislocation dynamics simulations of the interaction between a short rigid fiber and a glide dislocation pile–up. *Comp. Mater. Sci.*, 24:310–322, 2002a.

T. A. Khraishi and H. M. Zbib. Free surface effects in 3d dislocation dynamics: formulation and modeling. *ASME J. of Engineering Materials and Technology*, 124(3):342–351, 2002b.

T. A. Khraishi, H. M. Zbib, and T. Diaz de la Rubia. The treatment of traction–free boundary condition in three–dimensional dislocation dynamics using generalized image stress analysis. *Materials Science and Engineering*, A309–310:283–287, 2001.

T. A. Khraishi, H. M. Zbib, T. Diaz de la Rubia, and M. Victoria. Localized deformation and hardening in irradiated metals: Three–dimensional discrete dislocation dynamics simulations. *Metall. Mater. Trans.*, 33B: 285–296, 2002.

T. A. Khraishi, L. Yan, and Y. L. Shen. Dynamic simulations of dislocation–particle interaction in metal–matrix composites (mmcs). *International Journal of Plasticity*, in press, 2003.

U. F. Kocks, A. S. Argon, and M. F. Ashby. *Thermodynamics and kinetics of slip.* Pergamon Press, 1975.

T. J. Koppenaal and D. Kuhlmann-Wilsdorf. The effect of prestressing on the strength of neutron–irradiated copper single crystals. *Appl. Phys. Lett.*, 4:59, 1964.

L. P. Kubin. Dislocation patterning during multiple slip of fcc crystals. *Phys. Stat. Sol (a)*, 135:433–443, 1993.

L. P. Kubin and G. Canova. The modelling of dislocation patterns. *Scripta Metall*, 27:957–962, 1992.

K. C. Le and H. Stumpf. A model of elasticplastic bodies with continuously distributed dislocations. *Int. J. Plasticity*, 12:611–628, 1996.

J. Lepinoux and L. P. Kubin. The dynamic organization of dislocation structures: A simulation. *Scripta Metall.*, 21:833–838, 1987.

I. N. Mastorakos and H. M. Zbib. Shielding and amplification of a penny–shape crack due to the presence of dislocations. *Int. J. Fract.*, 142: 103–117, 2006.

I. N. Mastorakos and H. M. Zbib. Dd simulations of dislocation-crack interaction during fatigue. *Journal of ASTM International*, 4(8):1–9, 2007.

I. N. Mastorakos and H. M. Zbib. Dislocation-cracks interaction during fatigue: A discrete dislocation dynamics simulation. *Journal of Materials*, 60(4):59–63, 2008.

I.M. Mastorakos, A. Bellou, D.F. Bahr, and H.M. Zbib. Size dependent strength in nanolaminate metallic systems. *Journal Materials Research*, 26:1179–1187, 2011.

T. J. McKrell and J. M. Galligan. Instantaneous dislocation velocity in iron at low temperature. *Scripta Materialia*, 42:79–82, 2000.

O. Mügge. *Neues Jahrb.*, Min 13, 1883.

T. Ohashi, M. Kawamukai, and H. M. Zbib. A multiscale approach for modeling scale–dependent yield stress in polycrystalline metals. *Int. J. Plasticity*, 23:8970914, 2007.

C. Overman, N. R. Overman, D. F. Bahr, and H. M. Zbib. Yield and deformation in biaxially stressed multilayer metallic thin films. *Journal of Engineering Materials and Technology*, in press, 2009.

D. Raabe. Introduction of a hybrid model for the discrete 3d simulation of dislocation dynamics. *Comput. Mater. Sci.*, 11:1–15, 1998.

S. Rao, T. A. Parthasarathy, and C. Woodward. Atomistic simulation of cross–slip processes in model fcc structures. *Phil. Mag. A*, 79:1167, 1999.

M. Rhee, H. M. Zbib, J. P. Hirth, H. Huang, and T. D. de la Rubia. Models for long/short range interactions in 3d dislocation simulation. *Modeling & Simulations in Maters. Sci. & Enger*, 6:467–492, 1998.

D. Ronnpagel, T. Streit, and T. Pretorius. Including thermal activation in simulation calculation of dislocation glide. *Phys. Stat. Sol.*, 135:445–454, 1993.

R. Sandstrom. Subgrain growth occurring by boundary migration. *Acta Metall.*, 25:905–911, 1977.

K. W. Schwarz and J. Tersoff. Interaction of threading and misfit dislocations in a strained epitaxial layer. *Applied Physics Letters*, 69(9):1220, 1996.

M. A. Shehadeh, H. M. Zbib, and T. Diaz de la Rubia. Multiscale dislocation dynamics simulations of shock compressions in copper single crystal. *Int. J. Plasticity*, 21:2396–2390, 2005a.

M. A. Shehadeh, H. M. Zbib, T. Diaz de la Rubia, and V. V. Bulatov. Modeling the dynamic deformation and patterning in fcc single crystals at high strain rates: dislocation dynamic plasticity analysis. *Phil Mag.*, 85:1667–1684, 2005b.

M. A. Shehadeh, E. M. Bringa, H. M. Zbib, J. M. McNaney, and B. A. Remington. Simulation of shock–induced plasticity including homogeneous and heterogeneous dislocation nucleation. *Applied Physics Letters*, 89: 171918, 2006.

K. Shizawa and H. M. Zbib. Thermodynamical theory of strain gradient elastoplasticity with dislocation density: Part i–fundamentals. *Int. J. Plasticity*, 15:899–938, 1999.

N. Urabe and J. Weertman. Dislocation mobility in potassium and iron single crystals. *Mater. Sci. Engng*, 18:41, 1975.

H. Y. Wang and R. LeSar. $o(n)$ algorithm for dislocation dynamics. *Philosophical Magazine A*, 71:149–164, 1995.

W. Wasserbäch. Plastic deformation and dislocation arrangement of nb-34 at. ta alloy crystals. *Phil. Mag. A*, 53:335–356, 1986.

H. Yasin, H. M. Zbib, and M. A. Khaleel. Size and boundary effects in discrete dislocation dynamics: Coupling with continuum finite element. *Materials Science and Engineering*, A309–310:294–299, 2001.

H. M. Zbib and T. Diaz de la Rubia. A multiscale model of plasticity. *Int. J. Plasticity*, 18(9):1133–1163, 2002.

H. M. Zbib, M. Rhee, and J. P. Hirth. 3*d* simulation of curved dislocations: Discretization and long range interactions. In T. Abe and T. Tsuta, editors, *Advances in Engineering Plasticity and its Applications*, pages 15–20. Pergamon, 1996.

H. M. Zbib, M. Rhee, and J. P. Hirth. On plastic deformation and the dynamics of 3*d* dislocations. *International Journal of Mechanical Science*, 40:113–127, 1998.

H. M. Zbib, M. Rhee, J. P. Hirth, and T. Diaz de la Rubia. A 3*d* dislocation simulation model for plastic deformation and instabilities in single crystals. *J. Mech. Behavior of Mater.*, 11:251–255, 2000.

H. M. Zbib, S. Shehadeh, S. Khan, and G. Karami. Multiscale dislocation dynamics plasticity. *Int. J. for Multiscale Computational Engineering*, 1:73–89, 2003.

H. M. Zbib, S. Akarapu, F. Akasheh, C. Overman, and D. F. Bahr. Deformation and size effects in small scale structures. In A. Khan et al, editor, *proceedings of the 15th International Symposium on Plasticity and its Current Applications*, pages 220–222, 2009.

H.M. Zbib, C. Overman, F. Akasheh, and D.F. Bahr. Analysis of plastic deformation in nanoscale metallic multilayers with coherent and incoherent interfaces. *International Journal of Plasticity*, 27:1618–1638, 2011.